Mathematics and Visualization

More information about this series at http://www.springer.com/series/4562

Lars Linsen
Bernd Hamann
Hans-Christian Hege
Editors

Visualization in Medicine and Life Sciences III

Towards Making an Impact

With 161 Figures, 132 in color

Editors
Lars Linsen
Department of Computer Science and Electrical Engineering
Jacobs University
Bremen, Germany

Bernd Hamann
Department of Computer Science
University of California
Davis, California, USA

Hans-Christian Hege
Visualization and Data Analysis
Zuse Institute Berlin (ZIB)
Berlin, Germany

ISSN 1612-3786 ISSN 2197-666X (electronic)
Mathematics and Visualization
ISBN 978-3-319-79640-6 ISBN 978-3-319-24523-2 (eBook)
DOI 10.1007/978-3-319-24523-2

Mathematics Subject Classification (2010): 68-XX, 68-06, 68Uxx, 68U99

Cover illustration: "Extraction of Robust Voids and Pockets in Proteins" by R. Sridharamurthy et al. with kind permission

Printed on acid-free paper

This Springer imprint is published by Springer Nature
The registered company is Springer International Publishing AG Switzerland

Preface

Medicine has for a long time been a driving force for the development of visualization and visual analysis methods. Visualization has become an integral part of several clinical treatment and planning processes. Still, much more needs to be done. Many existing state-of-the art visualization approaches have not yet become part of daily clinical routine. It should be a major interest of the visualization community to strengthen the role and impact of visualization components in medical applications and, more generally, in health services. The topics are not limited to volume visualization, the traditional field of medical visualization, but include visual analysis of highly complex data and of large databases. With this broader focus, many challenging visualization problems remain to be solved.

Significant advances in data acquisition and new experimental techniques have boosted research in the life sciences in recent decades. The huge amount of data collected made it necessary to include sophisticated computational tools for data analysis, and a steady stream of new types of data requires continuous development of analysis tools. An important component to data analysis is provided by visual means, i.e., by visual encoding for intuitive and effective display of the acquired data and by interaction mechanisms for user-centered exploration. This lies in the core of visualization research.

The third international workshop on Visualization in Medicine and Life Sciences took place in 2013 (VMLS 2013). After two successful events in 2006 and 2009, VMLS 2013 was, for the first time, colocated with the EuroVis conference. The goal of the workshop was to discuss novel visualization techniques driven by the needs in medicine and life sciences as well as new application areas and challenges for visualization within these fields. VMLS 2013 generated ideas and concepts for visual analysis of data from scientific studies of living organs or to the delivery of healthcare services. Target scientific domains include the entire field of biology at all scales—from genes and proteins to organs and populations—as well as interdisciplinary research based on technological advances in other fields such as bioinformatics, biomedicine, biochemistry, and biophysics. Moreover, they comprise the field of medicine and the application of science and technology to healthcare problems.

Internationally leading experts came together at VMLS 2013 in Leipzig, Germany. This book reflects the outcome of the workshop. It contains research and survey articles, which were solicited and peer-reviewed after the conference. The research topics covered by the papers in this book address the following visualization themes:

- Segmentation and uncertainty
- Visualization of 3D medical images
- Visualization for diffusion-weighted imaging
- Cohort studies and time-varying phenomena
- Visualization in life sciences

The workshop was supported, in part, by the Deutsche Forschungsgemeinschaft (DFG) and the Eurographics Association. We would like to thank the organizers of EuroVis 2013 for their cooperation.

Bremen, Germany Lars Linsen
Davis, CA, USA Bernd Hamann
Berlin, Germany Hans-Christian Hege
January 2016

Contents

Part I
Segmentation and Uncertainty

Lung Segmentation of MR Images: A Review

Tatyana Ivanovska, Katrin Hegenscheid, René Laqua, Sven Gläser, Ralf Ewert, and Henry Völzke

Abstract Magnetic resonance imaging (MRI) is a non-radiation based examination method, which gains an increasing popularity in research and clinical settings. Manual analysis of large data volumes is a very time-consuming and tedious process. Therefore, automatic analysis methods are required. This paper reviews different methods that have been recently proposed for automatic and semi-automatic lung segmentation from magnetic resonance imaging data. These techniques include thresholding, region growing, morphological operations, active contours, level sets, and neural networks. We also discuss the methodologies that have been utilized for performance and accuracy evaluation of each method.

1 Introduction

Magnetic resonance imaging (MRI) has non-invasive, non-ionizing nature and provides a superior soft tissue contrast. This technique is rapidly developing over the last years. For example, continuous technological advances such as parallel imaging [16, 54] and integrated parallel acquisition techniques (iPAT) [30] significantly reduce the acquisition time and enable standardized imaging acquisition at a very high spatial and temporal resolution providing morphological and functional information of the highest detail. This makes MRI attractive in clinical practice and research settings.

T. Ivanovska (✉) • H. Völzke
Institute of Community Medicine, Ernst-Moritz-Arndt University, Greifswald, Germany
e-mail: tetyana.ivanovska@uni-greifswald.de; voelzke@uni-greifswald.de

K. Hegenscheid • R. Laqua
Institute of Diagnostic Radiology and Neuroradiology, Ernst-Moritz-Arndt University, Greifswald, Germany
e-mail: katrin.hegenscheid@uni-greifswald.de; laquar@uni-greifswald.de

S. Gläser • R. Ewert
Department of Internal Medicine B, Division of Pulmonary Medicine, Ernst-Moritz-Arndt University, Greifswald, Germany

L. Linsen et al. (eds.), *Visualization in Medicine and Life Sciences III*, Mathematics and Visualization, DOI 10.1007/978-3-319-24523-2_1

Lung imaging is one of the most challenging topics in MRI due to motion artifacts and low signal. Nevertheless, in recent years lung MRI is becoming increasingly popular in comparison to the pulmo-imaging "gold standard", computed tomography (CT). High quality MRI assists in detection of numerous pulmonary diseases. Moreover, MRI of the lung is important for specific clinical applications. For instance, it can contribute to decision making the case of such diseases as lung cancer [67], malignant pleural mesothelioma, and acute pulmonary embolism. The advantages of MR over CT are not limited to the lack of ionizing radiation, which is of particular interest for the assessment of lung disease in children, pregnant women, or in patients, who require frequent follow-up examinations (e.g., immunocompromised patients with fever of unknown origin). Chest wall invasion by a tumor and mediastinal masses are accepted indications benefiting from MRI superior soft tissue contrast. Dynamic examinations to study respiratory mechanics and contrast enhanced first pass perfusion imaging reach far beyond the scope of CT [25].

MRI of the lung has multiple application areas. For example, conventional (so-called proton or ^{1}H) MRI provides anatomical details and is standardly used for assessing lung volumes, boundaries, detection of nodules, infiltrates, and masses. There is a significant number of protocols designed for these purposes [5, 6]. In Fig. 1, two example slices of anatomical MR images from two different sequences, namely, T1-weighted VIBE (volume-interpolated breath hold examination) and T2-weighted HASTE (half-Fourier single shot turbo spin-echo), are shown. The VIBE sequence has a higher spatial resolution and a smaller slice thickness than the HASTE sequence (for instance, $1.8 \times 1.8 \times 3$ mm vs. $2.3 \times 1.8 \times 5$ mm on 1.5 Tesla Magnetom Avanto Siemens device) and is preferable for lung volumetry, whereas lung infiltrates serve as key pathologies for applying the HASTE sequence.

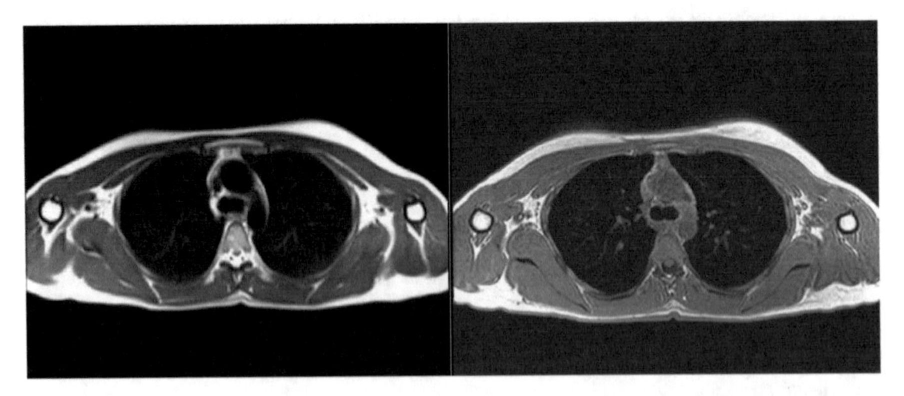

Fig. 1 An example of two slices from two anatomical MR thoracic scans from the same participant. *Left*: T2-weighted Half-Fourier Acquisition Single-Shot Turbo Spin-Echo sequence (HASTE), $256 \times 206 \times 44$; *Right*: T1-weighted volume-interpolated breath hold sequence (VIBE), axial orientation, $512 \times 416 \times 88$. Image courtesy of Study of Health in Pomerania, Germany [18]

MRI with non-radioactive noble gases, such as hyperpolarized ^{3}He (helium-3) or ^{129}Xe (xenon-129), is applied for detection and evaluation of functional, e.g., ventilation, abnormalities [40, 42, 62]. MR angiography (MRA) gains its popularity for assessment of pulmonary vascular disease [25].

(Semi-)automatic segmentation is an essential step in medical image analysis, medical visualization, and computer-aided diagnosis. Since manual segmentation of large volumes of MR data is a very laborious, observer-dependent, and time-consuming process, which is prone to inter- and intra-observer variability, automatized segmentation methods for extraction of different organs and structures are actively developed.

This paper aims to provide an overview of recently published literature on automatic and semi-automatic segmentation techniques for MR images of human lungs. Our main source of references was the Internet: we have searched for the terms *lung, segmentation, volumetry, pulmonary, magnetic resonance imaging (MRI)* on Google Scholar, PubMed, and IEEE-Xplore.

The paper is organized as follows. First, a short introduction to human lung anatomy is given in Sect. 2. In Sect. 3, we introduce the methods and the data, which they are applied to. The performance of the methods is discussed in Sect. 4. Section 5 concludes the paper.

2 Pulmonary Anatomy

In this section, we briefly present the human pulmonary system. For a more detailed description, we refer to specialized medical literature, for example [35].

Human lungs are located in the chest wall, consisting of the rib cage, diaphragm, and mediastinum, the area between the lungs, which includes all of the organs in the chest, such as heart, vessels, trachea, except the lungs. The right lung is composed of three lobes, while the left lung has only two lobes. The lobes are further divided into sub-lobar segments, which are defined by the branching pattern of the airway tree. The lobar fissures, space between the surface of the lobes, allow the lobes to rotate relative to one another to accommodate body posture-related changes in chest wall geometry.

Air travels into the lungs through the nose or mouth and through the pharynx, larynx, and into the tracheobronchial (airway) tree (cf. Fig. 2). The pulmonary blood circulation is provided by pulmonary arteries and veins.

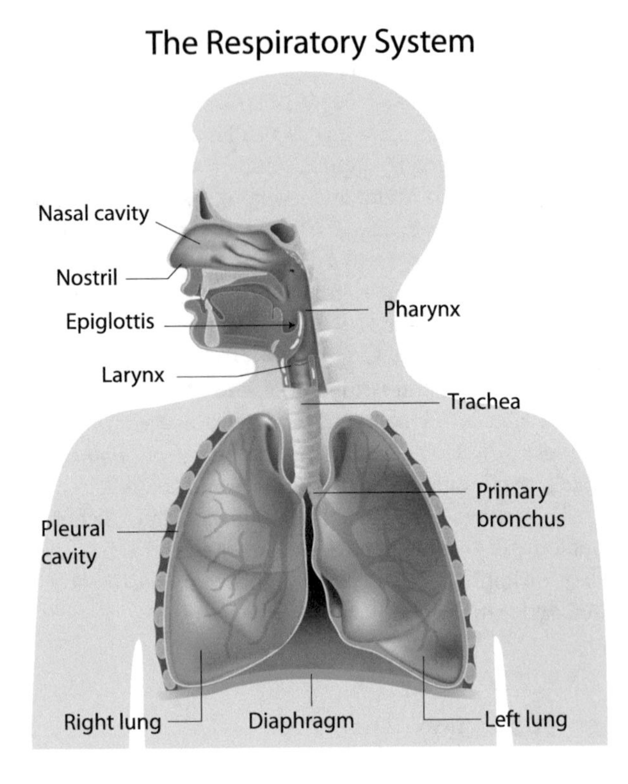

Fig. 2 A schematic description of the human respiratory system

3 Lung Segmentation

Segmentation is the first and essential step in computer analysis of medical imaging data. Segmentation results are further used, for instance, for volumetric measurements, visualization, and biomechanical modeling.

In thoracic MR images, various anatomical entities are presented, such as lungs (including lobes and fissures), bronchial tree, and pulmonary vessels. Whereas in high-resolution CT images all these structures are clearly visible and can be distinguished, lung imaging is a challenge in MRI. Here, usually only organ borders can be identified. Moreover, such fine structures as the complete bronchial tree or the fissures are hardly visible in anatomical MR images. Therefore, the task of lung segmentation from MR images usually consists of lung region detection, separation of the lungs from the large airways, and exclusion or inclusion of the pulmonary vessel regions.

Numerous approaches have been proposed for CT data [13, 38, 53], whereas there are only a few methods for MRI data presented in the literature. Here, we classify and discuss these methods for automatic and semi-automatic lung segmentation from MRI data. The literature can be roughly divided into the following categories: intensity-based, model-based, and active contour-based approaches. A summary of published works is presented in Table 1. The table rows are ordered as the techniques are discussed in the paper body. We leave table cells dashed, if the correspondent information is missing in the original papers.

Table 1 A summary of published works on MR lung segmentation

Authors	Main points	Data		
		Resolution	Orientation	Sequence
Sensakovic et al. 2006 [52]	*Intensity-based* 2D, slice based, Histogram thresholding, Morphological operations	256 × 256	Axial	T1 SPGR
Ivanovska et al. 2012 [22]	*Intensity-based* 3D, Clustering, CC Analysis, Trachea removal, Lung separation, Morphological operations	2.3 × 1.8 × 5 mm	Axial	T1 VIBE
Kullberg et al. 2009 [27]	*Intensity-based* 3D, Thresholding, Distance transform, Region growing, CC Analysis	2.1 × 2.1 × 8 mm	Coronal	Whole body multigradient
Lui et al. 2013 [33]	*Intensity-based* 2D, slice based, Semi-automatic Histogram thresholding, Clustering, Trachea removal, Morphological operations	256 × 256, 8–14 slices	Coronal	^{3}He Fast GRE

(continued)

Table 1 (continued)

Authors	Main points	Data		
		Resolution	Orientation	Sequence
Plathow et al. 2005 [45]	*Intensity-based* 2D, slice based, Semi-automatic Interactive region growing Morphological operations	$3.75 \times 3.75 \times 3.8$ mm	Coronal	3D FLASH
		$2.5 \times 1.6 \times 10$ mm	Coronal	2D trueFISP
Heydarian et al. 2012 [19]	*Intensity-based* 2D and 3D K-Means Region growing	256×128, 14 slices	Coronal	T1 SPGR
Kirby et al. 2012 [26]		128×128, 14 slices	Coronal	^{3}He
Tokuda et al. 2009 [58]	*Intensity-based* 2D, Semi-automatic Confidence connectedness Fuzzy connectedness Region growing (ITK)	$3.125 \times 3.125 \times 5$ mm	Coronal	3D FLASH + GRAPPA
Woodhouse et al. 2005 [66]	*Intensity-based* 2D, Semi-automatic Thresholding	256×256	Coronal	^{1}H SSFSE
		112×128	Coronal	^{3}He
Virgincar et al. 2012 [64]	*Intensity-based* 2D, Semi-automatic Region Growing Morphological operations	192×192	Coronal	SSFP
		192×192	Coronal	^{3}He SPGRE
Lelieveldt et al. 2000 [28, 29]	*Model-based* 3D, Multiorgan Energy potential Minimization	–	Thoracic scout	–

(continued)

Table 1 (continued)

Authors	Main points	Data		
		Resolution	Orientation	Sequence
Tustison et al. 2011 [60]	*Model-based* 3D, Automatic Multiorgan Shape prior, PCA Ventilation-based defects N4 correction	80 × 128, 19–28 slices	Axial	^3He
Ray et al. 2003 [48]	*Active contours* 2D, Automatic Contour merging Snakes Slice based	–	Coronal	^1H
Middleton and Damper, 2004 [39]	*Active contours* *Neural network* Binary classification Refinement with Snakes 3D	256 × 256 × 35	Axial	$T1$-weighted
Tetzlaff et al. 2010 [57]	*Semi-automatic* *2D* Graph Cuts	128 × 128	Coronal	2D FLASH
Böttger et al. 2007 [9]	*Semi-automatic* *2D and 3D* Parametric active contours Simplex mesh	–	Coronal	HASTE

The table rows are ordered as the techniques are discussed in the paper body

3.1 Intensity-Based Segmentation

Intensity-based segmentation steps are similar in MR and CT data. Namely, intensity-based approaches usually consist of the following steps. First, lungs are extracted and separated from the large airways. Second, if needed, anterior and posterior lung junctions are corrected. Third, the lung contours are smoothed and the cavities are filled.

Sensakovic et al. proposed a multi-step approach based on thresholding, shape descriptors, and morphological operations [50–52]. The method works on a slice-by-slice basis. In each slice, the thorax is separated from the background using a histogram-based thresholding and a boundary smoothing [17]. Once the thorax segmented image is computed, a histogram-based thresholding [17] is applied to it to create a lung-segmented images. Thereafter, the connected components are analyzed, and all regions that fail to satisfy the descriptor thresholds (area, compactness, center of mass, and perimeter) are classified as non-lung. If no lung region is detected, the threshold is decreased and the procedure is repeated. After this procedure, which is referred as the "core stage", a refinement step ("the correction stage") starts. This step is important especially in the presence of disease and acquisition artifacts. It aims to capture valid lung regions that may have been obscured by disease or acquisition artifacts. First, a gray scale erosion operator [55] is applied to the thorax-segmented image. It lowers the intensity values of the lung regions masked by acquisition artifacts or disease. Second, a thresholding procedure, similar to the one in the core stage, is applied. A series of rolling ball filters [2] is applied to the internal boundaries the lungs. Candidate lung regions from the correction stage lung-segmented image that satisfy the circularity criterion are combined with lung regions from the core stage, and the final lung-segmented image is constructed.

Ivanovska et al. implemented a 3D method for lung segmentation [22]. The approach consists of four steps: the main extraction and three refinement steps. First, the lung and trachea are segmented. Any automatic clustering technique (for example, K-Means [34]) is applicable here. Second, the trachea and main bronchi are found using a 3D connected component analysis [17]. The separation of the large airways from the lungs is done slice-wise using 2D Watershed [63] on the inverted distance map [55]. Third, the lungs are separated from each other using a 3D Watershed procedure on a subimage, which represents an anterior or posterior junction. The subimage is selected such that the computational costs of the watershed procedure stay relatively low. Fourth, the lung cavities are corrected with a morphological closing [55].

Kullberg et al. presented an automatic approach for lung segmentation as a part of the adipose tissue depots segmentation from the whole-body MRI datasets [27]. Segmentation of lungs is performed with thresholding and morphological operations [17]. To increase robustness to erroneous inclusion of intestinal gas, the lungs are segmented one by one. First, the right lung is segmented, since the position of the liver increases the distance between the right lung and the intestines. The segmentation is performed as follows. Initial thresholding is followed by a distance transform [55]. Then, a region growing [17] is applied to the distance transform image. The condition for the growing process is continuous decrease of the distance transform values. The result is dilated to ensure inclusion of the whole lung. An intensity threshold is applied in the dilated lung to determine the final result. The threshold value is computed from the fitted Gaussian function to the histogram from the dilated lung region.

Lui et al. proposed a semi-automatic segmentation algorithm to isolate areas of ventilation from hyperpolarized ^3He MRI [33]. First, the background is removed by determining an optimal threshold from a sampled background noise distribution located outside of the lung field. Second, the lung mask is refined with a four-class fuzzy C-Means [56] clustering. Third, the trachea is semi-automatically removed with a seeded region growing method [56]. Finally, morphological operators [55] are used to refine the result.

Plathow et al. applied a semi-automatic method based on an interactive region growing technique [56] to evaluate lung volumes during the breathing cycles using dynamic MRI [45]. First, displacement of the chest wall and the diaphragm as a surrogate of the lung volume and lung surface was measured in dynamic 2D MRI images during the breathing cycle. Thereafter, vital capacity (VC) was calculated using a model, consisting of a half ellipse, an ellipsoid cone, and a prisma [44]. An automatic segmentation in 3D dynamic MRI failed due to intensity inhomogeneities and motion artifacts. Therefore, the 3D data was segmented in a slice-by-slice manner using interactive region growing. For each dataset, the 2 segmentation results were combined into a voxel-based 3D description of the current breathing state. The voxel-based segmentation results were transformed into triangular meshes using the marching cubes algorithm [31]. The isosurfaces were then smoothed in a post-processing step. Total surface lung areas were computed by summing up the area of each individual triangle.

Heydarian et al. proposed a semi-automated method for generating ^{3}He measurements of individual slice (2D) or whole lung (3D) gas distribution [19]. The same technique was also used by Kirby et al. [26]. The authors applied hierarchical K-Means [34] clustering for the ^{3}He and a seeded region-growing algorithm [56] for anatomical MR images. Thereafter, the images were registered semi-automatically using landmark-based registration [41]. Finally, the segmented areas were postprocessed with a morphological closing algorithm with a structuring element disk [55]. The vertical central region was excluded from the processing to avoid connecting the right and left lung areas.

Tokuda et al. measured lung volumes in dynamic 3D lung images [58]. The segmentation was performed in a slice-by-slice manner. In each slice, the lung area was segmented using a combination of confidence-connectedness and fuzzy-connectedness region growing algorithms, implemented in the Insight Segmentation and Registration Toolkit [21]. First, a rough segmentation of the lung area was obtained with the confidence-connectedness region growing method. The algorithm extracted a connected set of pixels whose pixel intensities are consistent with the pixel statistics (the mean and variance across a neighborhood) of predefined seed points. The pixels, connected to the seed points, whose values are within the confidence intervals are grouped together. The width of the confidence interval is controlled by a multiplier parameter. Second, the mean and variance of intensities in the lung area were then used for the fuzzy connectedness algorithm [21] to compute an affinity map, which represents degrees of adjacency and the similarity of pairs of nearby voxels. Finally, the refined lung area was extracted by thresholding the affinity maps. The multiplier and threshold parameters were tuned such the segmented lung area included the blood vessels.

Woodhouse et al. used a combination of anatomical and ^{3}He MR images to compare the ventilated and thoracic lung volumes in groups of smokers and never-smokers [66]. The processing was done manually in a slice-by-slice manner. ^{3}He images were also segmented semi-automatically with adaptive thresholding [56].

The threshold value was derived from the signal-to-noise (SNR) [56] value of each image and the mean signal in a manually selected region of interest in each image.

Virgincar et al. analysed ^{129}Xe and SSFP (steady state free precession) ^{1}H images in a semi-automatic manner [64]. The anatomical MRI was segmented by a region growing method [56]. The seeds were placed manually in the lowest intensity areas of the right and left lung in the central slice of the thoracic cavity. Thereafter, a threshold range was computed from the intensities of the seeds. Then, the morphological closing [55] was applied to the extracted lung. The ^{1}H and ^{129}Xe were acquired over different breath-holds, they required registration. It was performed using either affine transform [3] or similarity transform [20]. After the registration, ventilation images were also segmented with the region growing method.

3.2 Model-Based Segmentation

Lelieveldt et al. developed a model-based method to simultaneously segment lungs as well as some other nearby organs [28, 29]. An anatomical model of thorax is built by modeling of individual organs with implicit surfaces from manually delineated training images and subsequent grouping of single organ models into a tree structure. The whole hierarchical scene is described by a boundary model, which characterizes the scene volume as a boundary potential (or energy) function. Thereafter, a model is matched to image data. First, an initial parameter set for the pose and scale parameters is selected. Second, an automatic thresholding based on empirical histogram evaluation discriminates air from tissue. A set of target boundary points is obtained. Third, a model is placed and the energy function is minimized with Levenberg-Marquardt nonlinear by fitting the minimization [32] by fitting the target boundary points to the model. The authors reported that their method was designed for a coarse pre-segmentation and a more accurate local segmentation process is further required.

Tustison et al. proposed an automated segmentation method for differentiation of the ventilated lung volumes on ^{3}He MRI [60]. The method consists of two main steps: template and statistical model construction and individual subject processing. All images in the database are registered to a normalized space. A normalized unbiased template is built from seven representative subjects with a symmetric diffeomorphic registration algorithm [4]. Thereafter, a principal component analysis (PCA) model [15] from an image database is built. Each image is transformed to the template using an affine transformation, so that the presence of any global shape differences in the statistical model is avoided. Then the processing of individual subjects starts. Each dataset is mapped to the template, the bias field is corrected [59], and then a shape-based level set procedure [43] starts to extract the lungs, where the PCA model is used.

3.3 Active Contours and Neural Network Segmentation

Ray et al. implemented a slice-based active contour approach for automatic segmentation of lungs from ^1H MRI [48]. Initial snakes are placed automatically within the lung cavities that are to be segmented. Then all snakes are evolved independently of each other. Finally, the union of the regions covered by all snakes is considered as the final results. For the snake evolution, the authors proposed to apply a modified gradient vector flow (GVF) with Dirichlet boundary conditions [47].

Middleton and Damper proposed a combined method, consisting of neural networks and active contours (snakes) to segment lungs in MRI [39]. First, a neural network was trained for binary classification of each pixel as a "boundary" or a "non-boundary". The inputs to the neural network are 7×7 image patches of the pixel to be classified. The weights were determined by training on MR sections with lung boundary pixels segmented by an expert observer. The resulting edge-point image was used as the external energy for the snake evolution.

Tetzlaff et al. analysed lung areas on 2D dynamic MR images [57]. The segmentation was performed semi-automatically, using Graph Cut algorithm [10]. The algorithm was initialized by adapting a bounding box to the size of the region of interest, and then the areas inside and outside of the lungs were roughly marked. If the segmentation leaked into the thoracic wall, an additional scribble had to be drawn at the point of leakage.

Böttger et al. presented a new segmentation approach [8, 9], based on parametric active contours [24]. Discrete surface meshes [12] were used here. The authors proposed to use an additional speed term, which was derived from the magnitude of the Gaussian gradient image. New attractor forces were introduced into the simplex mesh deformation scheme [7]. These forces were obtained from the user defined attractor points. To cope with complex surface geometries and avoid oscillating behavior during the evolution, the mesh was refined during the deformation automatically after every 50 iterations.

4 Discussion

4.1 Methodical Analysis

There are several aspects that define the relationships between the described algorithms. As it is shown above, most of the presented techniques are pipelines consisting of well-known algorithms, such as region growing, intensity clustering or thresholding, morphological operations [55, 56], combined together for a specific purpose. The novelty of the presented pipelines is in the application areas. Therefore, the algorithms are usually designed for a specific data type, i.e., MR sequence, and would often require substantial changes to be adapted to other data. Most of the presented techniques are developed for coronal or axial anatomical

MRI. Unfortunately, often no exact data protocols are given in the papers (cf. Table 1). Some techniques are designed for detection of ventilation defects in MR with noble gases (^3He and ^{129}Xe). Since ^3He images are often acquired concurrent with the traditional anatomical MR images, some techniques of ^3He segmentation are based on the prior segmentation of ^1H images and subsequent image registration, for instance, [19]. Moreover, some methods were designed only for $2D$ (slice-wise) processing [48, 51] and do not consider any $3D$ object consistency and connectivity. More complicated approaches, such as presented in [22, 60], were developed for a $3D$ organ extraction. Most of the presented techniques consist of one or two steps and include the user interaction. Usually, the more users are involved in the processing, the simpler the algorithmic pipeline is, since the complicated parts are accomplished by the users. For example, for many algorithms [19, 33] the mediastinum area is manually pre-excluded from the processing, and the user initiates the computations by placing the initial seed points [33, 58, 64] or adapting threshold values [66]. Of course, semi-automatic algorithms are less tedious for the users than the completely manual segmentation, but still can be inappropriate for processing of thousands of datasets. The fully automated pipelines, such as [22, 60], have a significant number of algorithmic steps and pre-selected parameters.

Therefore, the practitioners should take into account the following facts, while selecting, which technique to apply:

- What type of data is available?
- How many datasets need to be processed?
- Is $2D$ or $3D$ processing required?
- What level of user interaction is accepted?

For example, if the user interaction and $2D$ processing are acceptable, then slice-by-slice algorithmic solutions for ^1H or ^3He MR data, similar to the ones presented in [19, 26, 66], can be considered.

4.2 Performance Evaluation

All automatic and semi-automatic methods require some form of quantitative measure of the accuracy. Usually, ground truth masks are obtained manually and the results produced by the algorithm are compared to it. Here, region based metrics, such as DICE coefficient [14] or sensitivity and specificity, are often applied. Moreover, for certain applications the correlation to the clinical examinations, such as spirometry, is important. Hence, the correlation coefficients between these measurements are assessed.

However, there is no unique system of quality measures used in the community, which complicates the evaluation of the algorithms' efficiency. This section reviews the approaches, presented above, from the quantitative and qualitative perspectives.

A summary of the test sets and details of validation for each method is given in Table 2. The order of the rows in kept the same, as in Table 1.

Table 2 A summary of performance measurements of the segmentation methods on MR lung segmentation

Authors	Test set	Evaluation
Sensakovic et al. 2006 [52]	101 slices	Area of overlap (regions) $AOM \geq 82\,\%$
Ivanovska et al. 2012 [22]	10 datasets	Volume fractions (regions) $TPVF \geq 97\,\%$ Error fractions $\leq 5\,\%$
Kullberg et al. 2009 [27]	24 datasets	No evaluation
Lui et al. 2013 [33]	10 datasets	DICE's coefficient (regions) $DICE = 0.96 \pm 0.1$
Plathow et al. 2005 [45]	20 datasets	Correlation with spirometry Spearman's correlation $r > 0.83$ Significance level $P < 0.005$
Heydarian et al. 2012 [19]	4 datasets	Correlation with manual Pearson's correlation $r > 0.98$ Significance level $P < 0.0001$ DICE > 0.9
Kirby et al. 2012 [26]	15 datasets 3 pathologies	Ventilation Defect Volume (VDV) Ventilation Volume (VV) Correlation with manual Pearson's correlation $r \geq 0.84$ Significance level $P < 0.0001$ DICE ≥ 0.88
Tokuda et al. 2009 [58]	2 datasets	No evaluation
Woodhouse et al. 2005 [66]	18 datasets Smokers and never-smokers	Ventilation Volume (VV) Correlation with manual ICC ≥ 0.98
Virgincar et al. 2012 [64]	44 datasets	Ventilation Defect Percentage (VDP) Correlation with manual Pearson's $r = 0.97, P < 0.0001$
Lelieveldt et al. 2000 [28, 29]	15 Datasets	No evaluation

(continued)

Table 2 (continued)

Authors	Test set	Evaluation
Tustison et al. 2011 [60]	43 datasets for defect evaluation 18 datasets for ventilation based segmentation	Defect evaluation: $ICC = 0.85$ Segmentation: STAPLE method from 4 observers $Sens = 0.898, Spec = 0.905$ (regions, no ground truth)
Ray et al. 2003 [48]	10 datasets	$FOM = 69\,\%$ (edges) Error $= 6\,\%$ (regions)
Middleton and Damper 2004 [39]	13 datasets	Segmentation performance $F \geq 0.84$ (regions)
Tetzlaff et al. 2010 [57]	10 datasets	Correlation to spirometry $r \geq 0.97$
Böttger et al. 2007 [8, 9]	10 datasets	DICE ≥ 0.88 Surface distance ≤ 2 mm Hausdorff distance ≤ 20 mm

4.2.1 Region-Based and Edge-Based Metrics

Sensakovic et al. [51, 52] and Sensakovic and Armato [50] applied their technique to a random sample of 101 thoracic MR sections. True lung regions were manually delineated by two radiologists. 90 % of the patient data included abnormalities, such as mesothelioma, scarring, enlarged lymph node and others. To measure the segmentation quality, the authors used an area-of-overlap (AOM) measure, also known as the Jaccard coefficient [23]. The AOM measure of two regions, A and B, is defined as the number of pixels contained within the intersection of the regions divided by the number of the pixels contained within the union of the regions:

$$AOM = \frac{A \cap B}{A \cup B}, AOM \in [0, 1], \quad (1)$$

where 0 corresponds to disjoint regions and 1 is the complete overlap. The AOM is similar to the Dice coefficient [14], but the normalizing part (the denominator) is different. The reported AOM values were equal to 0.82 ± 0.16 and 0.83 ± 0.13, when to compared to the first and second observers, respectively.

Ivanovska et al. [22] tested the proposed method on ten randomly selected participants with normal lungs. The accuracy evaluation was done by the methodology proposed by Udupa et al. [61]. Namely, the following volume measures (VF: volume fractions) were computed: TPVF (true positives is the fraction of voxels in the intersection of the automatic and manual segmentation results), FPVF (false positives is the fraction of voxels falsely identified by the automatic segmentation), FNVF (false negatives is the fraction of voxels defined in manual segmentation, but missed by the automatic method). Additionally, TPVE (true positive volume error) was calculated. Let C_{exp} and C_{auto} denote the binary masks produced the expert and the automatic pipeline, correspondingly. Then, the measures are defined as

$$TPVF = \frac{|C_{auto} \cap C_{exp}|}{|C_{exp}|}, \tag{2}$$

$$FNVF = \frac{|C_{exp} - C_{auto}|}{|C_{exp}|}, \tag{3}$$

$$FPVF = \frac{|C_{auto} - C_{exp}|}{|C_{exp}|}. \tag{4}$$

Let V_{auto} and V_{exp} denote the volumes, computed by the algorithm and the expert, correspondingly. Then $TPVE$ is defined as

$$TPVE = \frac{|V_{auto} - V_{exp}|}{|V_{exp}|}. \tag{5}$$

The accuracy measures have been computed for left and right lungs separately. For the left lung: $TPVF = 97.06 \pm 1.36\,\%, FPVF = 2.97 \pm 1.08\,\%, FNVF = 2.94 \pm 1.36\,\%, TPVE = 1.58 \pm 1.16\,\%$. For the right lung: $TPVF = 97.33 \pm 1.57\,\%, FPVF = 3.96 \pm 1.69\,\%, FNVF = 2.66 \pm 1.57\,\%, TPVE = 2.37 \pm 1.74\,\%$. Moreover, the authors acquired the ground truth from two independent experts and observed that agreement about 93.5 %, which shows that the automatic result accuracy lies with the variation interval between the experts.

Lui et al. [33] validated their method on four healthy and six asthmatic subjects, calculating the ventilated lung volume (LVL). To assess the accuracy of the method, the authors calculated a Dice coefficient [14], comparing their results to the manually obtained ground truth. The Dice coefficient measures the overlap between the regions A and B:

$$DICE = \frac{2|A \cap B|}{|A| + |B|}, DICE \in [0, 1], \tag{6}$$

when 0 there is no overlap, 1 is the perfect match. The reported Dice value is 0.96 ± 0.01. Bland-Altman analysis [36] was used to determine the 95 % limits of agreement calculated from the mean and standard deviation of the volume difference between the segmentation results and the ground truth. The authors processed a total of 109 coronal slice for ten subjects. It showed that the means LVLs of the semi-automatic approach were 3.88 ± 0.75 L and 3.83 ± 1.11 L for healthy and asthmatics subject, respectively. The differences to the manual and standard spirometric measurements were statistically insignificant.

Ray et al. [48] applied their approach to ten different datasets. The authors measured Pratt's figure of merit (FOM) [1]. $FOM \in [0, 1]$. The FOM quantifies the comparison between ideal edges (ground truth) and detected edges of the image.

$$FOM = \frac{1}{max(I_A, I_I)} \sum_{i=1}^{IA} \frac{1}{1 + d_i \alpha^2}, \tag{7}$$

where I_A and I_I are the detected and ideal edge images, respectively. d_i is the distance between the actual and ideal edge i, and α is a penalty factor for displaced edges. The average obtained FOM is 0.69. The authors also utilized a region-based measure, namely, the percentage error , computed as

$$Error = \frac{\sum_i |Segm(i) - I_g(i)|}{\sum_i I_g(i)} 100\,\%, \tag{8}$$

where $Segm$ and I_g are the segmentation result and the ground truth, correspondingly. This metric is a combination of FP (false positives) and FN (false negatives) measures. The mean percentage error for ten datasets was about 6 %.

Tustison et al. [60] used data from seven random subjects to build the unbiased template. Experimentally, it was seen that such number of datasets provided a satisfactory compromise between quality of results and required computational time. For the PCA statistical model 156 images from normal subjects were used. For the evaluation two comparative analyses were performed. First, the number of defects scored by two human readers in 43 subjects. Here, the intraclass correlation coefficient (ICC) [37].This metric is a general measurement of agreement or consensus, where the measurements used are assumed to be parametric (continuous and has a Normal distribution). There was a high agreement between the algorithm and the readers ($ICC = 0.85$ and $ICC = 0.86$ for the first and second observers, correspondingly). Second, the simultaneous truth and performance estimation (STAPLE) [65] was performed on 18 subjects in which the ventilation defects were manually segmented by four human readers. Here, the sensitivity and specificity are defined as

$$Sens = \frac{TP}{TP + FN}, \tag{9}$$

$$Spec = \frac{TN}{TN + FP}, \tag{10}$$

where TP, FN, FP, TN denote the true positives, the false negatives, false positives, and true negatives, respectively. The STAPLE results yielded the best sensitivity and specificity combination for the algorithm ($Sens = 0.898, Spec = 0.905$).

Heydarian et al. [19] tested their method on two healthy subjects and two subjects with chronic obstructive lung disease (COPD). The results of hierarchical K-means 2D and 3D segmentation were compared to an expert observer's manual segmentation results using linear regression, Pearson correlations [11] and the Dice similarity coefficient. 2D hierarchical K-means segmentation of ventilation volume (VV) and ventilation defect volume (VDV) was strongly and significantly correlated with manual measurements (VV: $r = 0.98, P < 0001$; VDV: $r = 0.97, P < 0.0001$) and mean Dice coefficients were greater than 0.92 for all subjects. 3D hierarchical K-means segmentation of VV and VDV was also strongly and significantly correlated with manual measurements (VV: $r = 0.98, P < 0.0001$; VDV: $r = 0.64, P < 0.0001$) and the mean Dice coefficients were greater than 0.91 for all subjects.

Kirby et al. [26] applied the same approach for segmentation of ventilation defect volume (VDV) and ventilation volume (VV) on five patients with asthma, five patients with chronic obstructive pulmonary disease (COPD), and five patients with cystic fibrosis (CF). They compared semi-automatic and manual measurements and observed strong significant correlations between the VDV values generated by each method (asthma: $r = 0.89, P < 0.0001$; COPD: $r = 0.84, P < 0.0001$; CF:$r = 0.89, P < 0.0001$). The spatial agreement for VV values was measured with the Dice coefficient (asthma: 0.95; COPD: 0.88; CF: 0.9).

Middleton and Damper [39] applied their technique to 13 datasets. To evaluate the performance, they utilized the following region-based measures: precision (P), recall (R), effectiveness E [49], and segmentation performance $F = 1 - E$. The measures are defined as:

$$P = \frac{TP}{TP + FP}, \tag{11}$$

$$R = \frac{TP}{TP + FN}, \tag{12}$$

$$E = 1 - \frac{PR}{(1-\alpha)P + \alpha R}, \tag{13}$$

where $\alpha = 0.5$, i.e., the precision and recall are weighted equally. The mean of F was 0.866 and 0.844 for the left and right lungs, respectively.

Böttger et al. [9] applied their tool to 10 datasets and compared the semi-automatically obtained results to two expert readings. For evaluation, the average surface distance, Hausdorff distance, and the Dice coefficient of two compared segmentation were computed. For all segmentation, the average surface distance was significantly lower than 2 mm, and the Hausdorff distance was lower than 20 mm, apart from one outlier. The Dice coefficient was higher than 0.88.

4.2.2 Other Metrics

Plathow et al. [45] measured on a dynamic 2D MRI displacement of the chest wall and the diaphragm as a surrogate of the lung volume and lung surface in 20 healthy subjects. Moreover, a 3D volumetric evaluation of the breathing cycle was done, using a dynamic 3D MRI. The results were correlated to spirometry and the vital capacity (VC) [46] was measured. VC using spirometry was 4.3 ± 1.0 L; using the 2D model, VC was 4.9 ± 1.2 L. Using the 3D MRI VC was 4.65 ± 0.9 L. Correlation between spirometry and the 2D and 3D models was highly significant: the Spearman's correlation coefficient $r > 0.83$ and significance level $P < 0.005$ [11]. The differences of absolute VC values between spirometry, 2D and 3D MRI were insignificant.

Tokuda et al. [58] assessed lung motion sequences from two volunteers, the quality of segmentation was determined by visual observation of the original images.

Woodhouse et al. [66] measured the ventilated lung volumes from a combination of ^3He and proton single-shot fast spin echo (SSFSE) coronal MR images in groups of "healthy" smokers (five subjects), smokers with moderate COPD (five subjects), and never-smokers (eight subjects). The results (volume values) from slice-wise semi-automatic segmentation and manual segmentations from two observers were compared to each other. It showed high agreement (Pearson correlation coefficient $r \geq 0.94, P < 0.05$). The main source of disagreement occurred in the central five slices of each dataset, with the middle slice centered on the bifurcation of the trachea.

Virgincar et al. [64] developed their method for ^129Xe MR ventilation and anatomical MR images and tested in on a group of forty four participants: 24 healthy subjects, 10 subjects with COPD, 9 subjects with Global Initiative for Chronic Obstructive Lung Disease (GOLD), and 10 age-matched control (AMC) subjects. Ventilation images were quantified by two methods: the ventilation defect percentage (VDP), as the ratio between the thoracic cavity volume and ventilated volume, was computed from the semi-automatically segmented images; an expert computed the ventilation score percentage (VDS%). For the ventilation images, the intensity histograms from the thoracic cavity volume were analysed, and the coefficient of variation (CV) was computed there. The study showed that there was a correlation between VDS% and VDP ($r = 0.97, P < 0.0001$), and between VDS% and CV ($r = 0.82, P < 0.0001$).

Lelieveldt et al. [28, 29] tested their pipeline on 15 MR scans and assessed the total model matching. However, no separate evaluation of the lung segmentation results was done.

Kullberg et al. [27] used the segmented lungs only as a marker for further detection of adipose tissue in the body. No additional evaluation of the lung segmentation accuracy was done.

Tetzlaff et al. [57] tested their method on 10 datasets from healthy patients and compared the semi-automatically obtained volumes with the spirometric volumes. The comparison showed high agreement (mean Pearson correlation coefficient is ≥ 0.97).

5 Conclusion

The paper provides a review of existing automatic and semi-automatic methods for human lung segmentation from MR data. The categorization of the methods is done according to the main segmentation strategy. The approaches are related and compared to each other from the application point of view. Namely, such aspects as data types, which the techniques are designed for, the level of user involvement and the methods' complexity are discussed. Accuracy of the segmentation methods is crucial according to the nature of the work. The review of performance evaluation approaches is also done. Unfortunately, there is no standardized evaluation system, which would make the methods really comparable to each other. Moreover, for certain application areas there is no existing ground truth, which additionally complicates the evaluation.

References

1. Abdou, I.E., Pratt, W.K.: Quantitative design and evaluation of enhancement/thresholding edge detectors. Proc. IEEE **67**(5), 753–763 (1979)
2. Armato, S.G., MacMahon, H.: Automated lung segmentation and computer-aided diagnosis for thoracic ct scans. Int. Congr. Ser. **1256**, 977–982 (2003)
3. Avants, B.B., Tustison, N., Song, G.: Advanced normalization tools (ANTs). Insight J. **2**, 1–35 (2009)
4. Avants, B.B., Tustison, N.J., Song, G., Cook, P.A., Klein, A., Gee, J.C.: A reproducible evaluation of ants similarity metric performance in brain image registration. Neuroimage **54**(3), 2033–2044 (2011)
5. Biederer, J., Hintze, C., Fabel, M., Jakob, P., Horger, W., Graessner, J., Bolster, B., Heller, M.: MRI of the lung—ready ... get set ... go. Magnetom Flash **46**, 6–15 (2011)
6. Biederer, J., Beer, M., Hirsch, W., Wild, J., Fabel, M., Puderbach, M., Van Beek, E.: MRI of the lung (2/3). why ... when ... how? Insights Imaging **3**(4), 355–371 (2012)
7. Boettger, T., Kunert, T., Meinzer, H.P., Wolf, I.: Interactive constraints for 3d-simplex meshes. In: Proceedings of SPIE Medical Imaging 2005. Image Processing, vol. 5747, pp. 1692–1702. International Society for Optics and Photonics (2005)
8. Böttger, T., Grunewald, K., Schöbinger, M., Fink, C., Risse, F., Kauczor, H., Meinzer, H., Wolf, I.: Implementation and evaluation of a new workflow for registration and segmentation of pulmonary MRI data for regional lung perfusion assessment. Phys. Med. Biol. **52**(5), 1261 (2007)
9. Böttger, T., Kunert, T., Meinzer, H.P., Wolf, I.: Application of a new segmentation tool based on interactive simplex meshes to cardiac images and pulmonary MRI data. Acad. Radiol. **14**(3), 319–329 (2007)
10. Boykov, Y., Veksler, O., Zabih, R.: Fast approximate energy minimization via graph cuts. IEEE Trans. Pattern Anal. Mach. Intell. **23**(11), 1222–1239 (2001)
11. Cohen, J., Cohen, P.: Applied Multiple Regression/Correlation Analysis for the Behavioral Sciences. Lawrence Erlbaum, Hillsdale (1975)
12. Delingette, H.: General object reconstruction based on simplex meshes. Int. J. Comput. Vis. **32**(2), 111–146 (1999)
13. Devaki, K., MuraliBhaskaran, V.: Study of computed tomography images of the lungs: A survey. In: 2011 International Conference on Recent Trends in Information Technology (ICRTIT), pp. 837–842. IEEE (2011)

14. Dice, L.R.: Measures of the amount of ecologic association between species. Ecology **26**(3), 297–302 (1945)
15. Duda, R.O., Hart, P.E., Stork, D.G.: Pattern Classification. Wiley, New York (2000)
16. Fenchel, M., Requardt, M., Tomaschko, K., Kramer, U., Stauder, N.I., Naegele, T., Schlemmer, H.P., Claussen, C.D., Miller, S.: Whole-body MR angiography using a novel 32-receiving-channel MR system with surface coil technology: First clinical experience. J. Magn. Reson. Imaging **21**(5), 596–603 (2005)
17. Gonzalez, R., Woods, R.E.: Digital Image Processing. Prentice Hall, New York (2002)
18. Hegenscheid, K., Kühn, J., Völzke, H., Biffar, R., Hosten, N., Puls, R.: Whole-body magnetic resonance imaging of healthy volunteers: pilot study results from the population-based ship study. Rofo **181**(8), 748–759 (2009)
19. Heydarian, M., Kirby, M., Wheatley, A., Fenster, A., Parraga, G.: Two and three-dimensional segmentation of hyperpolarized ^{3}He magnetic resonance imaging of pulmonary gas distribution. In: Proceedings of SPIE Medical Imaging 2012, vol. 8317. International Society for Optics and Photonics (2012)
20. Horn, B.K.: Closed-form solution of absolute orientation using unit quaternions. J. Opt. Soc. Am. A **4**(4), 629–642 (1987)
21. Ibanez, L., Schroeder, W., Ng, L., Cates, J.: The ITK Software Guide. Kitware Inc., Clifton Park (2003)
22. Ivanovska, T., Hegenscheid, K., Laqua, R., Kühn, J.P., Gläser, S., Ewert, R., Hosten, N., Puls, R., Völzke, H.: A fast and accurate automatic lung segmentation and volumetry method for mr data used in epidemiological studies. Comput. Med. Imaging Graph. **36**(4), 281–293 (2012)
23. Jaccard, P.: The distribution of the flora in the alpine zone. 1. New Phytol. **11**(2), 37–50 (1912)
24. Kass, M., Witkin, A., Terzopoulos, D.: Snakes: Active contour models. Int. J. Comput. Vis. **1**(4), 321–331 (1988)
25. Kauczor, H.U.: MRI of the Lung. Springer, Berlin (2009)
26. Kirby, M., Heydarian, M., Svenningsen, S., Wheatley, A., McCormack, D.G., Etemad-Rezai, R., Parraga, G.: Hyperpolarized ^{3}He magnetic resonance functional imaging semiautomated segmentation. Acad. Radiol. **19**(2), 141–152 (2012)
27. Kullberg, J., Johansson, L., Ahlström, H., Courivaud, F., Koken, P., Eggers, H., Börnert, P.: Automated assessment of whole-body adipose tissue depots from continuously moving bed MRI: a feasibility study. J. Magn. Reson. Imaging **30**(1), 185–193 (2009)
28. Lelieveldt, B.P.F., van der Geest, R.J., Ramze Rezaee, M., Bosch, J.G., Reiber, J.H.C.: Anatomical model matching with fuzzy implicit surfaces for segmentation of thoracic volume scans. IEEE Trans. Med. Imaging **18**(3), 218–230 (1999)
29. Lelieveldt, B.P.F., Sonka, M., Bolinger, L., Scholz, T.D., Kayser, H.W.M., van der Geest, R.J., Reiber, J.H.C.: Anatomical modeling with fuzzy implicit surface templates: application to automated localization of the heart and lungs in thoracic MR volumes. Comput. Vis. Image Underst. **80**(1), 1–20 (2000)
30. Lichy, M.P., Mugler, B.M.W.J., Horger, W., Menzel, M.I., Anastasiadis, A., Siegmann, K., Niemeyer, T., Knigsrainer, A., Kiefer, B., Schick, F., Claussen, C.D., Schlemmer, H.P.: Magnetic resonance imaging of the body trunk using a single-slab, 3-dimensional, T2-weighted turbo-spin-echo sequence with high sampling efficiency (SPACE) for high spatial resolution imaging: initial clinical experiences. Investig. Radiol. **40**(12), 754–760 (2005)
31. Lorensen, W.E., Cline, H.E.: Marching cubes: a high resolution 3d surface construction algorithm. In: ACM Siggraph Computer Graphics, vol. 21, pp. 163–169. ACM, New York (1987)
32. Lowe, D.G., et al.: Fitting parameterized three-dimensional models to images. IEEE Trans. Pattern Anal. Mach. Intell. **13**(5), 441–450 (1991)
33. Lui, J.K., LaPrad, A.S., Parameswaran, H., Sun, Y., Albert, M.S., Lutchen, K.R.: Semi-automatic segmentation of ventilated airspaces in healthy and asthmatic subjects using hyperpolarized ^{3}He MRI. Comput. Math. Methods Med. **2013**, Article ID 624683 (2013)
34. MacQueen, J.B.: Some methods for classification and analysis of multivariate observations. In: Proceedings of 5-th Berkeley Symposium on Mathematical Statistics and Probability, vol. 1, pp. 281–297 (1967)

35. Marieb, E.N., Hoehn, K.: Human Anatomy and Physiology. Pearson Education, London (2007)
36. Martin Bland, J., Altman, D.: Statistical methods for assessing agreement between two methods of clinical measurement. Lancet **327**(8476), 307–310 (1986)
37. McGraw, K.O., Wong, S.: Forming inferences about some intraclass correlation coefficients. Psychol. Methods **1**(1), 30 (1996)
38. Memon, N.A., Mirza, A.M., Gilani, S.: Limitations of Lung Segmentation Techniques, vol. 27, chap. 76, pp. 753–766. Springer, Berlin (2009)
39. Middleton, I., Damper, R.I.: Segmentation of magnetic resonance images using a combination of neural networks and active contour models. Med. Eng. Phys. **26**(1), 71–86 (2004)
40. Mills, G., Wild, J., Eberle, B., Van Beek, E.: Functional magnetic resonance imaging of the lung. Br. J. Anaesth. **91**(1), 16–30 (2003)
41. Modersitzki, J.: Numerical Methods for Image Registration (Numerical Mathematics and Scientific Computation). Oxford university press, Oxford (2004)
42. Möller, H.E., Chen, X.J., Saam, B., Hagspiel, K.D., Johnson, G.A., Altes, T.A., de Lange, E.E., Kauczor, H.U.: Mri of the lungs using hyperpolarized noble gases. Magn. Reson. Med. **47**(6), 1029–1051 (2002)
43. Osher, S., Sethian, J.A.: Fronts propagating with curvature-dependent speed: algorithms based on hamilton-jacobi formulations. J. Comput. Phys. **79**(1), 12–49 (1988)
44. Plathow, C., Ley, S., Fink, C., Puderbach, M., Heilmann, M., Zuna, I., Kauczor, H.U.: Evaluation of chest motion and volumetry during the breathing cycle by dynamic MRI in healthy subjects: comparison with pulmonary function tests. Investig. Radiol. **39**(4), 202–209 (2004)
45. Plathow, C., Schoebinger, M., Fink, C., Ley, S., Puderbach, M., Eichinger, M., Bock, M., Meinzer, H.P., Kauczor, H.U.: Evaluation of lung volumetry using dynamic three-dimensional magnetic resonance imaging. Investig. Radiol. **40**(3), 173–179 (2005)
46. Pratt, J.H.: Long-continued observations on the vital capacity in health and heart disease. Am. J. Med. Sci. **164**(6), 819–831 (1922)
47. Ray, N., Acton, S.T., Altes, T., De Lange, E.E.: Mri ventilation analysis by merging parametric active contours. In: Proceedings. 2001 International Conference on Image Processing, vol. 2, pp. 861–864. IEEE (2001)
48. Ray, N., Acton, S.T., Altes, T., De Lange, E.E., Brookeman, J.R.: Merging parametric active contours within homogeneous image regions for MRI-based lung segmentation. IEEE Trans. Med. Imaging **22**(2), 189–199 (2003)
49. van Rijsbergen, C.J.: Information Retrieval. Butterworth, London (1979)
50. Sensakovic, W.F., Armato III, S.G.: Magnetic resonance imaging of the lung: automated segmentation methods. In: General Methods and Overviews, Lung Carcinoma and Prostate Carcinoma, pp. 219–234. Springer, Berlin (2008)
51. Sensakovic, W.F., Armato III, S.G., Starkey, A.: Automated lung segmentation in magnetic resonance images. In: Proc. SPIE **5747**, 1776–1781 (2005)
52. Sensakovic, W.F., Armato III, S.G., Starkey, A., Caligiuri, P.: Automated lung segmentation of diseased and artifact-corrupted magnetic resonance sections. Med. Phys. **33**, 3085 (2006)
53. Sluimer, I., Schilham, A., Prokop, M., van Ginneken, B.: Computer analysis of computed tomography scans of the lung: a survey. IEEE Trans. Med. Imaging **25**(4), 385–405 (2006)
54. Sodickson, D.K., McKenzie, C.A., Ohliger, M.A., Yeh, E.N., Price, M.D.: Recent advances in image reconstruction, coil sensitivity calibration, and coil array design for smash and generalized parallel MRI. Magn. Reson. Mater. Phys., Biol. Med. **13**(3), 158–163 (2002)
55. Soille, P.: Morphological Image Processing: Principles and Applications. Cambridge University Press, Cambridge (1999)
56. Sonka, M., Hlavac, V., Boyle, R.: Image Processing, Analysis, and Machine Vision. Thomson, Toronto (2008)
57. Tetzlaff, R., Schwarz, T., Kauczor, H.U., Meinzer, H.P., Puderbach, M., Eichinger, M.: Lung function measurement of single lungs by lung area segmentation on 2d dynamic MRI. Acad. Radiol. **17**(4), 496–503 (2010)

58. Tokuda, J., Schmitt, M., Sun, Y., Patz, S., Tang, Y., Mountford, C.E., Hata, N., Wald, L.L., Hatabu, H.: Lung motion and volume measurement by dynamic 3d MRI using a 128-channel receiver coil. Acad. Radiol. **16**(1), 22–27 (2009)
59. Tustison, N.J., Gee, J.C.: N4itk: Nick's n3 itk implementation for MRI bias field correction. Insight J. (2009)
60. Tustison, N.J., Avants, B.B., Flors, L., Altes, T.A., de Lange, E.E., Mugler, J.P., Gee, J.C.: Ventilation-based segmentation of the lungs using hyperpolarized ^{3}He MRI. J. Magn. Reson. Imaging **34**(4), 831–841 (2011)
61. Udupa, J.K., LaBlanc, V.R., Schmidt, H., Imielinska, C., Saha, P.K., Grevera, G.J., Zhuge, Y., Currie, L.M., Molholt, P., Jin, Y.: Methodology for evaluating image-segmentation algorithms. Med. Imaging 2002 Image Process. **4684**(1), 266–277 (2002)
62. van Beek, E.J., Wild, J.M., Kauczor, H.U., Schreiber, W., Mugler, J.P., de Lange, E.E.: Functional MRI of the lung using hyperpolarized 3-helium gas. J. Magn. Reson. Imaging **20**(4), 540–554 (2004)
63. Vincent, L., Soille, P.: Watersheds in digital spaces: An efficient algorithm based on immersion simulations. IEEE Trans. Pattern Anal. Mach. Intell. **13**(6), 583–598 (1991)
64. Virgincar, R.S., Cleveland, Z.I., Sivaram Kaushik, S., Freeman, M.S., Nouls, J., Cofer, G.P., Martinez-Jimenez, S., He, M., Kraft, M., Wolber, J., et al.: Quantitative analysis of hyperpolarized 129Xe ventilation imaging in healthy volunteers and subjects with chronic obstructive pulmonary disease. NMR Biomed. **26**(4), 424–435 (2012)
65. Warfield, S.K., Zou, K.H., Wells, W.M.: Simultaneous truth and performance level estimation (staple): an algorithm for the validation of image segmentation. IEEE Trans. Med. Imaging **23**(7), 903–921 (2004)
66. Woodhouse, N., Wild, J.M., Paley, M.N., Fichele, S., Said, Z., Swift, A.J., van Beek, E.J.: Combined helium-3/proton magnetic resonance imaging measurement of ventilated lung volumes in smokers compared to never-smokers. J. Magn. Reson. Imaging **21**(4), 365–369 (2005)
67. Wu, N.Y., Cheng, H.C., Ko, J.S., Cheng, Y.C., Lin, P.W., Lin, W.C., Chang, C.Y., Liou, D.M.: Magnetic resonance imaging for lung cancer detection: experience in a population of more than 10,000 healthy individuals. BMC Cancer **11**(1), 242 (2011)

Fast Uncertainty-Guided Fuzzy C-Means Segmentation of Medical Images

Ahmed Al-Taie, Horst K. Hahn, and Lars Linsen

Abstract Image segmentation is a crucial step of the medical visualization pipeline. In this paper, we present a novel fast algorithm for modified fuzzy c-means segmentation of MRI data. The algorithm consists of two steps, which are executed as two iterations of a fuzzy c-means approach: the first iteration is a standard fuzzy c-means (FCM) iteration, while the second iteration is our modified FCM iteration with misclassification correction. In the second iteration, we use the classification probability vectors (uncertainties) of the neighbor pixels found by the first iteration to confirm or correct the classification decision of the current pixel. The application of the proposed algorithm on synthetic data, simulated MRI data, and real MRI data show that our algorithm is insensitive to different types of noise and outperforms the standard FCM and several versions of modified FCM algorithms in terms of accuracy and speed. In fact, our algorithm can easily be combined with many modified FCM algorithms to improve their segmentation result while reducing the computation costs (using two FCM iterations only). An optional simple post-processing step can further improve the segmentation result by correcting isolated misclassified pixels. We also show that our algorithm reduces the uncertainty in the segmentation result, by using recently proposed uncertainty estimation and visualization tools.

A. Al-Taie (✉)
Jacobs University, Bremen, Germany

Baghdad University, Baghdad, Iraq
e-mail: a.altaie@jacobs-university.de

H.K. Hahn
Fraunhofer MEVIS, Bremen, Germany

Jacobs University, Bremen, Germany
e-mail: horst.hahn@mevis.fraunhofer.de

L. Linsen
Jacobs University, Bremen, Germany
e-mail: l.linsen@jacobs-university.de

L. Linsen et al. (eds.), *Visualization in Medicine and Life Sciences III*, Mathematics and Visualization, DOI 10.1007/978-3-319-24523-2_2

1 Introduction

Image segmentation plays an essential role in a variety of applications such as pattern recognition, geographical imaging, remote sensing, and medical imaging. In image segmentation, images are partitioned into disjoint regions of homogeneous properties. In the medical imaging context, it is an important step in medical research and clinical applications, e.g., in visualization and analysis of anatomical structures, multi-modality fusion and registration, pathology detection, and surgical planning.

Traditionally, domain experts were required to manually segment images. Even for 2D images, this is a time-consuming process. In recent years, a variety of semi- and fully automatic techniques have been developed to address the segmentation problem. Magnetic resonance imaging (MRI) is one of the most used medical imaging techniques due to its ability to distinguish soft tissues and due to not using x-rays. However, it suffers from artifacts that make the automatic segmentation task challenging. MR images of a brain have a particularly complicated structure and contain artifacts such as noise, intensity in-homogeneity, and partial volume effect. When segmenting brain images, the regions of interest are typically white matter (WM), gray matter (GM), and cerebrospinal fluid (CSF). Several studies show that changes in the composition of these tissues in the whole volume or within specific regions can be used to characterize physiological processes and disease entities [1] or to characterize disease severity [2].

Fuzzy c-means (FCM) and its modified versions are among the most used algorithms for image segmentation [3–10]. Their main advantages include the straightforward implementation, the applicability to multichannel data, its robustness in the absence of prior knowledge about cluster centers, its robustness when the number of classes is known (as in the problem of segmenting the human brain image), and the ability to model the uncertainty within the data. FCM algorithms assign pixels to fuzzy clusters without labels. Unlike the non-probabilistic clustering methods which classify pixels to belong exclusively to one class, FCM is more flexible by allowing pixels to belong to multiple cluster with varying degrees of membership. However, the existence of the different artifacts in MR images causes problems for the standard FCM approach. Therefore, many attempts have been made to modify the standard FCM in order to improve the results. These modification include replacing the Euclidian distance of the standard FCM with other measures [3, 6, 7], decomposing the observed MRI signal into two components containing the true values and bias field artifacts [4, 8], and enforcing spatial constraints on estimating the partition vector of the current pixel to reduce the effect of noise [5, 6, 9]. Tolias et al. [11] suggest to apply spatial constraints using fuzzy logic on the probabilistic segmentation. Although the modified FCM algorithms decrease the sensitivity to noise to some extent, they still lack enough robustness to noise or outliers (especially in the absence of prior knowledge of the noise) and require a large number of iterations to converge. Furthermore, recent studies showed that even the most sophisticated segmentation algorithms require further post-segmentation expert editing, which is usually done in a time-consuming manual process relying largely on visual inspection [12, 13].

In this paper, we present a novel fast modified FCM algorithm for segmenting brain MRI data. We propose a two-step algorithm that embeds fuzzy logic decisions with the structure of an FCM algorithm. Each of the two steps executes one FCM iteration, i.e., the entire algorithm consists of only two FCM iterations. While the first iteration is a standard FCM step, the second iteration exploits the fuzzy partition matrix produced by the first step to adaptively correct misclassifications, see Sect. 4. Hence, we call our variant of FCM fast adaptive FCM (FAFCM). Our approach is not limited to being based on the standard FCM approach. It can be used to improve most FCM variants. We combine our approach with several existing variants of the FCM approach such as the modified fuzzy c-means (mFCM) [3], the spatial FCM (SFCM) [9], the kernelized FCM (KFCM), and the spatial kernelized FCM approach (SKFCM) [7], see Sect. 3.

The experimental results of applying the proposed algorithm on synthetic data corrupted with different types of noise, simulated brain MRI data, and real brain MRI data show that our methods outperform the standard FCM and its variants in terms of segmentation accuracy and computational complexity, see Sect. 5 and 6. We also show that a simple postprocessing step can further improve the results by fixing misclassifications of the isolated pixels. Since the fuzzy logic we apply is related to the uncertainty in the classification, we further analyze the uncertainty in our segmentations. We use uncertainty measures that have recently been shown as being able to serve as segmentation quality measures in numerical and visual inspections [14]. We evaluate our segmentation results with respect to the uncertainty estimates numerically and visually. The latter uses color-coded uncertainty visualization techniques.

2 Related Work

Many FCM approaches have been proposed recently that tackle the issue of artifacts in medical imaging to improve the segmentation quality [3–10, 15]. Many of the paper address segmentation of MRI data, as MR imaging has become very popular.

We can group these FCM variants into the following categories: (1) Approaches that modify the used distance metric (e.g., [3, 6, 7]). (2) Approaches that modify the observed image model to compensate for bias field or noise artifacts (e.g., [4, 8]). (3) Approaches that include spatial constraints to adjust the observed feature (intensity), the measured distance, or the partition matrix probabilities of the considered pixel by the involvement of neighbor pixels with fixed or varying weights (e.g., [5, 9, 10, 16–18]). (4) Approaches that modify the minimization of the objective function domain by considering the frequencies of observed intensities (e.g., [10, 19]).

Mohammed et al. [3] use a modified version of the Euclidean-distance approach that involves the membership value and the spatial distance of the neighboring pixels to make the method less sensitive to noise. Ahmed et al. [4] proposed an algorithm to compensate for inhomogeneity and to label a pixel considering its immediate neighborhood. This algorithm is called bias-corrected fuzzy c-means (BCFCM). The BCFCM method is a common reference for most of the modified fuzzy c-means algorithms that were introduced later and its objective function is used as a basis for their modifications. Yuan et al. [8] enhanced the method by replacing the fixed value of the alpha parameter that controls the influence of the neighborhood voxels during optimization with a weighted function that varies according to the closeness of the intensity of each neighboring pixel to the considered pixel.

We want to point out a few important FCM approaches. Pham [5] generalizes the FCM objective function by including a spatial penalty term on the membership function to be more robust against noise and other imaging artifacts. Chuang et al. [9] follow up in utilizing the spatial information in the image to further improve the segmentation performance. They use a spatial function in the weighted summation of membership function of the neighborhood pixels to adjust the membership value of the considered pixel, which leads to more homogeneous regions and removes noisy spots.

Zhang and Chen [6] proposed two variants of the algorithm by Ahmed et al. [4] to simplify its computation and replace the Euclidean distance with a kernel-induced distance to construct a non-linear version of the linear algorithm [6, 7]. Furthermore, they added a spatial penalty term to act as regularizer. Szilágyi et al.[19] introduced the enhanced fuzzy c-means algorithm (EnFCM) that modifies the FCM algorithm such that the objective function is based on the number of intensity levels available in the image but not on the number of pixels. This modification led to a very fast algorithm that slightly improves the segmentation result. Cai et al. [10] proposed the fast generalized fuzzy c-means algorithm (FGFCM) as an improvement to the enhanced fuzzy c-means algorithm (EnFCM). They use a new factor S_{ij} as a local (both spatial and intensity) similarity measure aiming to for noise-immunity while preserving details. However, the algorithm is still inaccurate at segment boundaries.

Tolias et al. [11] suggest a non-iterative algorithm to apply spatial constraints using fuzzy logic on top of fuzzy segmentation to improve the segmentation result. This approach improves the segmentation results to some extent, but still it has drawbacks: First, it is applied to the fuzzy segmentation results, i.e., it introduces additional computations. Second, it inherits the drawbacks of the fuzzy segmentation. Third, it corrupts the probability vector condition of the fuzzy partition matrix, as the algorithm updates the membership values according to the applied constraints, i.e., the resulting probabilities do not sum up to 1 anymore. Recent studies [12, 14, 20, 21] show that this probability vector condition is important for estimating and visualizing the uncertainty associated with the probabilistic segmentation result, which is useful for drawing conclusions from the visualizations.

Li et al. proposed a new energy minimization method based on coherent local intensity clustering (CLIC) for simultaneous tissue classification and bias field estimation of MR images [16]. For each pixel, a clustering criterion function is defined with the clustering centers replaced by the product of the bias within a neighborhood and the clustering centers. In addition, a weight for each pixel is introduced to control its influence on the clustering criterion function. A pixel at a location far from the neighborhood center should have less influence on the clustering criterion function than the closer pixels. Integration of the clustering criterion function leads to an energy, whose minimization leads to simultaneous tissue classification and bias field estimation. The CLIC algorithm is sensitive to the choice of the parameter of the weighting function. In addition, the spatially coherent nature of the CLIC criterion function that omits the local gray level relationship makes the estimated membership functions inaccurate, and finally cause misclassifications, especially for the pixels around the boundaries. In order to reduce the noise effect during segmentation, Wang et al. [17] introduced a method that incorporates both the local spatial context and the non-local information into the standard FCM algorithm using a dissimilarity index instead of the usual distance metric. Ji et al. [18] follow up on Wang et al.'s approach by using the local spatial context and non-local information to develop the possibilistic fuzzy c-means clustering algorithm PFCM. Unfortunately, the two methods by Wang et al. and Ji et al. are computationally expensive, which makes them impractical.

Although the modified FCM algorithms reduce the sensitivity to noise to some extent, they still lack enough robustness to noise or outliers and they require, at the same time, a large number of iterations in order to converge. This can have different reasons: (1) A method is modeled to be robust against one type of noise and omit the other types. (2) A method treats all neighbor pixels in a similar way, i.e., it disregards whether some pixels are noisy or outliers with respect to the considered pixel or its neighbors. (3) A method treats all considered pixels in a similar way, i.e., it disregards whether the pixel is affected by noise or an outlier and thereby should be corrected before being treated.

To overcome such obstacles a method should be adaptive and dynamic to deal with the different situations of the considered pixels in different ways based on meaningful (smart) interpretations of the observed spatial information, especially in the presence of a partition matrix as a measure of the degree of membership which can be exploited to correct the misclassified decision. In addition to the above drawbacks, the FCM algorithms with spatial constraints typically require more time in each iteration for including the neighbor information in their computations and they may require more iterations to converge. Our proposed method always requires only two iterations. In the first iteration the normal standard fuzzy c-means iteration is used to produce an initial estimation of the fuzzy segmentation. Based on the uncertainty information (stored in the probability vectors of the partition matrix), we use an adaptive heuristic filter the partition matrix or adjust pixel intensity values. The heuristic considers spatial information in form of neighborhoods, probability values of the considered pixel and its neighborhood, and majority consideration of the classification within the neighborhood. Since our approach is based on the estimates found by the FCM algorithm, which can be interpreted as probabilities or uncertainties, respectively, we call our approach uncertainty-guided FCM. The uncertainties are used to guide the algorithm towards the optimal solution using a set of heuristic rules. In contrast to other modified FCM approaches, our approach distinguishes among neighborhood pixels that should be participating in correcting the initial decision based on partition matrix probabilities. This is following the ideas of Tolias et al. [11], but in contrast to them our approach incorporates the fuzzy logic of applying spatial constraints to work within the FCM algorithm structure and conditions. Our proposed approach can be easily combined with other modified FCM approaches for improving them in terms of speed and accuracy, which we document with our results in this paper.

3 FCM Methods

In this section the standard fuzzy c-means (FCM) algorithm with some related versions of modified FCM are described. These variants are described in some detail, as we use them for generating our results. All variants can be used in our approach to improve their performances.

3.1 Standard FCM

The standard fuzzy c-means (FCM) approach is an iterative, unsupervised algorithm initially introduced by Bezdek [22]. The FCM objective function for partitioning an image $\{x_k\}_{k=1}^{N}$ with N pixels or voxels into c clusters is given by

$$J_{FCM} = \sum_{i=1}^{c} \sum_{k=1}^{N} u_{ik}^{m} D_{ik},$$

where $\{v_i\}_{i=1}^{c}$ are the representatives of the clusters, $D_{ik} = \|x_k - v_i\|^2$ is the squared distance between v_i and the intensity of pixel k (x_k), and the matrix $[u_{ik}] = U$ represents a partition matrix of size $c \times N$ with $U \in \mathscr{U}$ and

$$\mathscr{U} = \{u_{ik} \in [0,1] | \sum_{i=1}^{c} u_{ik} = 1 \ \forall k \wedge 0 < \sum_{k=1}^{N} u_{ik} < N \ \forall i\}.$$

The entries u_{ik} can be interpreted as the probability that pixel x_k belongs to cluster i. This is referred to as fuzzy membership. The memberships are updated iteratively based on the distance D_{ik} according to

$$u_{ik} = \left(\sum_{j=1}^{c} \left(\frac{D_{ik}}{D_{jk}} \right)^{\frac{2}{(m-1)}} \right)^{-1}.$$

The representatives of the clusters are updated using the formula

$$v_i = \frac{\sum_{k=1}^{n} u_{ik}^m x_k}{\sum_{k=1}^{n} u_{ik}^m}.$$

The parameter m is a weighting exponent on each fuzzy membership and determines the amount of fuzziness of the resulting classification. The algorithm is initialized with some representatives v_i for each cluster and the memberships are updated to minimize the objective function J_{FCM}.

3.2 *Modified FCM (mFCM)*

Mohamed et al. [3] proposed a modified FCM algorithm that replace the distance D_{ik} with one that considers the membership values and the spatial distance of the neighbor pixels. The modified D_{ik} distance is computed by

$$D'_{ik} = D_{ik}\big(1 - \alpha \frac{\sum_{j \in neighbors} u_{ij} \cdot p_{kj}}{\sum_j p_{kj}}\big),$$

where α is a constant that satisfies the condition $0 \leq \alpha \leq 1$. p_{kj} measures the proximity of pixel k to its neighbor pixel j. The proximity here is measured with respect to the relative locations between the two pixels $p_{kj} = \|k - j\|_2$. This variant of FCM considers D_{ik} as the resistance of pixel x_k to be clustered with class i. The membership values u_{ij} of the neighboring pixels x_j with cluster i were incorporated in the D_{ik} computation to tolerate its resistance.

3.3 *Spatial FCM (SFCM)*

Chuang et al. [9] proposed a fuzzy c-means algorithm that incorporates spatial information into the membership function for clustering. To exploit the spatial information, the spatial function is defined as

$$h_{ik} = \sum_{j \in N(x_k)} u_{ij},$$

where $N(x_k)$ represents a square window centered on pixel x_k in the spatial domain. The spatial function is incorporated into the membership function by

$$u'_{ik} = \frac{u_{ik}^p h_{ik}^q}{\sum_{j=1}^{c} u_{ij}^p h_{ij}^q},$$

where p and q are parameters to control the relative importance of both functions. Incorporating the spatial information into the membership function aims at generating segmentations with more homogeneous regions and at removing noisy spots.

3.4 Simplified FCM with Spatial Information

Zhang and Chen [6] proposed two variants of a simplified version of Ahmed et al.'s [4] objective function (by omitting the log transform and the bias field estimation)

$$J_m = \sum_{i=1}^{c}\sum_{k=1}^{N} u_{ik}^{m}\|x_k - v_i\|^2 + \frac{\alpha}{|N_k|}\sum_{i=1}^{c}\sum_{k=1}^{N} u_{ik}^{m}\Big(\sum_{x_r \in N_k}\|x_r - v_i\|^2\Big).$$

The two variants FCMS1 and FCMS2 aim to simplify the computation of J_m by replacing the term $\frac{1}{|N_k|}\sum_{x_r \in N_k}\|x_r - v_i\|^2$ with $\|\bar{x}_k - v_i\|^2$ where $\bar{x}_k$ is the mean for FCMS1 and the median for FCMS2 of neighboring pixels lying within a window around x_k:

$$J_{FCMS} = \sum_{i=1}^{c}\sum_{k=1}^{N} u_{ik}^{m}\|x_k - v_i\|^2 \alpha \sum_{i=1}^{c}\sum_{k=1}^{N} u_{ik}^{m}\|\bar{x}_k - v_i\|^2.$$

3.5 Kernelized FCM (KFCM) and Spatially Constrained KFCM (SKFCM)

Zhang and Chen [6, 7] proposed another variant that replaced the Euclidean distance in the standard FCM algorithm by a kernel-induced distance to construct a nonlinear version of the linear algorithm called the Kernelized FCM (KFCM) [6, 7]:

$$J_{KFCM} = \sum_{i=1}^{c}\sum_{k=1}^{N} u_{ik}^{m}\|\Phi(x_k) - \Phi(v_i)\|^2,$$

where Φ is an implicit nonlinear map from data space to the mapped feature space, $\Phi : X \to F(x \to \Phi(x))$. Using a Gaussian radial basis function $K(x, y) = exp\Big(\frac{-\|x-y\|^2}{\sigma^2}\Big)$ as kernel, they simplified the objective function to

$$J_{KFCM} = 2\sum_{i=1}^{c}\sum_{k=1}^{N} u_{ik}^{m}(1 - K(x_k, v_i)),$$

and derive

$$u_{ik} = \frac{\left(1 - K(x_k, v_i)\right)^{-1/(m-1)}}{\sum_{j=1}^{c} \left(1 - K(x_j, v_j)\right)^{-1/(m-1)}},$$

as well as

$$v_i = \frac{\sum_{k=1}^{n} u_{ik}^m K(x_k, v_i) x_k}{\sum_{k=1}^{n} u_{ik}^m K(x_k, v_i)},$$

as the two necessary but not sufficient conditions for J_{KFCM} to be at its local extrema. Furthermore, they introduced the spatially constrained KFCM (SKFCM) by adding a spatial penalty term to act as regularizer and to bias the solution towards piecewise-homogeneous labeling. The modified objective function is given by

$$J_{SKFCM} = \sum_{i=1}^{c} \sum_{k=1}^{N} u_{ik}^m (1 - K(x_k, v_i)) + \frac{\alpha}{|N_k|} \sum_{i=1}^{c} \sum_{k=1}^{N} u_{ik}^m \sum_{r \in N_k} (1 - u_r)^m,$$

where N_k is the set of neighbor pixels in a window around x_k.

4 Fast Adaptive FCM (FAFCM)

Although the modified FCM algorithms improve noise handling to some extent, they still lack enough robustness to noise or outliers and require a large number of iterations in order to converge. The main conceptual drawback of modified FCM algorithms with spatial constraints is the similar treatment of all neighbor pixels even if they belong to different classes than the considered pixel or if they are outliers. The considered pixel itself may be an outlier and should be corrected to be ready for accurate classification. Typically, pixel intensities are being adjusted using the (weighted) mean of neighbor pixels intensities, which in certain constellations may create even new outlier classes. For example, the mean filter used in [6] when applied across segment boundaries may generate a new class that does not belong to any of the classes in the region. This effect is similar to the partial volume effect (PVE) (and, consequently, we refer to as the artificial PVE). The constellation may even be worse, if the pixel at the boundary is affected by true PVE artifacts, which is typically the case.

To overcome these drawbacks of FCM and its variants, we propose a two-step algorithm that combines FCM steps with fuzzy logic similar to that used by Tolias et al. [11] but we propose to use it within the framework of the FCM algorithm not in a post-processing step. This integration into the FCM algorithm keeps the fuzzy partition matrix condition of FCM uncorrupted (i.e., the membership values sum up to 1). The first step of the proposed method is the standard FCM iteration. The purpose of this iteration is to produce an initial classification and an initial fuzzy partition matrix to be used as the basis for an adaptive heuristic filter in the second step. The second step uses the heuristic within an adaptive FCM iteration. Hence, the entire algorithm only requires the execution of two FCM iterations. The purpose of the second iteration is to reduce the misclassified pixels or improve the membership values of the correctly classified pixels. The spatial constraints of our algorithm differ from others in using neighborhood pixels that are highly correlated to the considered pixel. If the set of correlated neighbor pixels is empty, the considered pixel is an outlier pixel and needs to be corrected. We correct it by setting its intensity to the mean of the neighborhood majority *Mj*, where the neighborhood majority is the largest set of pixels in the neighborhood that belong to the same class (the class with highest probability). If the set of correlated neighbor pixels is not empty and represents the neighborhood majority with the majority matches the class of the considered pixel, the correctness of the classification of the considered pixel is confirmed. However, the intensity of the considered pixel is replaced with the mean of the correlated set plus the pixel itself for more homogeneity.

Based on the level of degradation (regardless of the reason) on the pixel intensity, our approach distinguishes three kinds of observed pixels: the outlier, the noisy, and the normal pixels. Since the classification process of the FCM depends mainly on the distance $D_{ik} = \|x_k - v_i\|^2$, our approach adjusts the distance computation when treating each of the three cases according to an adaptive heuristic filter that applies fuzzy rules to the pixel and its neighborhood.

- **The outlier pixel:**We refer to an outlier pixel as a pixel whose intensity differs significantly from all neighborhood pixels. Such pixels may be produced by salt and pepper noise. To correct outlier pixels, their intensity is replaced by the mean of the neighborhood majority (see above).
- **The noisy pixel:** We refer to noisy pixels as those that are affected by a moderate level of degradation that deviate from their true cluster centroid towards the closest cluster centroid. The PVE area is a good example for such pixels. Such noisy pixels typically have a probability vector with two large entries, i.e., the uncertainty is high. Depending on which of the two large entries is the largest one, the pixel may or may not be correctly classified. Our heuristic solution is as follows: If the considered pixel has a high probability in a class that matches the neighborhood majority, we replace the intensity of the pixel by the mean intensity of the pixel itself and the pixels of the neighborhood majority. If the majority class matches the true class of the pixel, then there are two cases that can occur: If the pixel was correctly classified, our adjustment creates a higher homogeneity of the pixel with its neighborhood. If the pixel was misclassified, the classification will be corrected. If the majority class does not match the true class of the pixel, there is still the possibility that the pixel is misclassified after our step, but this happens rarely. Classes with low values in the probability vector do not affect the pixel. Now, we want to discuss the situation of a noisy misclassified pixel regarding its position, i.e., whether it lies inside the segment or at its boundary. For noisy pixels inside the segment, e.g., the ones

affected by Gaussian noise, we expect the true class to have the second-highest probability. If the neighborhood majority matches the class ranked second, which is assumed to be the true class, and if this class has a sufficiently high probability, our approach corrects the misclassification. For noisy pixels on the segment boundary, there are two cases. The first case is that the pixel belongs to one class at the boundary and is misclassified to belong to the class on the other side. If the majority neighborhood matches the true class, the misclassification is corrected. Otherwise, the misclassification remains. Hence, on average, about half of the cases can be fixed. The second case occurs when the pixel is misclassified to belong to a third class different from the main two classes at the boundary, e.g., because of PVE. Here the neighborhood majority belongs to one of the two classes next to the boundary and the pixel is moved to that class. Hence, on average, it will be corrected to the true class, again, in about half of the cases.

- **The normal pixel:** We refer to normal pixels as the one with a low level of degradation, which is usually surrounded by pixels with similar intensities. For normal pixels, we apply the same strategy as for the noisy pixels, i.e., we adjust the intensity to be the mean intensity of the considered pixel and the neighborhood majority.

In the proposed algorithm, we use two thresholds: the center pixel threshold θ_C and the neighbor pixel threshold θ_{N_p}. The pixel k has a high probability to belong to class i if its membership value U_{ik} is greater than the center pixel threshold θ_C. The neighbor pixel j has a high probability to belong to class i if its membership value U_{ij} is greater than the neighbor pixel threshold θ_{N_p}.

The Algorithm In order to improve the performance of FCM segmentation for outlier, noisy, and normal pixels, we modify the computation of distance D_{ik} by correcting the observed intensity value x_k for pixel k using the following adaptive rules: Let x_k be the observed intensity for pixel k, v_i the centroid for cluster i, N_p the set of neighbor pixels, and $|N_p|$ its cardinality. Let $N_{p,i} = \{x_j \in N_p : U_{ij} > \theta_{N_p}\}$. Let Mj represents the set of pixels that belong to the neighborhood majority, and $|Mj|$ is its cardinality.

The D_{ik} Modification

if ($U_{ik} > \theta_C$) then

/*rule 1:for outlier pixel */

if ($|N_{p,i}| = 0$) then

$$x_k = \frac{\sum_{x_m \in Mj} x_m}{|Mj|}$$

/*rule 2:for noisy pixel case1 and normal pixel */

else if ($|N_{p,i}| > \frac{1}{2}|N_p|$) then

$$x_k = \frac{\sum_{x_j \in N_p} x_j + x_k}{|N_{p,i}|+1}$$

end if

end if

end if

$D_{ik} = \|x_k - v_i\|^2$.

As described above, we only need two iterations to compute the overall result. Of course, we could apply further iterations steps, but those additional computations would not or only slightly improve the result. This is the case because our heuristic rules re-compute the membership value of the current pixel based on the mean of the majority class (or mean of the pixel and the majority class) in the neighborhood window. Since the majority class is rarely changed and consequently neither is its mean (because the intensities of pixels are not changed), conducting more iterations does not improve the solution anymore in most cases. On the other hand, our method is stable as further iterations do not increase the error. In fact, the method converges quite quickly to it solution. We can also apply our second iteration (the modified iteration) on top of any FCM algorithm, but then we inherit the drawback of these algorithms and consume more time for more iterations.

Furthermore, we combine several modified FCM techniques with our approach (FAFCM) for further improvements, where the combined algorithms always require two iterations only. The first iteration is just a modified FCM iteration, while the second iteration uses our uncertainty-guided heuristics as described above. We combine the modified fuzzy c-means (mFCM) approach [3] to produce the fast adaptive modified fuzzy c-means (FAmFCM) algorithm, the spatial FCM (SFCM) [9] to produce the fast adaptive spatial FCM (FASFCM) algorithm (using $p = 1, q = 1$), the kernelized FCM (KFCM) to produce the fast adaptive KFCM (FAKFCM), and the spatial kernelized FCM (SKFCM) [7] to produce the fast adaptive SKFCM (FASKFCM).

5 Results and Discussions

We present and compare the experimental results when applying the standard FCM, the proposed methods, and several versions of the modified FCM algorithms on synthetic images, simulated MRI brain images [23], and real MRI T1 weighted brain images [24]. The reasons for using simulated images are the knowledge of the ground truth and the control over image parameters such as mean intensity values, noise, and intensity inhomogeneities.

To somehow mimic the main brain tissues of MRI T1 and T2 images in a synthetic image (i.e., background Bg, white matter WM, gray matter GM, and cerebrospinal fluid CSF), we generate an example of four respective classes with complex structures as shown in Fig. 1a. We believe that our examples mimic the structures better than the two-classes synthetic image in [4] or [6] and the four-classes synthetic image in [9]. We corrupted our synthetic image with different types of noise that are common in medical data such as Gaussian, salt-and-pepper, or sinusoidal noise, or a combination of them.

As initialization for clusters' representatives in the synthetic example, we use the cluster centroids of the synthetic image before corrupting it with noise. For the experiments on the simulated and real brain MRI images, we initialize the representatives of clusters based on the segmentation result of the EnFCM algorithm

[19] which is very fast and gives good starting points. This choice is suitable for experiments assuming the absence of ground truth and prior knowledge.
In order to compare the performance of the proposed methods with the other methods, we employed the segmentation accuracy *SA* measure to evaluate the differences among them:

$$SA = \frac{\text{number of correctly classified pixels}}{\text{total number of pixels}}.$$

In the following experiments, we compare the proposed methods (FAFCM and its variants) with: the standard FCM [22], the modified fuzzy c-means (MFCM) [3], the spatial fuzzy c-means (SFCM) [9], the kernelized fuzzy c-means (KFCM) and its spatial version, the spatial kernelized fuzzy c-means (SKFCM) [6, 7] (using $p = 1, q = 1$), the two variants of simplified fuzzy c-means methods: the (FCMS1) using mean filter, and (FCMS2) using median filter [6], the (CLIC) algorithm [16], the enhanced fuzzy c-means algorithm (EnFCM) [19] and its variant the fast generalized fuzzy c-means (FGFCM) [10], and finally the fuzzy rule based system (FRBS) [11]. Since the FRBS is applied on top of the FCM result, we consider the number of iterations of the standard FCM as the number of iterations for FBRS.

In the first experiment, we use the synthetic image of Fig. 1a corrupted by a mixture of Gaussian noise with mean 0 and variance 64, 5 % of salt-and-pepper noise, and a sinusoidal noise that models bias field errors. The result of applying the proposed methods and the other methods to this image shows how our improvements outperform the standard fuzzy c-means and all other variants, see Fig. 1 and Table 1.

The numbers in Table 1 indicate that the segmentation accuracy is higher for all the proposed methods when compared to the other methods, although our algorithms need only two iterations. Figure 2a shows a visual version of the comparison in Table 1 that combine the segmentation accuracy with the number of iterations in one plot. To show that this is also true for other combinations of noise, the plots in Fig. 2b–e show both the segmentation accuracy and the number of iterations comparison for synthetic image corrupted with individual noise and a combination of mixed noise. With a few exceptions (when only applying salt-and-pepper noise), it is clear that the proposed methods outperform all other methods. Furthermore, the plots show that our methods are more robust against any combinations of mixed noise than the others and that the other methods almost only perform well when the FCM already performed well. We can see from this result that the proposed methods have some difficulties for the synthetic image corrupted by salt and pepper noise (SaP). For that reason, we developed a simple post-segmentation noise removal step. This step is applied for post-processing of any segmentation result for further correcting the remaining isolated noisy spots of pixels (or island pixels) that have been classified as outlier pixels. For each pixel, the postprocessing step simply surveys the neighborhood classes and corrects the outlier pixels by setting the classification of the neighborhood majority. Furthermore, the uncertainty vector (the partition matrix vector) of the corrected pixel is estimated by using the probabilities

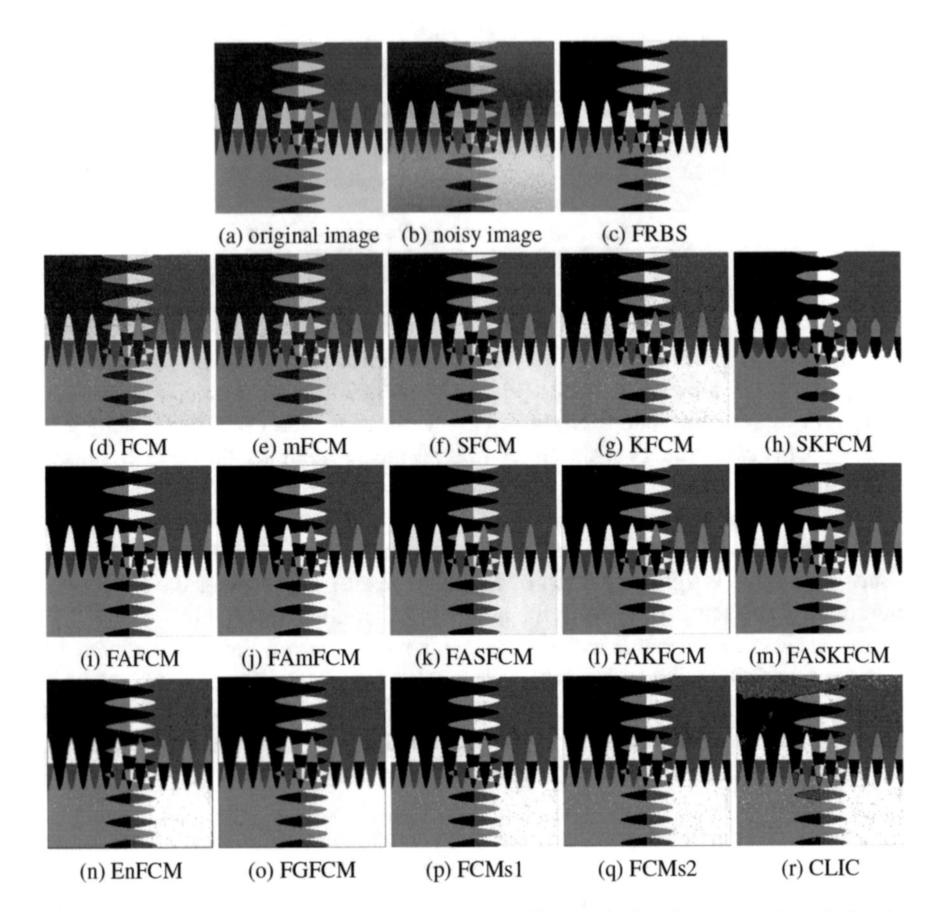

Fig. 1 Comparison of the synthetic image segmentation result for the proposed methods, the standard FCM, and several modified FCM methods. (**a**) The original image; (**b**) the original image corrupted with mixed noise (Gaussian(0,64), 5 % salt and pepper, and sinusoidal); (**c**) the FRBS result; (**d**)–(**h**) the original methods (i.e., the FCM, mFCM, SFCM, KFCM , and SKFCM) results; (**i**)–(**m**) the improved results with our proposed methods; (**n**)–(**r**) several other modified FCM results

of the classes in the neighborhood. The plot Fig. 2f shows how the application of the proposed post-segmentation step on the proposed methods, results for the synthetic image corrupted with salt-and-pepper noise improves the segmentation accuracy. With the postprocessing step, the overall result of our proposed method outperform the other methods also for the example with salt-and-pepper noise.

In the second group of experiments, we use the simulated MR brain images of different modalities (T1-, T2-, and PD-weighted) corrupted by different levels of Gaussian noise (3, 5, or 7 %) with and without 20 % intensity inhomogeneity. Figure 3 shows a comparison between the segmentation results when applying the methods on the simulated PD-weighted MR brain images with 5 % Gaussian noise and 20 % intensity inhomogeneity, while Figs. 4 and 5 show the comparison of the

Table 1 Segmentation accuracy (*SA*) in percentage and the number of iterations (Noit.) for (first column) our proposed methods, (second column) FCM and the original methods of our proposed variants, and (third column) several modified FCM methods on synthetic image of Fig. 1b corrupted by mixed noise

Figure 1b	SA%	Noit.	Figure 1b	SA%	Noit.	Figure 1b	SA%	Noit.
FAFCM	99.8199	2	FCM	97.0963	9	EnFCM	94.5251	10
FAmFCM	99.8413	2	mCFCM	98.7747	8	FGFCM	93.3868	8
FASFCM	99.8322	2	SFCM	98.5367	5	FCMS1	98.5565	9
FAKFCM	99.7314	2	SKFCM	94.0079	50	FCMS2	96.5958	9
FASKFCM	99.7375	2	KFCM	97.0963	11	CLIC	88.8062	120
						FRBS	98.5657	9

segmentation accuracies (SA) and the required number of iterations (Noit.) when applying all the methods to the three modalities corrupted with different levels of noise. In general, the outcome is similar to the one before, as our improved algorithms outperform all others in terms of accuracy and number of iterations or, in a few cases, produce similar accuracies with fewer number of iterations. This is true for all the given examples independent of noise level or imaging modality. Furthermore, these plots show that the accuracies of other methods degrade faster than our methods with increasing level of noise.

In addition to the synthetic images and simulated MRI brain images, we apply our methods to real T1-weighted MRI images [24]. We evaluate the segmentation accuracy based on the ground truth from manual segmentation. Figure 6 shows the comparison of the segmentation results when applying the proposed methods, the original FCM algorithm and all FCM variants mentioned above to the real MR image in Fig. 6a. Figure 7 shows the respective comparison of segmentation accuracy.

All experiments show that our improved methods outperform all other methods mostly in term of both the segmentation accuracy and the required number of iterations or, in a few cases, they give similar accuracies with fewer number of iterations. The experimental results show that the performance of the variants of our proposed method is approximately equivalent.

In conclusion, the results show that our approach is more robust against different types of noises and their mixture and reduces the possibilities of failure to a few cases. Hence, the convergence criterion which is always used by the FCM algorithm and its variants seems to not be enough to guarantee the improvement in the final results when compared to the initial or intermediate results. The results of the proposed approach shows that even the first iteration is informative enough to make the right decisions with high probability. The main drawback of the different versions of the modified FCM is in the application of blind filtering such as the (weighted) mean with spatial constraints, which may lead to the introduction of a new class that differs from the true classes in the region causing misclassification. We have shown that our adaptive rules improved results with less computational

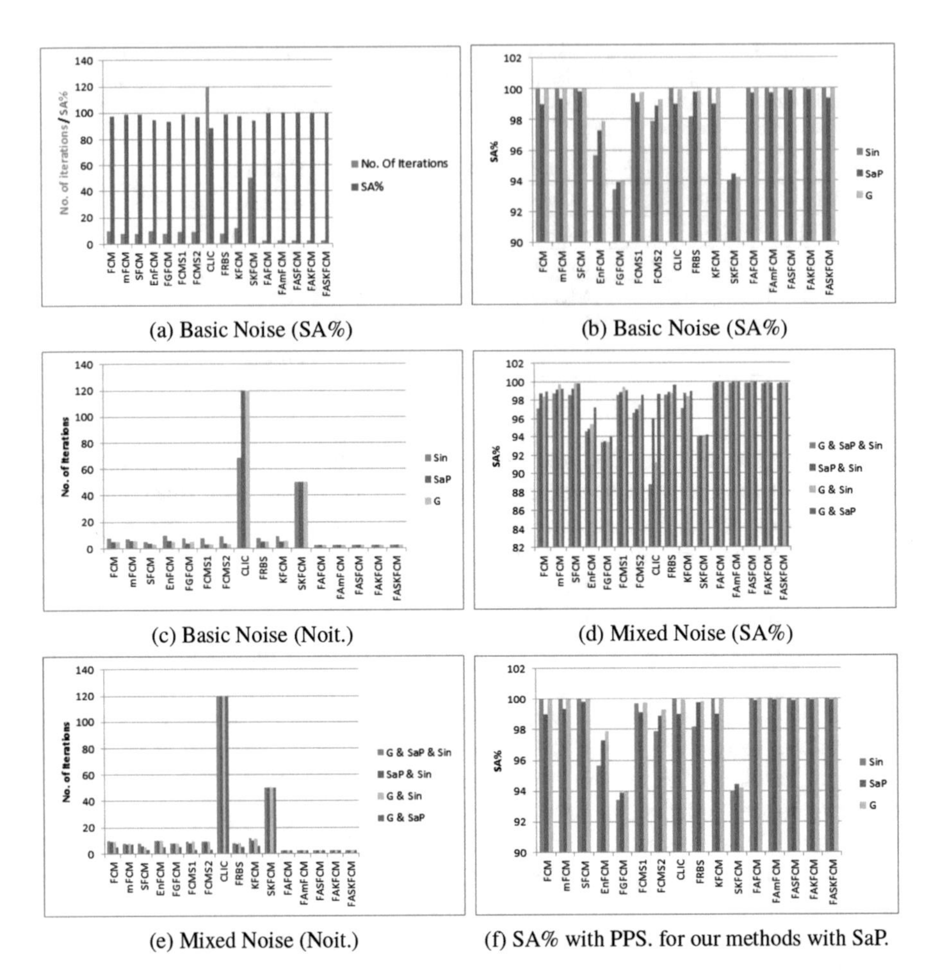

(a) Basic Noise (SA%)

(b) Basic Noise (SA%)

(c) Basic Noise (Noit.)

(d) Mixed Noise (SA%)

(e) Mixed Noise (Noit.)

(f) SA% with PPS. for our methods with SaP.

Fig. 2 Comparison of the synthetic image segmentation result for the proposed methods (last five groups of bars), the standard FCM (first group of bars), and several modified FCM methods (second to eleventh group of bars). (**a**) Segmentation accuracy (SA%) and number of iterations (Noit.) comparison of the synthetic image of Fig. 1a corrupted with mixed noise; (**b**)–(**e**) segmentation accuracy, and number of iterations comparison of synthetic image corrupted with basic and mixed noise; (**b**) and (**c**) corrupted with basic noises (G for Gaussian noise, SaP. for Salt and Pepper noise, and Sin for Sinusoidal noise) while (**d**) and (**e**) corrupted with a combination of mixed noise; (**f**) segmentation accuracy comparison similar to the plot in (**b**) but with applying the post-segmentation step (PSS) to the proposed methods result of the image with salt and pepper noise

effort. Still, for future work, we suggest to further tune the adaptive rules to cover more subtle differences in the cases.

We would like to discuss the selection of the two thresholds. The center pixel threshold θ_C should be chosen small enough in order to explore the second or even third highest probability classes, since they may be the true class, e.g., in PVE areas. Theoretically, the best choice is to use the maximum uncertainty probability level as

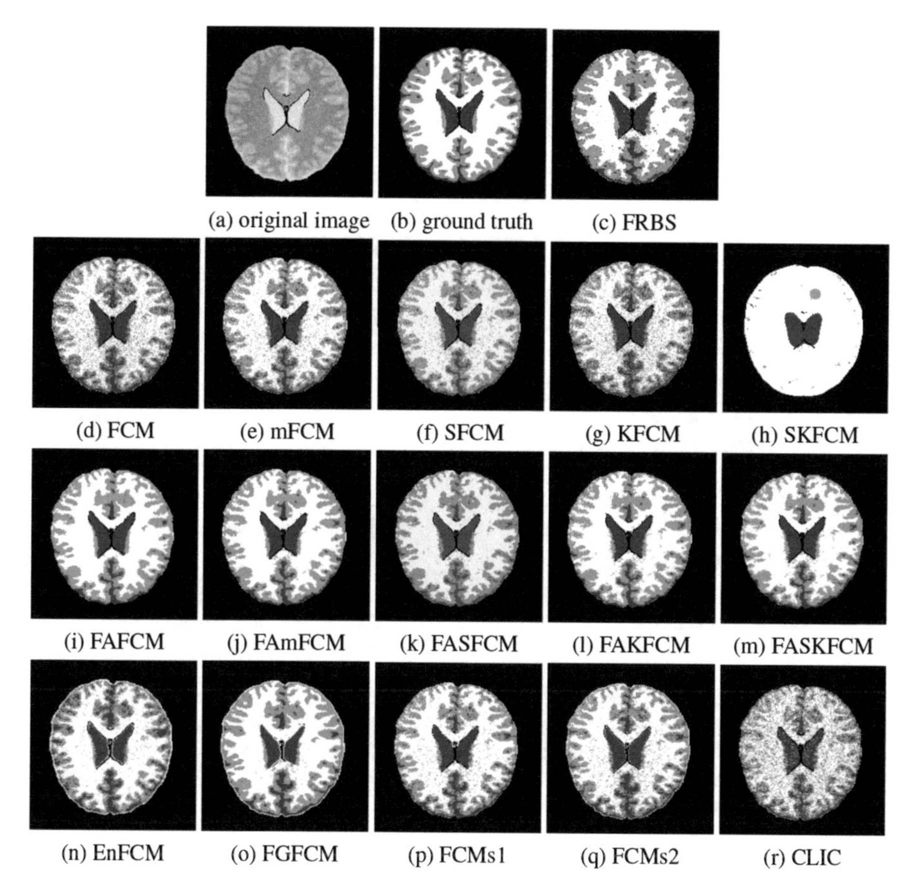

Fig. 3 Comparison of the segmentation result of simulated PD-weighted MR image for the proposed methods, the standard FCM, and several modified FCM methods; (**a**) original simulated PD-weighted MR image corrupted with 5 % Gaussian noise and 20 % intensity inhomogeneity; (**b**) The ground truth image; (**c**) the FRBS result; (**d**)–(**h**) the original methods (i.e., the FCM, mFCM, SFCM, KFCM , and SKFCM) results; (**i**)–(**m**) the improved results with our proposed methods; (**n**)–(**r**) several other modified FCM results

threshold when all classes have the probability $1/c$ where c is the number of classes. The reasoning behind this choice is that the high probability of a class to be the true class starts above this level, i.e., when the level of uncertainty starts declining. If we consider the effect of artifacts on the observed intensity, then 80–90 % of this level should be considered. For the neighbor threshold θ_{N_p}, a higher probability level, such as a probability equal to or higher than 0.5 is recommended to have a high confidence in the membership. However, in practice, the experimental results (for c=4) show that the thresholds can range from $\frac{1}{2c}$ to $\frac{1.6}{c}$ for the center pixel threshold θ_C and from 0.5 to 0.825 for the neighbor threshold θ_{N_p} with approximately similar results. The choice depends on the type of modality used and to a lower extent on the level of noise. For example, the intensities of the different tissues in the

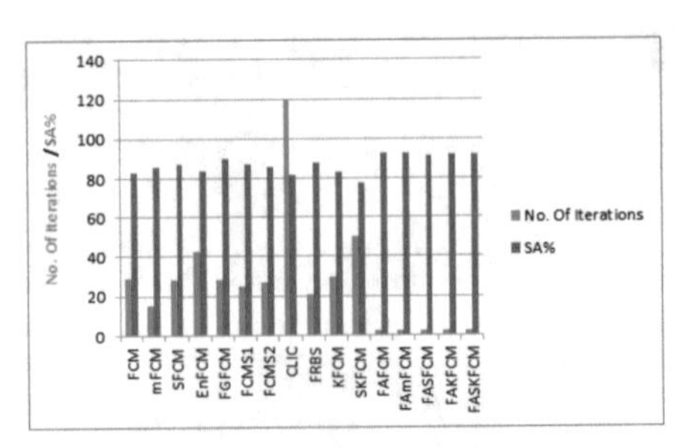

(a) 7% G 20% IIH. (SA% & Noit.)

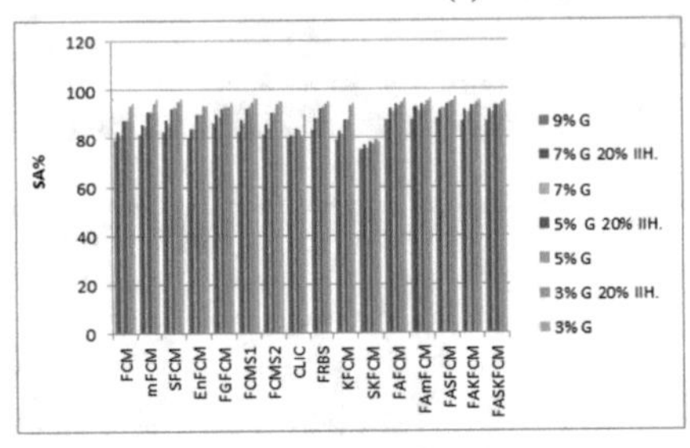

(b) All levels of Noise (SA%)

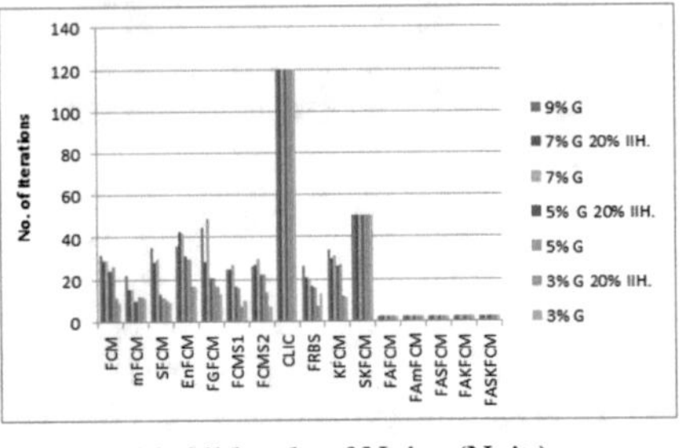

(c) All levels of Noise (Noit.)

Fig. 4 Comparison of the simulated PD-weighted MRI segmentation result for the proposed methods (last five groups of bars), the standard FCM (first group of bars), and several modified FCM methods (second to eleventh groups of bars). (**a**) segmentation accuracy (SA), and number of iterations (Noit.) comparison of simulated PD-weighted MRI corrupted with 7 % Gaussian noise (G) and 20 % intensity inhomogeneity (IIH.); (**b**) segmentation accuracy (SA), and (**c**) number of iterations (Noit.) comparison of the simulated PD-weighted MR image corrupted with different level of noise

simulated T2-weighted and PD-weighted MR images are close to each other and even overlap when adding a small amount of noise. So, the best threshold choices were $\theta_C = 0.16$ and $\theta_{N_p} = 0.78$ for simulated PD-weighted MRI and $\theta_C = 0.186$ and $\theta_{N_p} = 0.82$ for simulated T2-weighted MRI. For simulated T1-weighted MRI, the best threshold choices were $\theta_C = 0.31$ and $\theta_{N_p} = 0.75$. For real T1-weighted MRI, the best thresholds choices were $\theta_C = 0.381$ and $\theta_{N_p} = 0.51$. For the synthetic image, the best thresholds choices were $\theta_C = 0.181$ and $\theta_{N_p} = 0.61$.

6 Uncertainty-Based Segmentation Performance Comparison

In this section we compare the proposed segmentation methods with the other FCM variants using the uncertainty-based visual and numerical tools recently proposed by Al-Taie et al. [14]. They showed that the better segmentation algorithm is the one that produces fewer uncertain decisions (with lower uncertainty level) in the

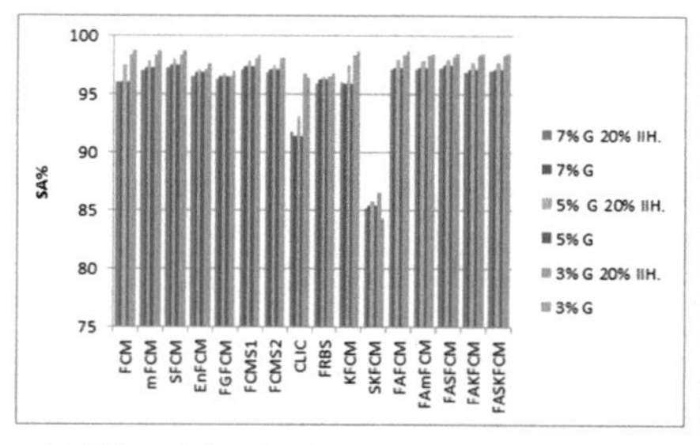

(a) T1-weighted - All levels of noise (SA%)

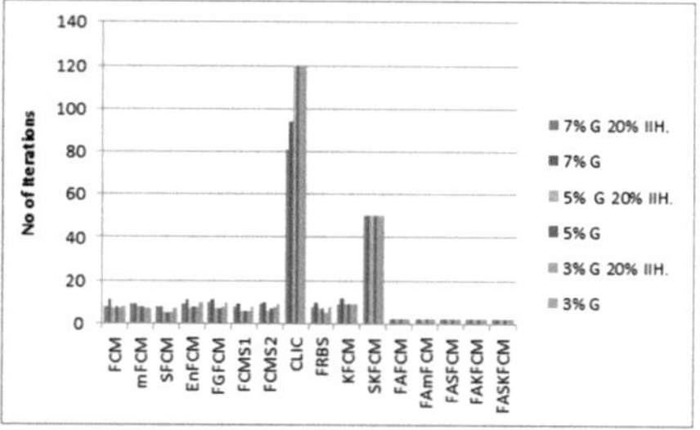

(b) T1-weighted - All levels of noise (Noit.)

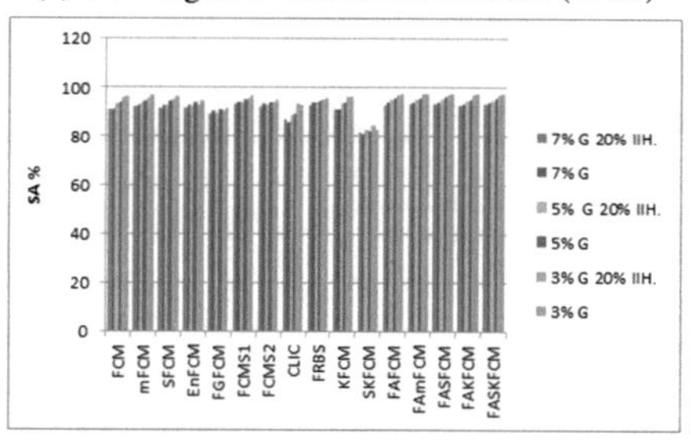

(c) T2-weighted - All levels of noise (SA%)

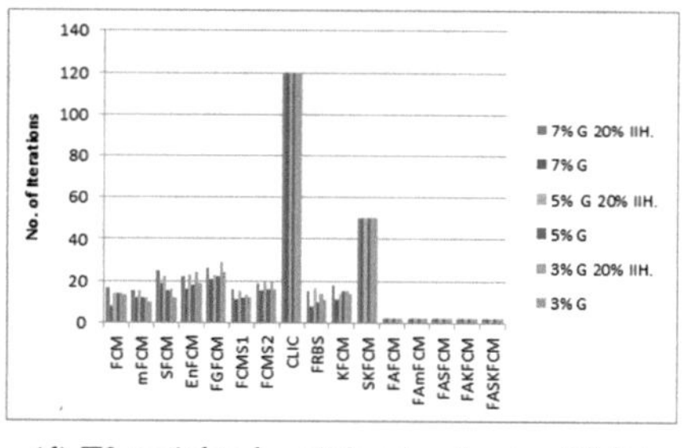

(d) T2-weighted - All levels of noise (Noit.)

Fig. 5 Comparison of the simulated T1- and T2-weighted MRI segmentation result for the proposed methods (last five groups of bars), the standard FCM (first group of bars), and several modified FCM methods (second to eleventh groups of bars). Segmentation accuracy (SA), and number of iterations (Noit.) comparison of the simulated T1-weighted (**a** and **b**) and T2-weighted (**c** and **d**) MR images corrupted with different level of Gaussian noise G with or without 20 % intensity inhomogeneity (IIH.)

uncertain area. Furthermore, the better algorithm tends to concentrate the high uncertainty decisions inside the problematic area (i.e., the misclassified area). While the first rule does not require the ground truth for comparison, the second one requires the ground truth for determining the misclassified area. The second rule is used as a second level of comparison, if the difference in the first rule is not large enough to decide for the better algorithm. In the first investigation, we use one of the visual tools proposed by Al-Taie et al. [14], which visualizes the measured uncertainty of the segmentation result over the original image after classifying the pixels into certain and uncertain. The algorithm that produces smaller uncertainty areas with lower uncertainty values is the better algorithm. Figure 8 shows the result of this visual comparison, where the color map for uncertainty values is shown next to the images. It can be observed that our proposed methods FASFCM and FAmFCM produce the best results and that our methods (third row) inherit difficulties from their original methods (second row). Moreover, these results show that the visualization tool is not only useful for comparing the performance of different algorithms, but it is also able to show how different algorithms differ in their behavior. For example, while the segmentation accuracy comparison in Fig. 7 shows that SKFCM outperforms KFCM, the uncertainty comparison reveals that

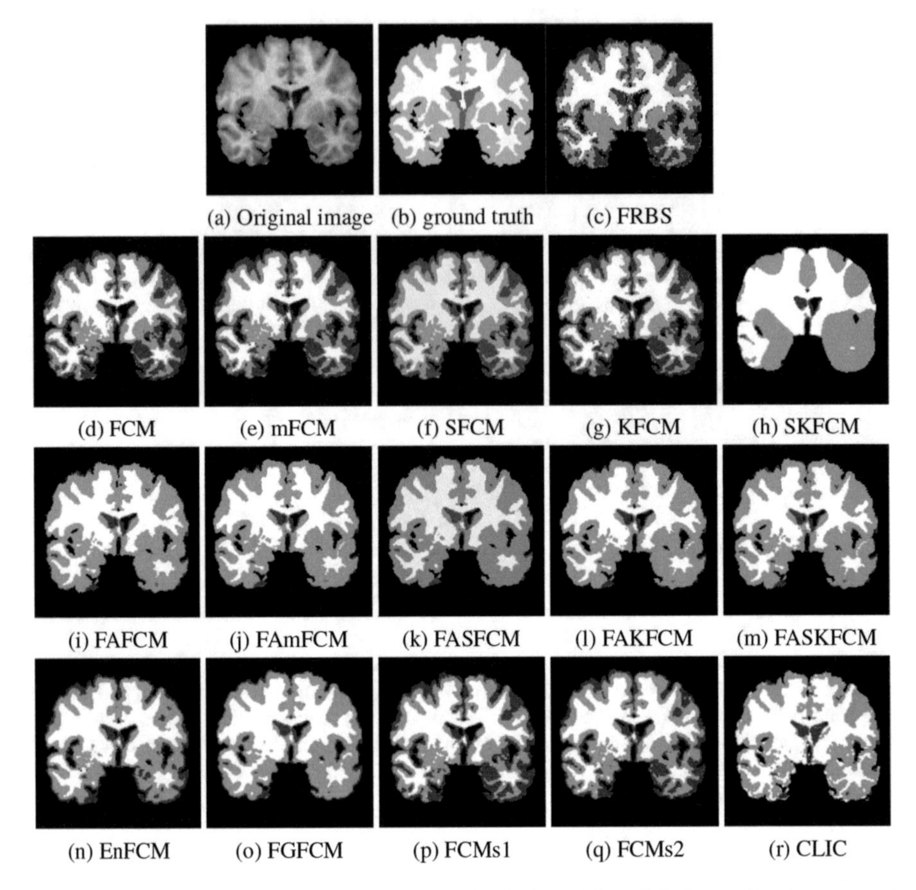

Fig. 6 Comparison of the segmentation result of real T1-weighted MR image for the proposed methods, the standard FCM, and several modified FCM methods. (**a**) original real T1-weighted MR image; (**b**) The ground truth image; (**c**) the FRBS result; (**d**)–(**h**) the original methods (i.e., the FCM, mFCM, SFCM, KFCM , and SKFCM) results; (**i**)–(**m**) the improved results with our proposed methods; (**n**)–(**r**) several other modified FCM results

KFCM performs better in modeling the uncertainty of the segmentation solution. In addition, the visualizations show how the effect of spatial constraints that are applied by the kernel of SKFCM (and in FASKFCM correspondingly) can corrupt the uncertainty model of the FCM algorithm. On the other hand, the spatial constraints applied by a non-kernelized FCM version (such as SFCM and mFCM) manage to improve the uncertainty model. Such comparisons can highlight the points of weaknesses and strengths of each algorithm.

We further investigate the behaviors of the different algorithms by examining how they distribute the uncertainty over the image. We compare the uncertainty area and the uncertainty density within the entire image, inside the misclassified area, and outside the misclassified area. The goal was to explore which algorithm

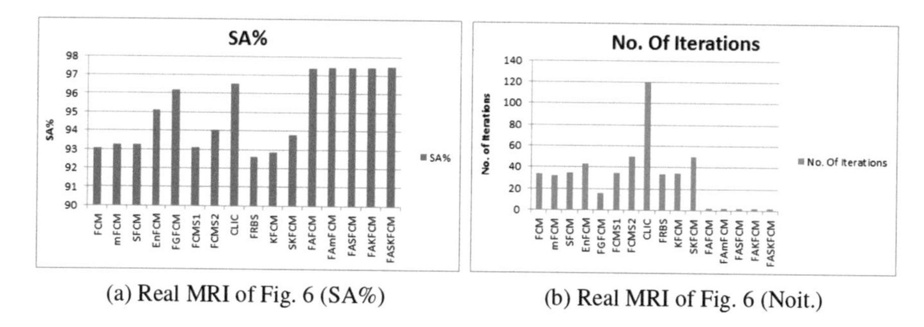

(a) Real MRI of Fig. 6 (SA%) (b) Real MRI of Fig. 6 (Noit.)

Fig. 7 Comparison of the real T1-weighted MRI segmentation result for the proposed methods (last five bars), the standard FCM (first bar), and several modified FCM methods (second to eleventh bar). (**a**) segmentation accuracy (SA%), and (**b**) number of iterations (Noit.) comparison of the real T1-weighted MR image of Fig. 6

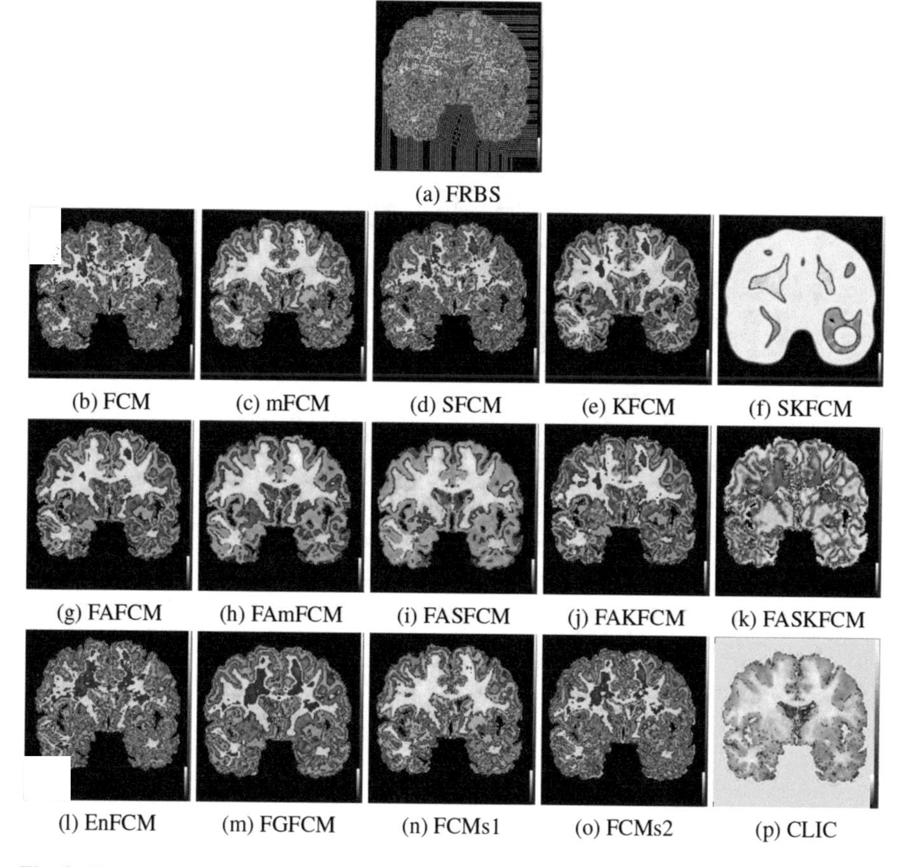

(a) FRBS

(b) FCM (c) mFCM (d) SFCM (e) KFCM (f) SKFCM

(g) FAFCM (h) FAmFCM (i) FASFCM (j) FAKFCM (k) FASKFCM

(l) EnFCM (m) FGFCM (n) FCMs1 (o) FCMs2 (p) CLIC

Fig. 8 Uncertainty-based visual comparison of the segmentation result of real T1-weighted MR image for the proposed methods, the standard FCM, and several modified FCM methods. (**a**) FRBS result (**b**) FCM result (**c**)–(**f**) the results of the other original methods (**g**)–(**k**) the results of our proposed methods, (**l**)–(**p**) several other modified FCM results

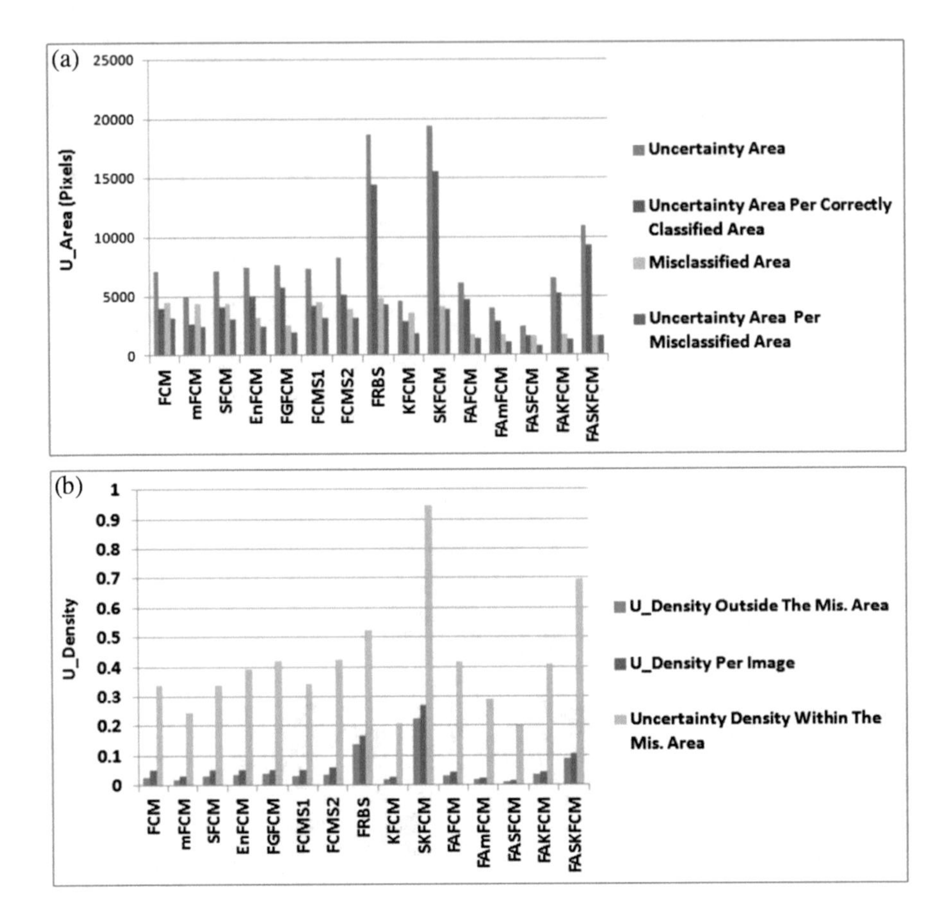

Fig. 9 Uncertainty-based numerical comparison. (**a**) Uncertainty area comparison. (**b**) Uncertainty density comparison

maximizes the uncertainty area and the uncertainty density inside the misclassified area, and, at the same time, minimizes the uncertainty area and the uncertainty density outside the misclassified area. Figure 9 shows that four of our proposed methods are among the best five algorithms, considering that our methods already have smaller misclassified areas.

7 Conclusions

In this paper, we propose a fast adaptive fuzzy c-means algorithm that is able to overcome several of the drawbacks of previous FCM variants with only two iterations. The first iteration is a standard FCM iteration, which gives the initial fuzzy segmentation. Based on strong assumptions, the second iteration exploits the initial membership value estimations using fuzzy rules for correcting the possible misclassifications. Moreover, we combine our approach with several existing modified FCM approaches for further improvements. In addition, we propose a post-segmentation noise removal step that survey the neighbor pixels' classes for correcting the outlier classes with the respective majority classes. This post-segmentation noise removal step is specially designed for pixels affected by salt-and-pepper noise.

The application of the proposed methods for synthetic images, simulated T1-, T2-, and PD-weighted brain MR images, and real T1-weighted MR images show that our methods outperform several FCM variants in term of both the segmentation accuracy and the number of iterations. Hence, the effect of different kinds of artifacts is reduced with low effort (i.e., computation time). Finally, the recently proposed uncertainty-based tools for visual and numerical segmentation algorithms evaluation [14] are used for further comparison. The uncertainty-based comparison shows that our methods also perform better than other FCM variants from an uncertainty point of view.

References

1. Guttmann, C., Jolesz, F.A., Kikinis, R., Killiany, R., Moss, M., Sandor, T., Albert, M.: White matter changes with normal aging. Neurology **50**, 972–978 (1998)
2. Heindel, W.C., Jernigan, T.L., Archibald, S.L., Achim, C.L., Masliah, E., Wiley, C.A.: The relationship of quantitative brain magnetic resonance imaging measures to neuropathologic indexes of human immunodeficiency virus infection. Arch. Neurol. **51**, 1129–35 (1994)
3. Mohamed, N.A., Ahmed, M.N., Farag, A.: Modified fuzzy c-mean in medical image segmentation. In: Proceedings IEEE International Conference on Acoustics, Speech, and Signal Processing, Piscataway, NI, USA, vol. 6, pp. 3429–3432 (1999)
4. Ahmed, M.N., Yamany, S.M., Mohamed, N., Farag, A.A., and Moriarty, T.: A modified fuzzy c-means algorithm for bias field estimation and segmentation of MRI data. IEEE Trans. Med. Imaging **21**(3), 193–199
5. Pham, D.L.: Fuzzy clustering with spatial constraints. In: 2002 Proceedings International Conference on Image Processing, vol. 2, pp. II–65–II–68 (2002)
6. Chen S., Zhang, D.: Robust image segmentation using FCM with spatial constraints based on new kernel-induced distance metric. IEEE Trans. Syst. Man Cybern. B **34**(4), 1907–1916 (2004)
7. Zhang, D., Chen, S.: A novel kernelised fuzzy c-means algorithm with application in medical image segmentation. Artif. Intell. Med. **32**(1), 37–50 (2004)
8. Yuan, K., Wu, L., Cheng, Q.S., Bao, S., Chen, C., Zhang, H.,: A novel fuzzy c-means algorithm and its application. Int. J. Pattern Recognit. Artif. Intell. **19**(8), 1059–1066 (2005)
9. Chuang, K.-S., Tzeng, H.-L., Chen, S., Wu, J., Chen, T.-J.: Fuzzy c-means clustering with spatial information for image segmentation. Comput. Med. Imaging Graph. **30**(1), 9–15 (2006)
10. Cai, W., Chen, S., Zhang, D.: Fast and robust fuzzy c-means clustering algorithms incorporating local information for image segmentation. Pattern Recogn. **40**(3), 825–838 (2007)
11. Tolias, Y.A., Panas, S.M.: On applying spatial constraints in fuzzy image clustering using a fuzzy rule-based system. IEEE Signal Process Lett. **5**(10), 245–247 (1998)
12. Saad, A., Möller, T. and Hamarneh, G.: Probexplorer: uncertainty-guided exploration and editing of probabilistic medical image segmentation. Comput. Graphics Forum **29**(3), 1113–1122 (2010)

13. Olabarriagaa S.D., Smeuldersb A.W.M.: Interaction in the segmentation of medical images: a survey. Med. Image Anal. **5**, 127–142 (2001)
14. Al-Taie, A., Hahn, H.K., Linsen, L.: Uncertainty estimation and visualization in probabilistic segmentation. Comput. Graph. **39**(0), 48–59 (2014); Available online: 26 October 2013
15. El-Melegy, M.T., Mokhtar, H.: Incorporating prior information in the fuzzy c-mean algorithm with application to brain tissues segmentation in MRI. In: International Conference on Image Processing (ICIP), pp. 3393–3396. IEEE (2009)
16. Li, C., Xu, C., Anderson, A.W., Gore, J.C.: MRI tissue classification and bias field estimation based on coherent local intensity clustering: a unified energy minimization framework. In Prince, J.L., Pham, D.L., Myers, K.J. (eds.) Information Processing in Medical Imaging. Lecture Notes in Computer Science, vol. 5636, pp. 288–299. Springer, Berlin/Heidelberg (2009)
17. Wang, J., Kong, J., Lu, Y., Qi, M., Zhang, B.: A modified FCM algorithm for MRI brain image segmentation using both local and non-local spatial constraints. Comput. Med. Imaging Graph. **32**(8), 685–698 (2008)
18. Ji, Z.-X., Sun, Q.-S., Xia, D.-S.: A modified possibilistic fuzzy c-means clustering algorithm for bias field estimation and segmentation of brain mr image. Comput. Med. Imaging Graph. **35**(5), 383–397 (2011)
19. Szilágyi, L., Benyo, Z., Szilágyi, S.M., Adam, H.S.: Mr brain image segmentation using an enhanced fuzzy c-means algorithm. In: Proceedings of the 25th Annual International Conference of the IEEE, vol. 1, pp. 724–726. Engineering in Medicine and Biology Society (2003)
20. Praßni, J.S., Ropinski, T., Hinrichs, K.: Uncertainty-aware guided volume segmentation. IEEE Trans. Vis. Comput. Graph. **16**(6), 1358–1365 (2010)
21. Potter, K.C., Gerber, S., Anderson, E.W.: Visualization of uncertainty without a mean. IEEE Comput. Graph. Appl. **33**(1), 75–79 (2013)
22. Bezdek J.C.: Pattern Recognition with Fuzzy Objective Function Algorithms. Plenum, New York (1981)
23. MNI. Brainweb, Simulated Brain Database: Available since 1997. Available at http://www.bic.mni.mcgill.ca/brainweb/, access time: on November 2012, 1997
24. IBSR. The Internet Brain Segmentation Repository (IBSR): Available since 1996. Available at http://www.cma.mgh.harvard.edu/ibsr/, access time: on October 2012, 1996

muView: A Visual Analysis System for Exploring Uncertainty in Myocardial Ischemia Simulations

Paul Rosen, Brett Burton, Kristin Potter, and Chris R. Johnson

Abstract In this paper we describe the Myocardial Uncertainty Viewer (muView or μView) system for exploring data stemming from the simulation of cardiac ischemia. The simulation uses a collection of conductivity values to understand how ischemic regions effect the undamaged anisotropic heart tissue. The data resulting from the simulation is multi-valued and volumetric, and thus, for every data point, we have a collection of samples describing cardiac electrical properties. μView combines a suite of visual analysis methods to explore the area surrounding the ischemic zone and identify how perturbations of variables change the propagation of their effects. In addition to presenting a collection of visualization techniques, which individually highlight different aspects of the data, the coordinated view system forms a cohesive environment for exploring the simulations. We also discuss the findings of our study, which are helping to steer further development of the simulation and strengthening our collaboration with the biomedical engineers attempting to understand the phenomenon.

1 Introduction

Myocardial ischemic is a disease caused by the restriction of blood flow to a region of the heart, most frequently due to the narrowing of one or more coronary vessels. Left untreated, cardiac cells will gradually weaken and die; in many cases, leading to heart attack. These consequences make ischemic heart disease

P. Rosen (✉)
Department of Computer Science and Engineering, University of South Florida, 4202 E. Fowler Avenue, ENB 118, Tampa, FL 33620, USA
e-mail: prosen@usf.edu

B. Burton • C.R. Johnson
Scientific Computing and Imaging Institute, University of Utah, 72 S. Central Campus Dr., Salt Lake City, UT 84112, USA
e-mail: bburton@sci.utah.edu; crj@sci.utah.edu

K. Potter
University of Oregon, 460 Mckenzie Hall, 5246 University of Oregon, Eugene, OR 97403, USA
e-mail: kpotter@cas.uoregon.edu

L. Linsen et al. (eds.), *Visualization in Medicine and Life Sciences III*, Mathematics and Visualization, DOI 10.1007/978-3-319-24523-2_3

the leading cause of death for men and women in the U.S. and most industrialized countries [34]. Detection of cardiac ischemia often requires inspecting the results of an electrocardiogram (ECG) and looking for abnormalities, particularly within the ST segment of the ECG trace. However, the relationship between cardiac ischemia and abnormalities in the ST segment is still unclear [7, 52].

To better understand the underlying physiology of cardiac ischemia, mathematical models are created to study the effect of ischemic regions on cardiac electrical signals, such as the electrical potentials external to the cell. These models have input parameters whose values cannot be practically obtained, and thus a series of parameter perturbations are used to cover a range of possible values. The result of these multi-run simulations is an ensemble of realizations that explores numerous possibilities within the domain and range of the simulation, but also introduces uncertainty within the output. We are currently developing the Myocardial Uncertainty Viewer (muView or μView) tool for visualizing the output of the cardiac model runs and, more specifically, the uncertainty present within the ensemble.

The goal of μView is to both directly explore the simulation results, helping scientists design and troubleshoot experiments, and to help understand the relationship of conductivity uncertainties to size and shape estimates of the ischemic zone. The challenges to this goal stem mainly from the complexity of the data; we are given multiple simulation outputs for each voxel. The structure of the data is inherently difficult; the spatial domain of the data is 3D, so simply displaying the data causes occlusion and clutter. Indicating further attributes within the 3D context is a formidable challenge. To address this issue, we have created μView to experiment with the collection of visualization techniques, including traditional two-dimensional and three-dimensional spatial displays, as well as the incorporation of information visualization approaches, to find a meaningful visual representation. The broader goal of this work is to develop visualization techniques that can concisely express the nature of the uncertainty within this type of complex data for domain scientists and health care professionals alike.

The main contributions of this work stem from the use of multiple visualization approaches to get a sense of the uncertainty in the data. Due to the complexity of the data and the domain, multiple views are employed to allow the user to explore different characteristics of the data. Each visual interface is designed to highlight specific aspects of the uncertainty within the data display, and interactions within a specific display are linked, as appropriate, to the other views. While each specific view is limited in its novelty for displaying this type of uncertainty information, the strength of our technique lies in the combination of our selected views to extract aspects of the uncertainty that are complimentary to the other views and appropriate for the specific needs of our domain scientists. We feel that this approach contributes knowledge to the field by demonstrating a collection of visualization techniques appropriate for uncertainty information within a 3D spatial domain that addresses the needs of scientists working with this specific type of complex data.

2 Related Work

Interest in uncertainty visualization is steadily increasing [10, 17, 32, 35], and the topic has been identified as a top research problem [16]. Related to this work are techniques aimed at incorporating uncertainty information into volume rendering and isosurfaces, using linked multiple windows, the visual representation of probability distribution functions (PDFs), and displaying the results of parameter-space explorations.

Volume rendering and isosurfacing are techniques designed to convey spatial characteristics of volumetric scalar data. The challenge for these methods is that uncertainty often accompanies data as a multi-valued attribute and applying volume rendering and isosurfacing is not straightforward. Approaches to add uncertainty information to these displays include pseudo-coloring, overlay, transparency, glyphs and animation [4, 23, 30, 44]. Fout and Ma [8] propose a computational model that computes a posteriori bounds on uncertainty propagated through the entire volume rendering algorithm and developed an interactive tool to inspect the resulting uncertainty. Pfaffelmoser et al. derive a mathematical framework for computing confidence levels of gradients of uncertainty parameters to allow for analysis such as the stability of features in an uncertain scalar field [38].

Rather than using isosurfaces to directly convey uncertainty in data, they can be used to show shape and extent of clusters, for example, in the exploration of a supervised fuzzy classification of 3D feature space plots [29]. Probabilistic formulations of marching cubes [40] and isocontours [39] allow for the display of positional uncertainty of isosurfaces colored by their distance from a mean [37].

While these three-dimensional representations are quite useful for conveying geometric structure and providing context, the complexity of the data often requires multiple presentation types to enable full understanding. For this reason multi-window linked-view systems are popular for addressing uncertainty—uniting a collection of visualization modes. Examples of such systems for understanding uncertainty have been used in magnetic resonance spectroscopy (MRS) [5], multi-dimensional cosmological particle data [11], and weather and climate modeling [41, 49].

Another way to look at uncertainty is to consider the multiple values as PDFs and to use statistical methods for characterizing them [42, 43]. Initial work in the area began by extending existing techniques to work with PDFs [31]. Clustering [2] and slice planes [25] can be used to reduce the dimensionality of the data for visualization, while colormaps, glyphs, and deformations have been used to express summaries and clusters [24, 26].

Finally, the type of data we are analyzing can be thought of in terms of parameter-space exploration in which the effect of perturbations of input parameters is related visually to outcomes through techniques such as parallel coordinates [1], preattentive highlighting [6], and high-dimensional boxplots [53].

3 Background

The electrocardiography forward problem aims to describe torso and cardiac electrical potentials that result from electrically active sources on or within the heart [15, 18, 19]. To this end, computational simulations have been developed to numerically approximate electrical outcomes like cardiac activation times, epicardial potentials and body surface potentials. However, several parameters are required for the simulations, many of which cannot be practically obtained through experimentation or other methods. The estimation of these parameters has the potential to introduce uncertainty into forward cardiac simulations, thereby reducing their potential for clinical applications.

Our parameter of interest is the conductivity of cardiac tissue. Cardiac muscle fibers (known as myocytes) are anisotropic, electrically active cells. Currents pass preferentially along the longitudinal axis of these fibers, investigating action potentials that ultimately cause contraction of the heart muscle. Experimental approaches have been used in an effort to extract appropriate conductivity values, but there are large discrepancies in the literature with regard to these values, which can differ in magnitude from each other by as much as five times [3, 20, 21, 45, 46]. Table 1 shows the range of conductivity values found in the literature that were used for our simulation with $\overline{\sigma}_i$ and $\overline{\sigma}_e$ representing intracellular and extracellular conductivities, respectively. The significance of these conductivity values are explained below.

3.1 Simulation

To solve the forward problem of electrocardiography, three components are linked together for a complete simulation. We defined an ischemic region with reduced transmembrane potential to mimic the ischemic condition, an anatomical cardiac geometry with associated conductivity properties, and a representation of the electrical activity through the cardiac tissue.

The ischemic region was obtained through experimental methods. Ischemia was induced in a canine model by restricting blood flow to the left anterior descending artery (LAD) of the heart. Plunge, or needle, electrodes with a 1.6 mm transmural resolution were positioned within the LAD vascular bed. Recordings of electrical potentials through the thickness of the myocardial wall were recorded at 1 KHz

Table 1 Conductivity ranges

	Longitudinal		Transmural	
	σ_i	σ_e	σ_i	σ_e
Min	0.00174	0.0012	0.000193	0.0008
Max	0.0034	0.00625	0.0006	0.00236
Ischemic scaling	1/10	1/2	1/1000	1/4

and potential values during the ST segment of the cardiac cycle were extracted. As ischemia progressed, the ST segment potentials were depressed within the ischemic region. Values that were greater than one standard deviation from baseline values were extracted and used to define the ischemic region.

The same heart in which ischemia was induced was used to extract accurate, cardiac geometry and fiber structure by way of MRI and diffusion weighted tensor imaging (DTI). The heart was scanned in a 7 tesla MRI with 0.31 mm/pixel resolution. These images were segmented and meshed to represent the cardiac anatomy. DTI was used to determine the preferred diffusion direction within the cells, thereby defining the anisotropic direction of the striated cardiac muscle.

Electrical activation of the heart was defined by the bidomain equations [12]. This model was adapted to generate cardiac potentials at a single time step, under the influence of ischemia, by reducing it to the passive current flow bidomain (Eq. (1)) [13]. The bidomain equations represent cardiac tissue by defining, on each node of a simulation, two continuous regions, or domains, that are coupled together by a membrane. The intracellular domain represents the region within cardiac myocytes while the extracellular region defines the extracellular space. Both of these regions are represented by respective conductivity tensors ($\overline{\sigma}_i$ and $\overline{\sigma}_e$). The extracellular and transmembrane potentials, V_e and V_m, are also represented as shown in the equation below. For the purposes of our study, V_m is assumed to have a constant potential difference of -35 mV.

$$\nabla \cdot (\overline{\sigma}_i + \overline{\sigma}_e)\nabla V_e = -\nabla \cdot \overline{\sigma}_i \nabla V_m \tag{1}$$

3.2 *Conductivity Values*

Given the variability in reported cardiac tissue conductivities [3, 20, 21, 45, 46], we selected and explored a range of conductivity values determined by the minimum and maximum values for longitudinal and transmural conductivities for both $\overline{\sigma}_i$ and $\overline{\sigma}_e$, as observed in the literature. However, when considering evaluation of high-dimensional, randomized parameters, it is necessary to consider both parameter distribution and computational complexity. To address both, we applied *generalized polynomial chaos with stochastic collocation* (gPC-SC) [54]. We used the gPC-SC method to reduce the amount of stochastic collocation points required to accurately compute statistical measures, thereby reducing computational complexity. We treated the conductivity ranges as uniformly distributed, stochastic process. Subsets of values within the range were selected based on Smolyak construction [9, 50], a linear combination of tensor products that span the subspace, while considered the four randomized conductivity parameters and the desired level of representative points. We selected a first level (9 representative conductivity value combinations), second level (41 representative conductivity value combinations), and a third level (137 conductivity combinations) Smolyak representations with which to run the deterministic solutions of the above mentioned bidomain equations.

4 Visualization System

The data input into the visualization system consists of three required components and one optional component. The required data components are a set of vertices, a solid mesh of tetrahedral or hexahedral elements that connect the vertices and form the geometric anatomy of the heart, and finally, an ensemble of simulation results attached to each vertex in the mesh. Optionally, a supply of directional tensor information can be added.

μView is an interactive, 6-way linked visual analysis system consisting of displays designed to highlight various aspects of the data in two main contexts. The first context is the anatomic or spatial context. In this context, the first view is a three-dimensional visualization of the data (Fig. 1 A), while three additional views contain orthogonal two-dimensional slices through the volume (Fig. 1 B-D). The second context is the PDF or uncertainty context. To explore this context, a

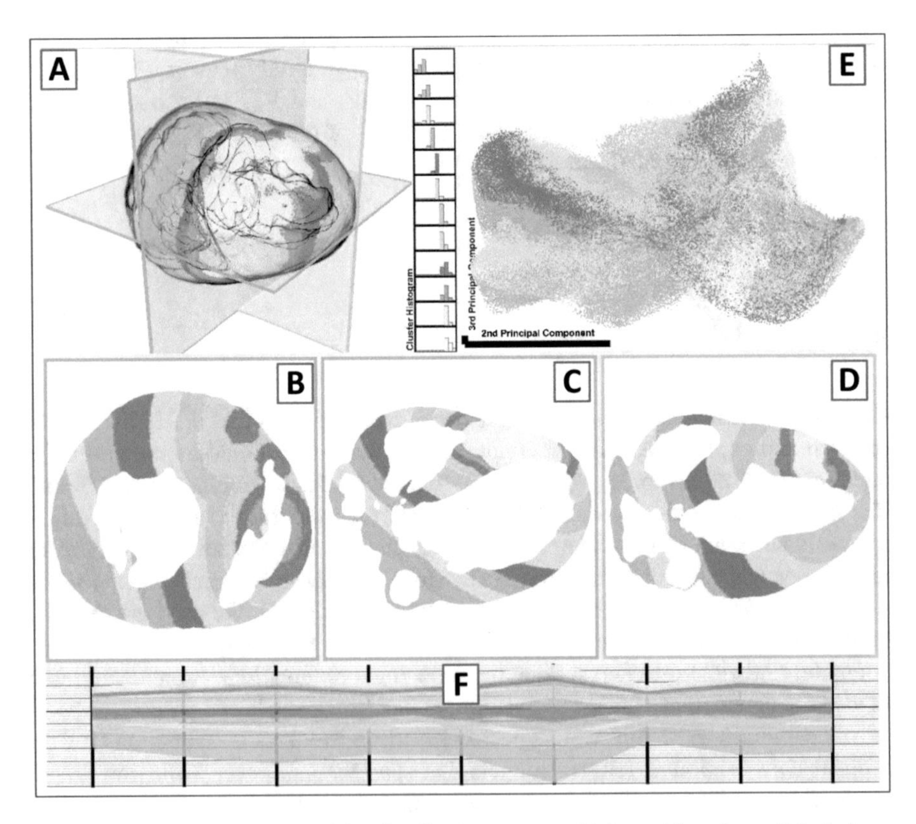

Fig. 1 *K9-2 dataset*. Overview of the visualization system, which combines 6-way linked views. *A*: Three-dimensional view, *B-D*: Two-dimensional views, *E*: Feature space view of the principal component analysis of the PDFs, *F*: Parallel coordinates. The data points are colored categorically using k-means clustering with L2-norm distance metric

feature space view that depicts the result of principal component analysis (PCA) applied to the PDFs (Fig. 1 E), and a parallel coordinates view of the PDFs (Fig. 1 F) shows each dimension of the data. Data point colors are held consistent across all visualizations for visually cross-correlating data between views and contexts (i.e. points receive the same coloring as their counterparts in other views). The interfaces are manipulated through mouse interactions and a small menu system (not shown).

4.1 Visual Interfaces

Three-Dimensional View (Fig. 1 A) Anatomic context is important in many medical applications, and thus we provide a three-dimensional rendering of the geometry. We extract a surface mesh from the volume by selecting tetrahedral faces that are not shared. Using the surface mesh, we darken triangle faces that are near perpendicular to the view direction to help contextualize the data within the three-dimensional shape, as seen in Fig. 3 left. This method has the advantage of showing both the surface and cavities of the heart. However, it can be difficult to determine the orientation and shape of the heart without interacting with it. Other more illustrative rendering methods [28] could be inserted in its place.

Within this interface, the data can be visualized either through a series of isosurfaces as discussed in Sect. 4.2 or by rendering the data points colored via a transfer function as discussed in Sect. 4.3. The three colored planes serve as a cross-reference mechanism to the two-dimensional views and can be switched on or off as desired.

Two-Dimensional View (Fig. 1 B-D) For a view of the data without perspective distortion, two-dimensional visualization slices of the volume are extracted by intersecting the solid elements with a plane and linearly interpolating vertices, triangles, and PDFs. The three slice planes, axial, coronal, and sagittal, are a more natural way for health care professionals to view the data, and are displayed using transfer functions to color the mesh with isolines to help highlight the variations in value. Isolines are extracted from the data with their frequency controlled by the user (i.e. every 0.5 step in data value). The orientation planes, optionally visible in the three-dimensional visualization, are color coordinated with the borders surrounding the views. These assist users in identify the three-dimensional spatial location of the slices and cross-referencing phenomena of interest.

Feature Space View (Fig. 1 E) A wide-variety of dimensionality reduction methods [27, 36, 48] allow for the conversion from a difficult to visualize and interpret high-dimensional space, to an easier to understand low-dimensional space. We have employed this approach in the feature space view with a goal of preserving as many features as possible.

We treat our PDFs as high-dimensional points and reduce them down to two-dimensional points. Principle component analysis (PCA) [36] is used as our dimensionality reduction approach. PCA works by extracting a vector representing

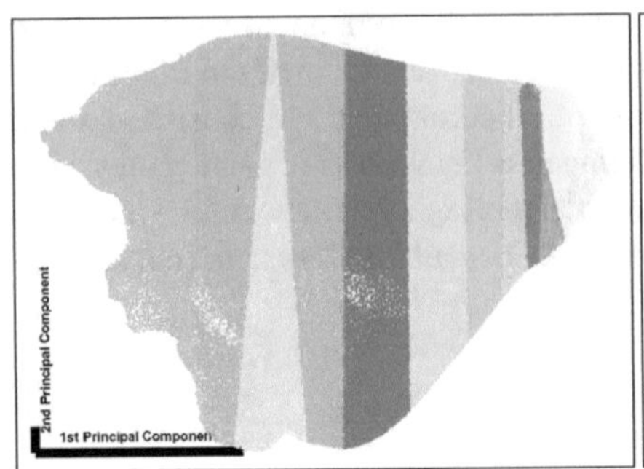

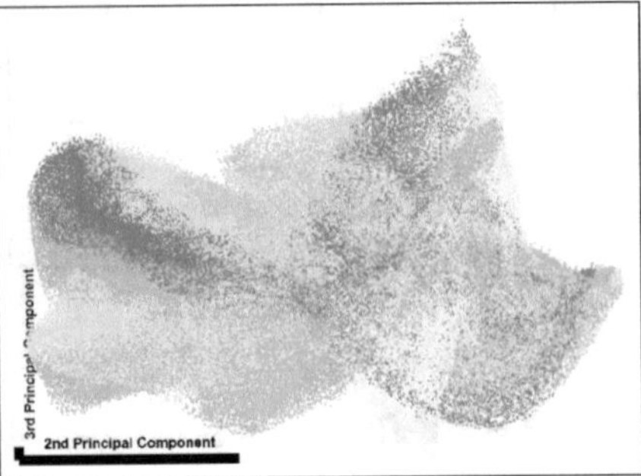

Fig. 2 *K9-2 dataset*. Visualization of PCA using the first and second principal components (*left*) and the second and third principal components (*right*). The data points are colored categorically using k-means clustering with L2-norm distance metric

the strongest component of the dataset. It then finds the next strongest vector which is orthogonal to the first. This process can continue to any number of dimensions. Our interface allows selecting any pair of principle components (not just the first two, see Fig. 2) to generate the display. This allows exploring the high-dimensional space for features which might not otherwise be visible in the spatial domain.

The coordinate system is placed in the lower left, with the length of the axes indicating the scaling of that principal component. For example, in Fig. 2 right, the second principal component (horizontal component) is significantly larger than the third principal component (vertical component). Therefore, the vertical axis is stretched, indicating that the natural dimensionality of the data may be reached.

Parallel Coordinate View (Fig. 1 F) Parallel coordinates are an alternative method for exploring the full high-dimensional space of the data. We supply an interactive parallel coordinates interface where each dimension represents a single simulation. Again, the data points (represented as lines in parallel coordinates) are colored using the same transfer function as in the other linked-views for easy cross-correlation. For example, in Fig. 1 it can be seen in the parallel coordinates that the yellow colored clusters are likely enclosing the ischemic zone. Similar observations are true of the other clusters as well, though some are difficult to see because of their limited size in the parallel coordinate view. We also enable interactive selection of the data points and reordering of axes within the parallel coordinates view for further exploration.

4.2 *Isosurfaces Over the PDFs*

The range of values for an individual data point makes isosurface location unclear [14]. Each dimension may maintain its own isosurface for a given isovalue, meaning the isosurface for the PDF could exist anywhere within a range of locations. To account for this, we reduce the PDF to a single dimension by applying an

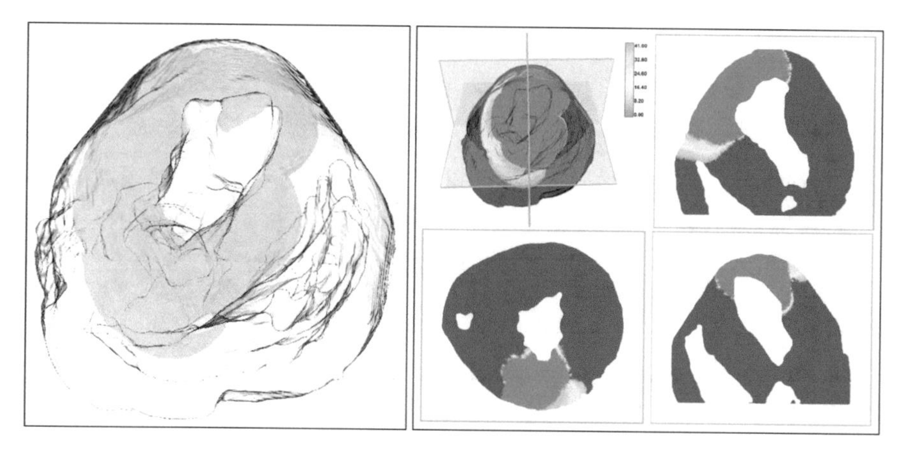

Fig. 3 *K9-1 dataset*. *Left*: Isovalue visualization is used to describe the range of potential isosurfaces of a given value with the minimum operator (in *blue*) and maximum operator (in *yellow*), combined with silhouettes (in *black*) for anatomic context. *Right*: Isovalue color mapping shows regions above (*blue*), below (*red*), and in-between (*yellow*) an isovalue

operator, such as minimum, maximum, or mean, to the data samples at each point. Isosurfaces are then extracted from the single dimensional field. Figure 3, left, shows an example where the minimum isosurface is blue and the maximum isosurface is yellow implying an envelope containing the range of possible surfaces.

4.3 Transfer Functions Over the PDFs

Because of the complexity of the data, we have adopted a number of transfer functions to color the data, each designed to aid understanding in a unique way.

Value-Based Coloring The first transfer function simply assigns a single value to each data point and applies an intensity-based sequential color map. The values can be related to individual dimensions, or derived values such as the mean (Fig. 4 left) or standard deviation (Fig. 4 right).

Coloring by Isovalue We have also explored coloring points by isovalue. This method takes each PDF and counts the number of dimensions above and below the isovalue. In this scheme, we choose solid colors to represent data points where all dimensions were above (red) or below (blue) the isovalue. The remaining points are colored using a sequential color map (orange to blue-green) which partially indicates how many dimensions fall above or below the isovalue. An example is seen in Fig. 3, right.

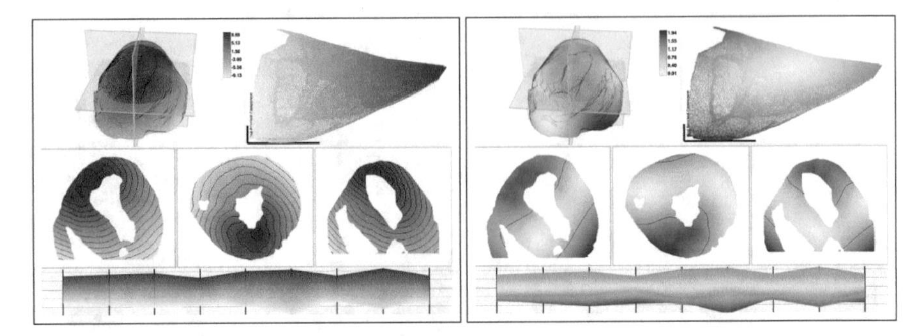

Fig. 4 *K9-1 dataset*. Single value, sequential color mapping of mean value (*left*) and standard deviation (*right*)

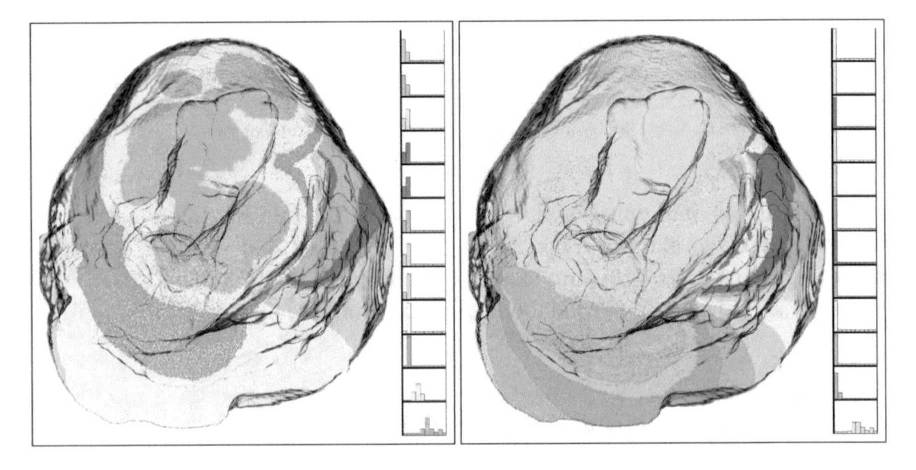

Fig. 5 *K9-1 dataset*. K-means clustering applied to the data using the L2-norm (*left*) and Pearson correlation coefficient (*right*) as distance metrics

Coloring by Clustering Clustering can reduce the set of data under investigation by grouping similar data together, such as points that respond similarly to variations in initial conditions. As points are placed into clusters, they are colored using a categorical color map. A collection of histograms showing the mean of each cluster is placed to the right.

We use k-means clustering [33] to exploring this space. We employ multiple distance metrics for comparing the underlying PDFs. The L2-norm (Fig. 5 left) groups points that are similar in a Euclidean sense and is defined by $d(X, Y) = \sqrt{\sum(X_i - Y_i)^2}$, where X and Y each contain the ensemble of simulation dimensions of the data points.

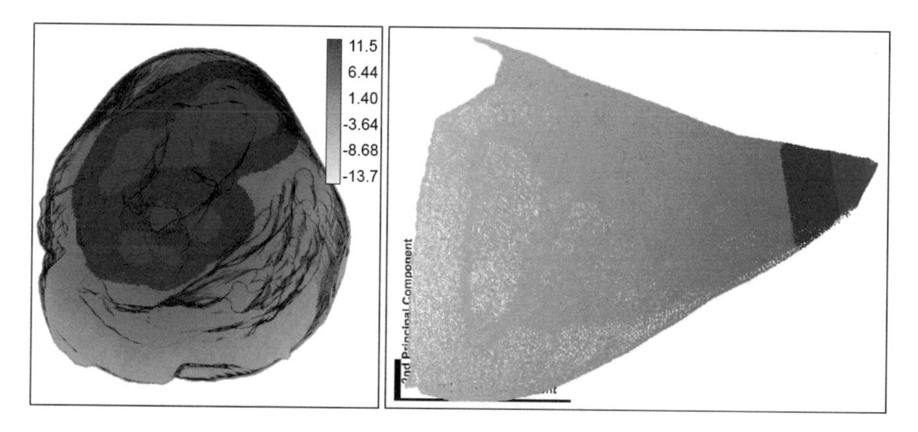

Fig. 6 *K9-1 dataset*. The result of painting the feature space (*right*) correlates a structure to its spatial counterpart (*left*). In this case, the ischemic zones are identified in the first principal component

Pearson correlation coefficient [47] (Fig. 5 right) clusters points that respond similarly to changes in input and is defined as $d(X, Y) = 1 - \frac{\sum(X_i - \overline{X})(Y_i - \overline{Y})}{\sqrt{\sum(X_i - \overline{X})}\sqrt{\sum(Y_i - \overline{Y})}}$, where $\overline{X}$ and $\overline{Y}$ are the means of the sets.

Selecting the number of clusters is a well known challenge to the k-means approach. Initially, the number of clusters is selected using an information-theoretic approach [51] where $k = log(n/2)$. The user is then given the opportunity to adjust the number of clusters based upon their intuition about the data.

Painting Over Feature Space We give the user the opportunity to interact indirectly with the high-dimensional data through a painting interface included in the feature space view. The interface allows users to select a paint color, brush over a region of interest, and see the resulting color change in both the feature and other domain visualizations. An example is seen in Fig. 6, where an interesting structure in feature space has been selected in various shades of purple. In this case, it turns out that the structures are related to the ischemic zone of the heart.

Coloring by Fiber Direction To better understand the impact of input versus output, we have included a common method of visualizing input fiber directions. In this visualization, red, green, and blue are assigned to x, y, and z, respectively. This gives the opportunity to correlate fiber directions with structures in either the feature or spatial domains. Figure 7 shows an example of this visualization.

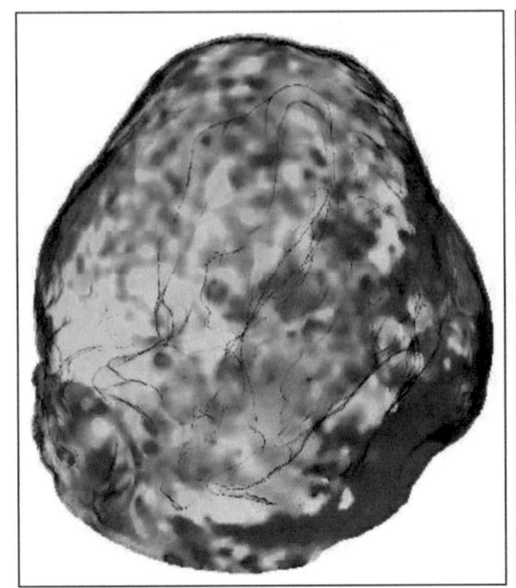

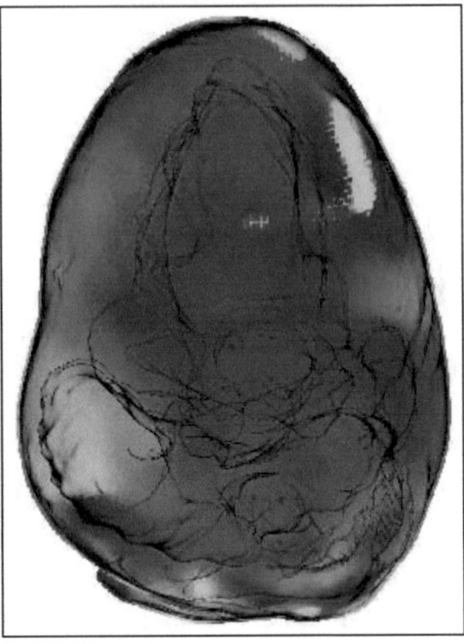

Fig. 7 Coloring by fiber directions in the three-dimensional view shows noisy MR diffusion tensor data used as input for the *K9-1 dataset* (*left*). A more atomically realistic model would be expected to vary more smoothly across the heart such as that of *K9-2 dataset* (*right*)

5 Results

Simulations were performed using two separate data sets. The first, "*K9-1*", was created by MRI segmentations of an excised canine heart with atrial tissue removed. The mesh consists of 1.4 M tetrahedra and 350 K vertices. In like manner, the second, "*K9-2*", was created by using the same imaging techniques from a different canine dataset. This mesh contains 2.5 M tetrahedra connecting 435 K vertices. Ischemic regions in both meshes were determined by thresholding measured potential values observed within the heart under experimental, ischemic conditions. Thresholds were determined to be within 1 standard deviation of control, baseline ECG values observed in the ST segment. As part of our experiments, each mesh was used to calculate first, second, and third order gPC-SC which produced 9, 41, and 137 realizations, respectively.

Performance is reported on Windows-based system with a 3.2GHz Intel Core i7 CPU, 8 GB of RAM, and an NVIDIA GeForce GTX 580. It is important to note, we were also able to run all of these experiments on a MacBook Air. The performance was slightly slower but still respectable. Our software takes a few moments to load data (15–30 s for smaller data and a few minutes for large data) and a few seconds to apply some operators (1–60 s; clustering being the slowest). OpenMP is used where possible to take advantage of the multicore environment. The individual OpenGL contexts can run at 50–100 frames per second, though there is a slowdown when all 6 contexts are rendered simultaneously. Nevertheless, the system remains interactive.

We developed this tool as part of a collaboration between visualization and biomedical researchers to better understand the physiology of cardiac ischemia. μView was actively developed simultaneously with the development of the simulation model, allowing: results from the simulation to be explored within μView; insights gleaned from μView to be incorporated back into the simulation; and μView revised and refined based on feedback from collaborators.

5.1 Identifying Ischemic Zones

Identifying the ischemic zones is one of the first tasks we engaged in when exploring the datasets. Many of the visualizations do not immediately enable access to this type of information. For example, in Fig. 8 top, the mean and standard deviation visualizations do not clearly delineate the ischemic zones. However, by exploring the isovalues of the data we can better understand the location of diseased tissue. Thresholding is a common approach used by biomedical engineers for capturing the shape of the ischemic zone. By selecting an isovalue, this visualization performs

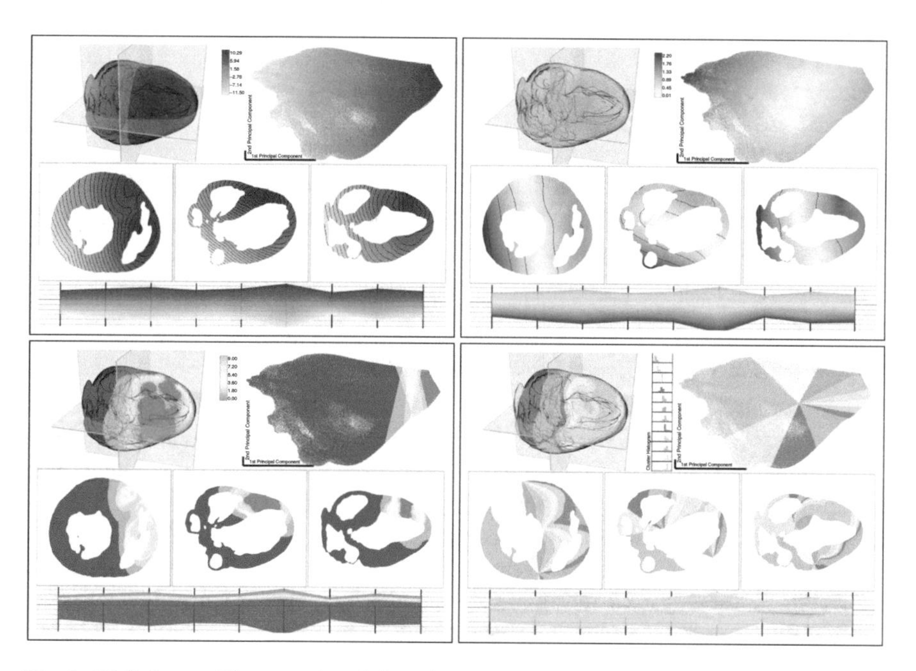

Fig. 8 *K9-2 dataset*. The mean (*top left*) and standard deviation (*top right*) visualizations often challenge the user searching for ischemic zones, while the isovalue (*bottom left*) and L2-norm clustering (*bottom right*) can enable quick identification

a pseudo thresholding while providing additional uncertainty information (Fig. 8 bottom left). It turns out that the L2-norm is also very useful in identifying ischemic zones. Figure 8 bottom right shows an example where the light green cluster represents the ischemic zones. The clustering approach also has the advantage of being almost entirely automatic. Finally, feature space painting can be used to potentially identify the ischemic region. As seen in Fig. 6, the first principal component primarily encodes a value that when explored, enables selecting the ischemic region. Our hypothesized ischemic regions can then be confirmed using the parallel coordinates view to note that the cluster or region has the highest voltage value, a known indication of the ischemic zone.

5.2 Detecting Noisy Fiber Orientations

The conductivities of the heart are highly dependent on the fiber directions across the tissue. The fiber direction data can be created any number of ways, such as rule-based methods or, as in our case, using DTI. Part of our study is understanding the impact of conductivity to fiber direction.

As we began our study, we noticed a bulge in the *K9-1* dataset for many isovalues (most prominently visible as the yellow area in Fig. 3 right) that we could not easily explain. We then dove into the data by directly visualizing some of the input, such as fiber direction. This lead us to discover that the fiber directions from our DTI imaging were noisy and poorly aligned for the first few millimeters of depth from the heart surface (see Fig. 7 left).

This finding drove us to obtain a secondary model, the *K9-2* dataset, which has much smoother fiber directional data (see Fig. 7 right). In this new model, no such irregular bulge is visible, leading us to believe that the feature was caused by the noisy tensors in the original data.

5.3 Comparing Higher-Order Methods

For our experiments, we ran first- through third-order gPC-SC with the goal of determining if there were significant advantages in higher-order methods over lower-order ones. Figure 9 show the results for the *K9-1* dataset, and Fig. 10 shows the results for the *K9-2* dataset. Both compare first- and third-order gPC-SC.

Both the first- and third-order versions of the data appear virtually identical. In *K9-1* (Fig. 9), there is a mirroring in the second principal component in the feature space visualization. However, the shape of the feature space is virtually identical. Otherwise, only the slightest of differences can be seen with close inspection.

In the *K9-2* dataset (Fig. 10), the only easily identifiable difference lies in the L2-norm clustering (row 3). However, this difference may be attributed as an artifact of initial mean selection in the k-means clustering algorithm.

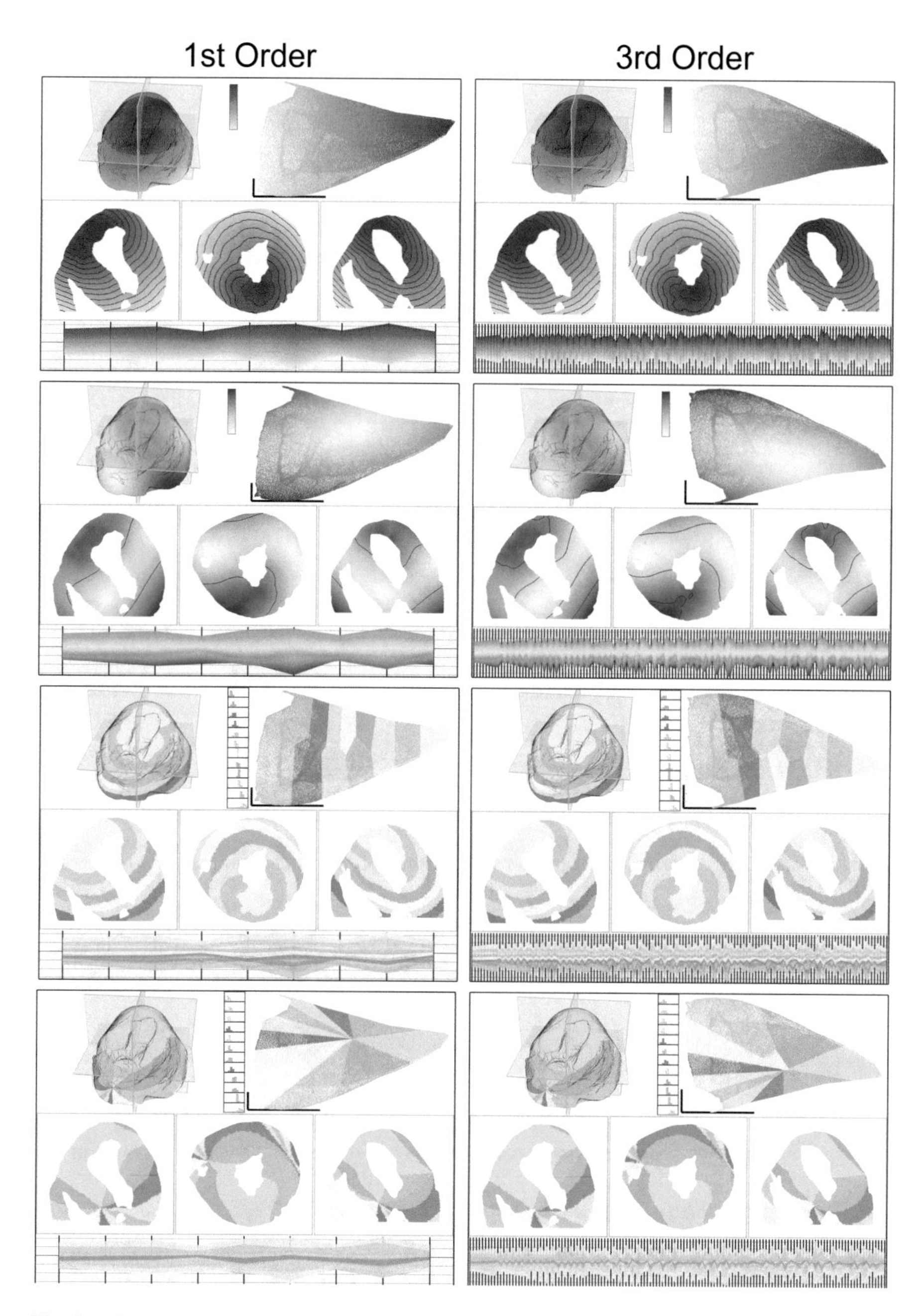

Fig. 9 *K9-1 dataset*. Comparison of the first- (*left*) and third-order (*right*) gPC-SC through mean (row 1), standard deviation (row 2), L2-norm clustering (row 3), and Pearson correlation clustering (row 4) visualizations

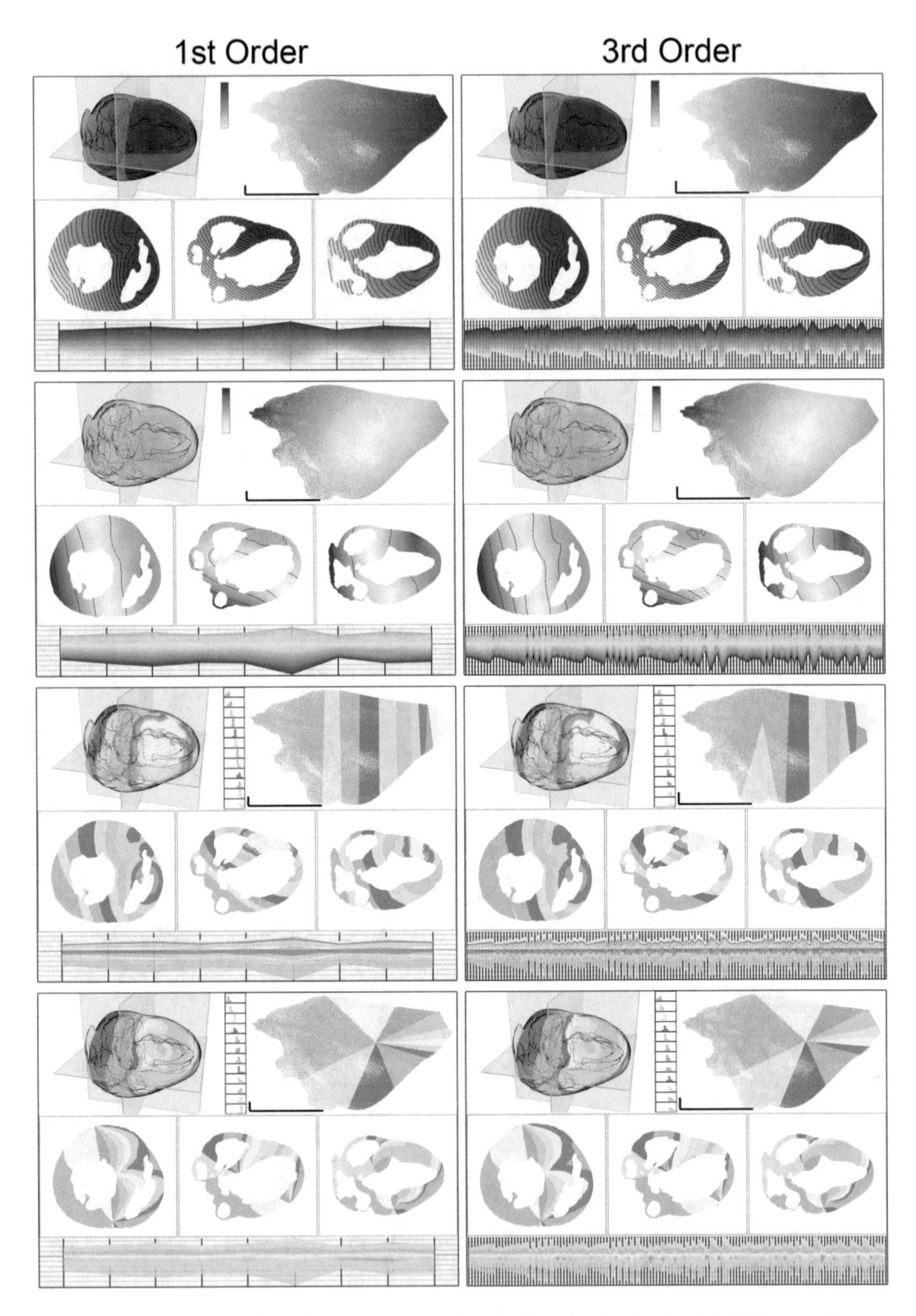

Fig. 10 *K9-2 dataset*. Comparison of the first- (*left*) and third-order (*right*) gPC-SC through mean (row 1), standard deviation (row 2), L2-norm clustering (row 3), and Pearson correlation clustering (row 4) visualizations

The results of both of these experiments convinced our team that, for conductivity-based uncertainty, higher-order gPC-SC does not provide a significant variation or advantage over the lower-order method. This assertion might not hold for other ischemia related uncertainties, such as positional uncertainty of the ischemic zone.

5.4 Uncertainty from Conductivity

The final issue we addressed was whether the uncertainty is in some way meaningful. Our conclusion after exploring the data using many visualization methods over a significant period of time is that, no, the uncertainty in this data is not particularly meaningful. We believe that the results can be explained primarily through scale and bias. This is most apparent in the parallel coordinates view of Fig. 1. The fact that the clustered colors maintain constant tracks is a clear indicator of the scale and bias effect. The other indication is that in principal component analysis, the first principal component almost entirely summarizes the voltage. Additionally, the secondary principal components are of significantly smaller scale, indicating that the underlying dimensionality of the data is probably just one or two (see Fig. 2).

6 Discussion and Future Work

The ultimate clinical goal of this work is to be able to assess the uncertainty in determining the ischemic zone from an inverse simulation linked to ST segment waveform changes. If we better understand how the uncertainty in the conductivities affect the size and shape of the ischemic zone, it will help determine what levels of uncertainty will be of consequence clinically and how much confidence can be assigned to the understanding of the ischemic zone size, shape, and location.

From a scientific point of view, these studies can also give us a better understanding of the relationship of conductivity uncertainty to both forward and inverse simulations of cardiac ischemia. In the future, we aim to provide confidence criteria of the simulation results as a function of both conductivity uncertainty and the problem we are trying to solve. It may be that for some problems, the level of uncertainty will not greatly effect the results, while the uncertainty may invalidate other applications. It may also indicate that uncertainty levels in the conductivities would have to be reduced in order to use such a method for a particular application, which could then spark research into generating better conductivity values.

6.1 Sources of Uncertainty

We are able to identify a wide variety of features associated with the location and magnitude of uncertainty. However, the source of the uncertainty remains largely a mystery (i.e. the classical correlation vs. causation problem). In Sect. 5.2, we suspected that the DTI noise was the source of the error, but were only able to find indirect evidence to validate our suspicion. Such a discovery encourages further investigation of the importance of fiber direction in simulation, which we plan to do through techniques such as those proposed by Jones [22] for visualizing uncertainty in fiber orientation. More generally, visualizations that indicate correlation and causation of uncertainty features will greatly improve the efficiency of these types of studies.

6.2 Repeated Features

We have presented a number of visualization approaches, many of which highlight the same features. Each visualization has its own advantages and disadvantages. Clustering has the advantage of highlighting multiple features simultaneously. However, it requires significant effort in visual search to wade through less important features. Using the isovalue visualization limits the number of features visible, making concentration easier, but requiring additional interaction. It is increasingly important to find visualizations which balance these modes of operation and identify which types of visualizations are most efficient from the perspectives of speed, accuracy, and cognitive load. Providing users with choice in visualization is valuable, but too much choice will overwhelm.

6.3 When Have We Found All Features?

The last issue arising from this work that often plagues visualizations is identifying when all of the data features have been located. We have presented visualizations that find a wide variety of features, but for a long time we lacked the confidence to claim the search had been exhaustive and all interesting features had been found. Making such claims demands further validation of which types of features are and are not identifiable by each visualization method.

7 Summary

This work is an initial exploration of uncertainty data obtained from the forward simulation of cardiac ischemia. We believe our close collaboration with the biomedical simulation scientists will greatly guide the choices we make regarding visualization, particularly in light of how our system has, to date, helped improve our understanding of the simulation. Both clinical problems and scientific exploration provide opportunities for improvement in uncertainty visualization techniques, and we look forward to extending μView to have greater research and clinical impact.

Acknowledgements This project was supported by grants from the National Center for Research Resources (5P41RR012553-14), National Institutes of Health's National Institute of General Medical Sciences (8 P41 GM103545-14), DOE NETL, and King Abdullah University of Science and Technology (KUS-C1-016-04).

References

1. Berger, W., Piringer, H., Filzmoser, P., Gröller, E.: Uncertainty-aware exploration of continuous parameter spaces using multivariate prediction. Comput. Graph. Forum **30**(3), 911–920 (2011)
2. Bordoloi, U.D., Kao, D.L., Shen, H.W.: Visualization techniques for spatial probability density function data. Data Sci. J. **3**, 153–162 (2004)
3. Clerc, L.: Directional differences of impulse spread in trabecular muscle from mammalian heart. J. Physiol. **255**, 335–346 (1976)
4. Djurcilov, S., Kim, K., Lermusiaux, P., Pang, A.: Visualizing scalar volumetric data with uncertainty. Comput. Graph. **26**, 239–248 (2002)
5. Feng, D., Kwock, L., Lee, Y., Taylor II, R.M.: Linked exploratory visualizations for uncertain mr spectroscopy data. SPIE Vis. Data Anal. **7530**(4), 1–12 (2010)
6. Feng, D., Kwock, L., Lee, Y., Taylor II, R.M.: Matching visual saliency to confidence in plots of uncertain data. IEEE Trans. Vis. Comput. Graph. **16**(6), 980–989 (2010)
7. Fleischmann, K.E., Zègre-Hemsey, J., Drew, B.J.: The new universal definition of myocardial infarction criteria improves electrocardiographic diagnosis of acute coronary syndrome. J. Electrocardiol. **44**, 69–73 (2011)
8. Fout, N., Ma, K.L.: Fuzzy volume rendering. IEEE Trans. Vis. Comput. Graph. **18**(12), 2335–2344 (2012)
9. Geneser, S., Hinkle, J., Kirby, R., Wang, B., Salter, B., Joshi, S.: Quantifying variability in radiation dose due to respiratory-induced tumor motion. Med. Image Anal. **15**(4), 640–649 (2011)
10. Griethe, H., Schumann, H.: Visualization of uncertain data: methods and problems. In: Proceedings of Simulation and Visualization, pp. 143–156 (2006)
11. Haroz, S., Ma, K.L., Heitmann, K.: Multiple uncertainties in time-variant cosmological particle data. In: IEEE Pacific Visualization, pp. 207–214 (2008)
12. Henriquez, C.: Simulating the electrical behaviour of cardiac tissue using the bidomain model. Crit. Rev. Biomed. Eng. **21**(1), 1–77 (1993)
13. Hopenfeld, B., Stinstra, J., Macleod, R.: Mechanism for st depression associated with contiguous subendocardial ischemia. J. Cardiovasc. Electrophysiol. **15**(10), 1200–1206 (2004)
14. Jiao, F., Phillips, J., Gur, Y., Johnson, C.: Uncertainty visualization in HARDI based on ensembles of ODFs. In: IEEE Pacific Visualization, pp. 193–200 (2012)

15. Johnson, C.: Numerical methods for bioelectric field problems. In: Bronzino, J. (ed.) The Biomedical Engineering Handbook, pp. 161–188. CRC, Boca Ratan (1995)
16. Johnson, C.R.: Top scientific visualization research problems. IEEE Comput. Graph. Appl. Mag. **24**(4), 13–17 (2004)
17. Johnson, C., Sanderson, A.: Next step: visualizing errors and uncertainty. IEEE Comput. Graph. Appl. Mag. **23**(5), 6–10 (2003)
18. Johnson, C., MacLeod, R., Matheson, M.: Computer simulations reveal complexity of electrical activity in the human thorax. Comput. Phys. **6**, 230–237 (1992)
19. Johnson, C., MacLeod, R., Matheson, M.: Computational medicine: bioelectric field problems. IEEE Comput. **26**(26), 59–67 (1993)
20. Johnston, P.R.: A cylindrical model for studying subendocardial ischaemia in the left ventricle. Math. Biosci. **186**, 43–61 (2003)
21. Johnston, P.R., Kilpatrick, D.: The effect of conductivity values on st segment shift in subendocardial ischaemia. IEEE Trans. Biomed. Eng. **50**, 150–158 (2003)
22. Jones, D.K.: Determining and visualizing uncertainty in estimates of fiber orientation from diffusion tensor MRI. Magn. Reson. Med. **49**, 7–12 (2003)
23. Jospeh, A.J., Lodha, S.K., Renteria, J.C., Pang, A.: Uisurf: Visualizing uncertainty in isosurfaces. In: Computer Graphics and Imaging, pp. 184–191 (1999)
24. Kao, D., Dungan, J.L., Pang, A.: Visualizing 2d probability distributions from eos satellite image-derived data sets: a case study. In: IEEE Visualization Conference, pp. 457–561 (2001)
25. Kao, D., Luo, A., Dungan, J.L., Pang, A.: Visualizing spatially varying distribution data. In: Information Visualisation, pp. 219–225 (2002)
26. Kao, D., Kramer, M., Luo, A., Dungan, J., Pang, A.: Visualizing distributions from multi-return lidar data to understand forest structure. Cartogr. J. **42**(1), 35–47 (2005)
27. Kruskal, J., Wish, M.: Multidimensional scaling. Sage University Paper series on Quantitative Application in the Social Sciences, vol. 07-011. Sage Publication, Beverly Hills/London (1978)
28. Lawonn, K., Moench, T., Preim, B.: Streamlines for illustrative real-time rendering. Comput. Graph. Forum **32**(3), 321–330 (2013)
29. Lucieer, A.: Visualization for exploration of uncertainty related to fuzzy classification. In: IEEE International Conference on Geoscience and Remote Sensing, pp. 903–906 (2006)
30. Lundström, C., Ljung, P., Persson, A., Ynnerman, A.: Uncertainty visualization in medical volume rendering using probabilistic animation. IEEE Trans. Vis. Comput. Graph. **13**(6), 1648–1655 (2007)
31. Luo, A., Kao, D., Pang, A.: Visualizing spatial distribution data sets. In: Symposium on Data Visualization, pp. 29–38 (2003)
32. MacEachren, A.M., Robinson, A., Hopper, S., Gardner, S., Murray, R., Gahegan, M., Hetzler, E.: Visualizing geospatial information uncertainty: what we know and what we need to know. Cartogr. Geogr. Inf. Sci. **32**(3), 139–160 (2005)
33. MacQueen, J., et al.: Some methods for classification and analysis of multivariate observations. In: Berkeley Symposium on Mathematical Statistics and Probability, vol. 1, p. 14 (1967)
34. Mathers, C.D., Fat, D.M., Boerma, J.T.: The global burden of disease: 2004 update. Technical Report, World Health Organization (2004)
35. Pang, A., Wittenbrink, C., Lodha., S.: Approaches to uncertainty visualization. Vis. Comput. **13**(8), 370–390 (1997)
36. Pearson, K.: On lines and planes of closest fit to systems of points in space. Philos. Mag. **2**(6), 559–572 (1901)
37. Pfaffelmoser, T., Reitinger, M., Westermann, R.: Visualizing the positional and geometrical variability of isosurfaces in uncertain scalar fields. Comput. Graph. Forum **30**(3), 951–960 (2011)
38. Pfaffelmoser, T., Mihai, M., Westermann, R.: Visualizing the variability of gradients in uncertain 2d scalar fields. IEEE Trans. Vis. Comput. Graph. **19**(11), 1948–1961 (2013)
39. Pöthkow, K., Heg, H.C.: Positional uncertainty of isocontours: condition analysis and probabilistic measures. IEEE Trans. Vis. Comput. Graph. **PP**(99), 1–15 (2010)

40. Pöthkow, K., Weber, B., Hege, H.C.: Probabilistic marching cubes. Comput. Graph. Forum **30**(3), 931–940 (2011)
41. Potter, K., Wilson, A., Bremer, P.T., Williams, D., Doutriaux, C., Pascucci, V., Johhson, C.R.: Ensemble-vis: a framework for the statistical visualization of ensemble data. In: IEEE Workshop on Knowledge Discovery from Climate Data: Prediction, Extremes, pp. 233–240 (2009)
42. Potter, K., Kniss, J., Riesenfeld, R., Johnson, C.: Visualizing summary statistics and uncertainty. In: Computer Graphics Forum (Proceedings of Eurovis 2010), vol. 29, pp. 823–831 (2010)
43. Potter, K., Kirby, R., Xiu, D., Johnson, C.: Interactive visualization of probability and cumulative density functions. Int. J. Uncertain. Quantif. **2**(4), 397–412 (2012)
44. Rhodes, P.J., Laramee, R.S., Bergeron, R.D., Sparr, T.M.: Uncertainty visualization methods in isosurface rendering. In: Eurographics Short Papers, pp. 83–88 (2003)
45. Roberts, D.E., Scher, A.M.: Effects of tissue anisotropy on extracellular potential fields in canine myocardium in situ. Circ. Res. **50**, 342–351 (1982)
46. Roberts, D.E., Hersh, L.T., Scher, A.M.: Influence of cardiac fiber orientation on wavefront voltage, conduction velocity and tissue resistivity in the dog. Circ. Res. **44**, 701–712 (1979)
47. Rodgers, J.L., Nicewander, W.A.: Thirteen ways to look at the correlation coefficient. Am. Stat. **42**(1), 59–66 (1988)
48. Roweis, S.T., Saul, L.K.: Nonlinear dimensionality reduction by locally linear embedding. Science **290**(5500), 2323–2326 (2000)
49. Sanyal, J., Zhang, S., Dyer, J., Mercer, A., Amburn, P., Moorhead, R.J.: Noodles: a tool for visualization of numerical weather model ensemble uncertainty. IEEE Trans. Vis. Comput. Graph. **16**(6), 1421–1430 (2010)
50. Smolyak, S.: Quadrature and interpolation formulas for tensor products of certain classes of functions. Sov. Math. Dokl. **4**, 240–243 (1963)
51. Sugar, C.A., Gareth, James, M.: Finding the number of clusters in a data set: An information theoretic approach. J. Am. Stat. Assoc. **98**, 750–763 (2003)
52. Toyoshima, H., Ekmekci, A., Flamm, E., Mizuno, Y., Nagaya, T., Nakayama, R., Yamada, K., Prinzmetal, M.: Angina pectoris vii. The nature of st depression in acute myocardial ischaemia. Am. J. Cardiol. **13**, 498–509 (1964)
53. Whitaker, R.T., Mirzargar, M., Kirby, R.M.: Contour boxplots: a method for characterizing uncertainty in feature sets from simulation ensembles. IEEE Trans. Vis. Comput. Graph. **19**(12), 2713–2722 (2013)
54. Xiu, D.: Efficient collocational approach for parametric uncertainty analysis. Commun. Comput. Phys. **2**, 293–309 (2007)

Part II
Visualization of 3D Medical Images

Combined Volume Registration and Visualization

Arlie G. Capps, Robert J. Zawadzki, John S. Werner, and Bernd Hamann

Abstract We describe a method for combining and visualizing a set of overlapping volumetric data sets with high resolution but limited spatial extent. Our system combines the calculation of a registration metric with ray casting for direct volume rendering on the graphics processing unit (GPU). We use the simulated annealing algorithm to find a registration close to optimal and allow the user to closely monitor the optimization progress. The combined calculation reduces memory traffic, increases rendering frame rate, and makes possible interactive-speed, user-supervised, semi-automatic combination of many component volumetric data sets.

1 Introduction

Volumetric imaging modalities have become centrally important in biological and medical applications. These modalities can produce a volumetric data set ("volume image") composed of samples or voxels commonly arranged in a 3D Cartesian grid of resolution $I \times J \times K$. Advances in technology and engineering have resulted in

A.G. Capps (✉)
Physical and Life Sciences, Lawrence Livermore National Laboratory, Livermore, CA 94550, USA

Vision Science and Advanced Retinal Imaging Laboratory (VSRI), Department of Ophthalmology and Vision Science, University of California, Davis, Sacramento, CA 95817, USA

Department of Computer Science, Institute for Data Analysis and Visualization (IDAV), University of California, Davis, CA 95616, USA
e-mail: agcapps@ucdavis.edu

R.J. Zawadzki • J.S. Werner
Vision Science and Advanced Retinal Imaging Laboratory (VSRI), Department of Ophthalmology and Vision Science, University of California, Davis, Sacramento, CA 95817, USA
e-mail: rjzawadzki@ucdavis.edu; jswerner@ucdavis.edu

B. Hamann
Institute for Data Analysis and Visualization (IDAV), Department of Computer Science, University of California, Davis, CA 95626, USA
e-mail: hamann@cs.ucdavis.edu

L. Linsen et al. (eds.), *Visualization in Medicine and Life Sciences III*, Mathematics and Visualization, DOI 10.1007/978-3-319-24523-2_4

rapidly increasing image resolution, data acquisition speed, and data set size and dimensionality. However, with some clinical in vivo imaging techniques such as optical coherence tomography (OCT) the coverage provided by individual high resolution volume images is smaller than the extent of the imaged structures. Multiple volume images that partially overlap must be captured, registered and combined ("stitched") to allow a single larger region to be analyzed as a single data set. Gaps or holes, areas of the larger region not covered by any volume image, must be easily identifiable as such by the user. Overlaps, areas covered by more than one volume image, must contain enough information for the stitching operation to unambiguously determine the volume images' correct relative position and are combined through an averaging operation to reduce noise and improve contrast.

We have developed methods and a prototype system for stitching multiple high-resolution volume data sets which combines on-screen visualization with registration metric computation. After the user places individual volume images into rough alignment, our system automatically refines the registration of the component images (see Fig. 1) using rigid-body transformations. Our system also supports the combination of scanned data sets in which correction for motion artifacts has changed a deformed, regular image into a correct image with sampling irregularities and gaps. Successive motion-corrected scans of the same region contain different

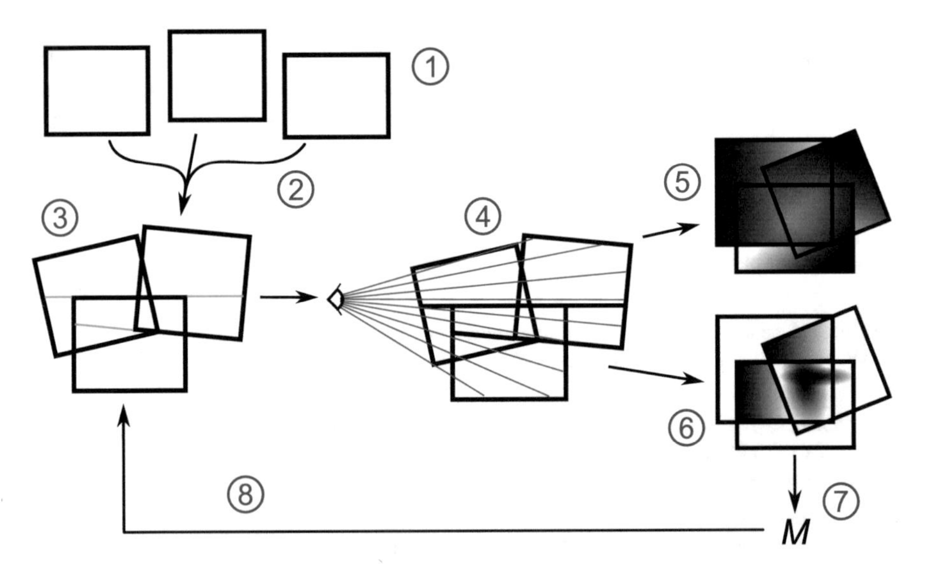

Fig. 1 Overview of the algorithm. (*1*) Input volumes. (*2*) User adjusts position. (*3*) Build a binary space partition (BSP) tree, cutting volumes into non-overlapping convex polyhedra. (*4*) Perform ray casting. (*5*) Each ray produces a color value for its pixel. The resulting image is displayed on the screen. (*6*) Each ray that pierces the intersection of multiple input volumes also produces an intermediate metric result. The resulting metric image is stored in a texture. (*7*) The intermediate metric results are reduced to a single value M. (*8*) An optimization step adjusts volume positions. Note that the image texture and the metric intermediate values shown at (*5*) and (*6*) represent the view from the eye point, a quarter turn from the representation at (*4*)

motion and hence different sampling gaps, which are filled in when the scans are properly registered and combined.

The procedures for visualizing multiple overlapping data sets and calculating a metric for registration are closely linked. Both tasks require sampling throughout the regions of image space where multiple volumes overlap followed by combination of the sample values to a summary, either to an image of the combined data or to a metric for goodness-of-fit. Since both tasks require intensive and wide-ranging data sampling, an efficient data access pattern directly benefits the speed of the system. We combine computation of the registration metric with the volume ray casting algorithm running on a workstation GPU.

We developed our system for use with non-invasive in vivo cellular resolution human retinal imaging modalities. Specifically, we applied our system to two varieties of optical coherence tomography, which acquires a series of consecutive 2D cross-sectional tomograms that when stacked define a 3D image. The first system [12] scans an area of 3 mm×3 mm on the retina in 4 s to produce an image of 375 × 360 steps with an axial resolution of 10 μm. The second system [14, 16] uses adaptive optics to achieve much finer resolution and scans a retinal patch of 300 μm × 250 μm in 512 × 100 steps over about 4 s, with an axial resolution of 3 μm. This fine resolution provides the potential for morphological analysis of microscopic structure in the human eye. The value of this analysis, already great, increases with the coverage of the image and provides a powerful motivation for volume image stitching. However, without some kind of correction, involuntary eye motion that is a significant nuisance in images from the first system becomes prohibitive to stitching ultra-high-resolution data sets acquired using adaptive optics with the second system. Our system should also be applicable to images from other volumetric modalities such as computed tomography (CT), magnetic resonance imaging (MRI), and confocal scanning optical microscopy.

In Sect. 2, we discuss work related to visualization and registration of volume data sets. We discuss volume partitioning, ray casting to produce images and calculate the registration metric, and optimization using simulated annealing in Sect. 3. In Sect. 4 we present examples demonstrating our method and the results of its application to OCT data sets, and we conclude with Sect. 5.

The key contributions made in this paper are (1) an algorithm combining visualization and computation of a registration metric for overlapping volumes and (2) a system using that algorithm to assist users in stitching multiple retinal OCT volumes to extend high-resolution coverage.

2 Related Work

Early work on multiple volume rendering includes that of Jacq and Roux [5], whose method casts rays through combined volumes and accumulated a derived value (the minimum, maximum, or average). Cai and Sakas [2] addressed the use of more general functions for data intermixing and established a useful taxonomy for the stage in the pipeline where data from the two volumes are combined. *Image*-level combines a 2D rendering of each individual volume, *accumulation*-level fusion combines colors from individual volumes' transfer functions during ray casting, and *illumination*-level fusion combines raw data from each volume at each step into one value which is transformed to color. The choice of data fusion level has consequences for the design of the rendering system, requiring different levels of flexibility. Recent work on multi-volume rendering has focused on facilitating arbitrary alignment and increasing the number of overlapping volumes supported. Shader programs are kept manageable by splitting up rays into sections that do not cross volume boundaries. The approach of Rößler et al. [13] treats overlapping volumes as a scene graph and auto-generates the shader programs necessary to render homogeneous regions of space. Kainz et al. [6] introduced a system that transforms the boundary polygons of each input volume to screen space and sorts the resulting fragments in depth to produce homogeneous ray segments. These techniques, while powerful, require expensive depth-sorting techniques. To avoid this cost, solutions suggested by Lindholm et al. [9] and Lux and Fröhlich [10] use a binary space partitioning (BSP) tree to produce homogeneous volume fragments which are rendered in depth order provided by traversal of the BSP tree. We base our system on the approach of Lindholm et al., which is simpler in its treatment of large volumes. The systems discussed so far and several other recent works [4, 8] perform multi-volume visualization, requiring registered volumes as input.

Many capable systems have been introduced that accomplish automatic or semi-automatic registration of volume data. Bria et al. [1] described a system for stitching 3D confocal ultramicroscopy (CU) data sets, with no function for visualization. Unlike OCT, the CU data sets are sparse; like OCT, the initial position of each volume is approximately known and is used as the starting point for registration. Dalvi et al. [3] introduced a multi-step process starting with recorded sensor position, performing feature extraction using a wavelet transform and finishing with intensity-based volume registration. This system performs non-interactive stitching of ultrasound data sets, with no visualization component. Ultrasound, like retinal OCT, is a coherent detection process and ultrasound data sets contain pervasive speckle noise. However, the examples in [3] show features that are much larger and less complicated than the range of feature scales present in retinal OCT images. Specific to OCT data sets, Zawadzki et al. [15, 16] presented a system that allows the user stitch multiple high-resolution adaptive optics OCT data sets using axis-aligned translation in whole-voxel increments. Image combination is accomplished by choosing the maximum intensity of overlapping voxels.

3 Methods

Our system is summarized in Fig. 1. The user supplies several input volumes (1) and adjusts their positions (2). The overlapping input volumes are divided using a BSP tree (3) on the CPU, using the boundaries of the input volumes as partition planes.

Interior nodes of the BSP tree represent cut planes that divide the overlapping input volumes; leaf nodes represent the subvolumes or BSP "cells" (4) defined by the cut planes. We produce a volume rendering by casting rays through the leaf node cells in order of increasing distance from the eye point. In addition to calculating an image pixel, each ray also calculates a partial metric value. Image pixels are displayed on the screen (5), and intermediate metric results (6) are stored in a graphics texture. The partial metric values are reduced to a final metric value M (7); an optimizer process uses the final metric to adjust the input volume positions (8). The optimizer repeatedly runs steps (3) through (8), using the simulated annealing algorithm [7] to choose successive relative positions.

BSP tree construction (3) and ray casting through the resulting volume fragments (4) to produce a display image (5) are adopted from the approach of Lindholm et al. [9] In our system, the ray casting algorithm produces not only an image for display but also calculates a registration metric. We also integrated the simulated annealing algorithm as the optimizer for this application.

The oriented plane in 3D space is a concept we use throughout our system. The plane pl with equation $ax + by + cz = d$ divides 3D space into its positive side, the open half-space pl^+ with equation $ax + by + cz > d$, and its negative side, pl^-, with equation $ax + by + cz < d$. The vector $\mathbf{n} = (a, b, c)$ is normal to the plane and points into pl^+.

3.1 Binary Space Partition

After the user selects and places input volumes, our system partitions the region occupied by the input volumes into a BSP tree (Fig. 1, step (3)). Algorithm 1 accomplishes the partition. The input to the algorithm is the list of the input volumes, which must all be convex polyhedra. The algorithm uses the oriented planes of the polyhedral faces to recursively divide the polyhedra until all the face planes have been *used*. Hence, all face planes of all input polyhedra are initially marked *unused*. The output is a binary tree whose nodes have the following fields:

- *pl*, the oriented cut plane (set on interior nodes, unset on leaf nodes),
- *polys*, a list of polyhedra (empty on interior nodes, populated on leaf nodes),
- *left*, the left child node, representing pl^+, the positive side of the cut plane
- *right*, the right child node, representing pl^-, the negative side of the cut plane.

Figure 2 illustrates the first iterations of applying Algorithm 1 to a 2D scene. Although here we show the application in two dimensions, BSP tree generation is generalizable to any dimensionality, which for our system is 3D. Implementers of the BSP tree algorithm must be careful to avoid errors in the geometric operations at lines 7, 9, and 13, which can result from floating-point inaccuracies.

The binary tree resulting from Algorithm 1 has properties useful for speedy ray casting. Each leaf node l contains a list of one or more polyhedra p. All p in the same leaf node have the same boundaries, and each p refers to one of the original

Algorithm 1 Generate BSP tree

Input: Convex polyhedra $v_1 \dots v_n$, empty node r.
Output: BSP tree rooted at node r.

1: $r.polys \leftarrow$ input volumes
2: Node $n \leftarrow r$
3: **Recursive section** on n:
4: $n.pl \leftarrow$ an unused face plane from $n.polys$
5: Mark $n.pl$ as *used*
6: **for all** p within $p.polys$ **do**
7:     **if** p lies entirely in pl^+ **then**
8:         Move p to $n.left.polys$
9:     **else if** p lies entirely in pl^- **then**
10:         Move p to $n.right.polys$
11:     **else**
12:         Remove p from $n.polys$
13:         Cut p with $n.pl$ into p^+ and p^-
14:         Add p^+ to $n.left.polys$
15:         Add p^- to $n.right.polys$
16:     **end if**
17: **end for**
18: **if** $n.left.polys$ contains anything **then**
19:     **Recur** on $n.left$
20: **end if**
21: **if** $n.right.polys$ contains anything **then**
22:     **Recur** on $n.right$
23: **end if**
24: **return** Node r, root of a new BSP tree

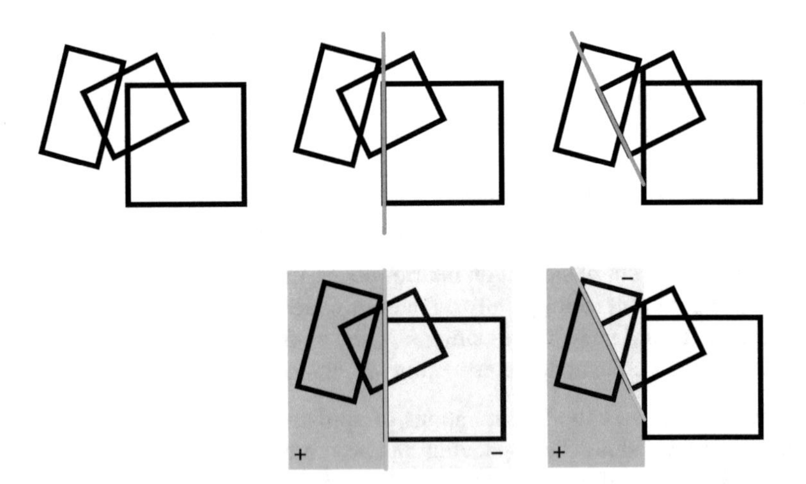

Fig. 2 First iterations of Algorithm 1 applied to a 2D scene. *Left*: initial state. *Middle*: first cut (*top*) and resulting partition (*bottom*). *Right*: second cut (*top*) and resulting partition (*bottom*)

input volumes. This means that we iterate over the list of p to get the list of input volumes that l intersects, and also that ray casting only needs to refer to the first p in each leaf node to determine ray entry and exit points. Since all leaf nodes have been constructed using cut planes chosen from the face planes of the input volumes, the interior of no p intersects any input volume boundary. This method of construction also assures that any ray cast through p must start precisely as it enters and stop precisely as it exits, and will never re-enter p. Finally, even with arbitrary viewpoints the Z-ordering of leaf volumes can be obtained without rebuilding the tree.

3.2 *Drawing the BSP Tree*

After constructing the BSP tree, the system executes the ray casting algorithm on the volume cells in order of increasing distance from the eye point i (Fig. 1, step (4)). To determine the depth ordering, the program traverses the BSP tree starting from its root. At each internal node n, if i lies on the positive side of cut plane $n.pl$, the program traverses first the left then the right child node. Otherwise it traverses the right then the left child. The side of each cut plane that contains i is always visited first, so that leaf nodes (which contain the cells) are visited in increasing depth order.

The program performs ray casting as it visits each leaf node l. This is done in several passes, using code on the CPU and several GPU shaders, as described below. Textures are used to pass intermediate results between rendering steps.

1. Sort faces of l's first polyhedron into two lists: *fronts*, with normals pointing toward i, and *backs*, with normals pointing away from i.
2. Draw each face in *backs* using shader **RecordDepth**, which returns the fragment's distance from i. We store this in a buffer d. We render computed distance to a buffer rather than use the depth buffer commonly provided by rendering toolkits because we need the Euclidean distance from i to the fragment lying on the back-face, not the Z-buffer distance resulting from the perspective transform.
3. Draw each face in *fronts* using shader **RayCast** to perform the ray casting algorithm:
    a. Record the fragment world coordinates as ray start point s.
    b. Stop point t is calculated from the value in buffer d, the distance from i to the ray exit point from l.
    c. For each input volume $v_1 \dots v_k$ that l overlaps, calculate texture-coordinate step vectors.
    d. Step through all input volumes in parallel from s to t. At each step p,
        i. Accumulate color from the average of $v_1[p] \dots v_k[p]$
        ii. Accumulate metric from the sum of the squares of all pairwise differences
    e. On ray termination, output three values: the color accumulated along each ray, stored in the buffer lc, the ray's partial metric value, stored in lm, and the number of sample steps p, stored in lp.

4. Draw each face in *fronts* using shader **Finalize** to accumulate the local color, metric, and sample count buffers into the overall buffers $C, Mbuf$, and P.

The heart of the rendering process is line 3d. Shader **RayCast** runs in parallel for each pixel of the final image that portrays the polyhedron l, stepping through l along that pixel's ray from its entry point s to exit point t. Here we explain line 3(d)i in more detail. At each step p, the shader finds the average of all k input volumes overlapping l, then looks up the corresponding color (with r, g, b and a components) in the transfer function table tf:

$$raw = \frac{1}{k}\sum_{i=1}^{k} v_i[p]$$

$$t.rgba = tf[raw]$$

The shader accumulates the color into the pixel color lc using the ray casting formula:

$$lc.rgb = lc.rgb + lc.a \cdot t.rgb$$

$$lc.a = lc.a \cdot t.a$$

To compute the metric over volume l, we compute the sum of the square of image differences over all sample points p within P, the set of ray casting sites lying within l (here $v_n[p]$ means the value located at world space point p in volume n):

$$M = \frac{1}{|P|}\sum_{i=1}^{k-1}\sum_{j=i+1}^{k}(v_i[p] - v_j[p])^2. \tag{1}$$

When a ray traverses a volume l that intersects more than one input volume, at each ray cast step p the shader **RayCast** calculates the contribution at p to the registration metric (line 3(d)ii) and accumulates the partial metric value along the ray. The subtotal for all p in the ray are stored in that ray's pixel in metric texture lm.

After completing **RayCast**, shader **Finalize** accumulates color and metric values into the final buffers (line 4):

$$C.rgb = C.rgb + C.a \cdot lc.rgb$$

$$C.a = C.a \cdot lc.a$$

$$Mbuf = Mbuf + lm$$

$$P = P + lp$$

After visiting all leaf nodes in order of increasing distance from i, the program displays color buffer C on screen (Fig. 1, (5)). Then the system divides the sum of the partial metrics stored in *Mbuf* (Fig. 1, (6)) by the sum of the ray sample counts stored in P to produce a final registration metric value M (Fig. 1, (7)).

3.3 Optimization

The final metric value (Eq. (1)) has a minimum when pixels from one volume overlap pixels in other volumes having equal value. In other words, the metric has a global minimum when volumes are perfectly registered. The position of one volume relative to another can be expressed by six values, three orthogonal translations and three orthogonal rotations. Given k input volumes, we use the $6(k-1)$-tuple θ to represent the rigid transformations of volumes $2 \ldots k$ with respect to volume 1. The metric function can be written as $M = f(\theta)$ to emphasize that the value of M depends on the rigid transformation (rotation and translation) of the input volumes. In Eq. (1), θ was implicit in the statement that overlapping volumes are sampled at points in world space.

3.3.1 Simulated Annealing

To search for the optimum arrangement of input volumes, we implement an iterative optimization routine. The optimizer calculates the metric M for a series of θ, choosing successive θ to converge to a global minimum for M. Optimizers exist which choose new θ based on predicted function behavior [11]. *Newton* and *gradient descent* optimization methods use the function value, first and/or second derivative to predict $f(\theta_{i+1})$ from θ_i based on the assumption that the derivatives exist and the function can be predicted over the interval from θ_i to θ_{i+1}. However, the OCT input volume images are full of high-frequency signal as well as speckle noise, making the metric function unpredictable even if the derivatives were available. The optimizer for the metric function must not require derivative information, and should not rely on function predictions to choose successive values for θ.

To satisfy these requirements, we implemented the simulated annealing (SA) optimizer process which executes a random walk in the parameter space, successively choosing values for θ_{i+1} that are "close" to θ_i. The step is "accepted" when it results in an improvement, and the random walk continues from θ_{i+1}. The step is "rejected" with gradually increasing probability when it results in a worse metric, and the random walk continues from θ_i. SA is well-suited to a noisy registration metric which must be calculated for each new value of θ: the possibility to accept suboptimal steps lets the algorithm escape from local minima, and in contrast to some other methods SA does not need an estimate of the function first or second derivative to choose θ_{i+1}. SA is modeled on the physical process of annealing, where a piece of hot metal is cooled slowly. When hot, the chance of random internal structure change is high. As the temperature drops, the chance of a random change in microstructure drops as well. When the temperature drops too quickly, regions of internal stress and irregular structure remain because the metal is too cold to allow the shifts that would relieve the weak spot. But in cases where the temperature drops slowly enough, the internal structure of the metal becomes homogeneous because regions of stress have a chance to relax in one of the random changes that occurs as part of the cooling process.

3.3.2 Details of Simulated Annealing

The SA algorithm tracks state with several variables: θ_i is the current rigid transformation of all input volumes, M_i is the metric calculated for θ_i using Eq. (1), and temperature T controls the likelihood of accepting a step with a worse metric than M_i. SA uses several parameters: T_0, the initial value of T; dT, the factor by which T periodically diminishes; and *freeze*, the value of T which determines the algorithm stopping point. SA has five steps:

1. $\theta_{i+1} = \textbf{randstep}(\theta_i, T)$
2. $M_{i+1} = \textbf{eval}(\theta_{i+1})$
3. If $\textbf{accept}(T, M_i, M_{i+1})$, then let $i = i + 1$
4. Let $T = T \cdot dT$
5. If $T >$ *freeze*, go to line 1.

The **randstep** function chooses θ_{i+1} to be "not far" from θ. This keeps the SA random walk from jumping far away from the user's initial placement. The function randomly chooses to translate a volume transversely or axially or to rotate the volume around one of its axes, then randomly chooses an amount by which to move the volume. We rely on the user to place the input volumes close to their desired final registration, within the diameter of the dominant features in the image, and we want the optimizer to refine the user's placement, not to discard it. Thus, we constrain the translation steps to be on the order of ten pixels and rotation steps to be less than 2° at the beginning of the annealing process. We multiply the random translation or rotation by $(T_0 + T)/2T_0$ to decrease the random step along with T.

The **eval** function renders the superimposed input volumes as described in Sects. 3.1 and 3.2 to find the metric M_{i+1} that results from θ_{i+1}.

The **accept** function implements the annealing behavior. It accepts all random steps that give an improved metric (where $M_i > M_{i+1}$), and also probabilistically allows random steps that don't give an improvement (where $M_i \leq M_{i+1}$). The probability of accepting a step from θ_i to θ_{i+1} is:

$$p(\text{step } \theta_i \rightarrow \theta_{i+1}) = \min\left(1, \exp\left(-\frac{M_{i+1} - M_i}{T}\right)\right). \quad (2)$$

Figure 3 shows a graph of Equation (2). When $M_i > M_{i+1}$ (to the right of the origin), the proposed step will improve the registration so **accept** always returns

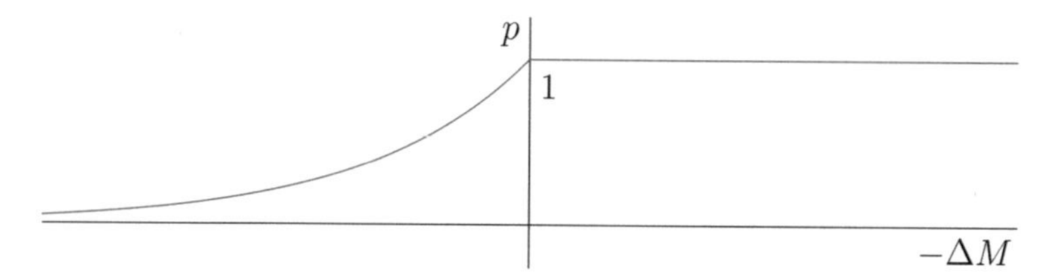

Fig. 3 Graph of Equation (2). The probability of accepting a random step is based on the change in metric

true. The further to the left of origin, the worse the effect on registration, so the lower the probability of accepting the step. As T decreases, the exponential curve becomes sharper and the probability of accepting a bad step decreases. Near freezing, the curve approaches a step from 0 to 1 at the origin, making the behavior of SA approximate a gradient-descent random walk.

3.3.3 Parameters and Starting Conditions

Successful optimization with SA requires good starting parameters. Together, the parameters T_0, dT and *freeze* determine the *annealing schedule*. The temperature T is closely linked to the random walk. When T_0 is too high, the algorithm will make many random steps with a very low penalty for a bad step and will tend to wander away from the user's specified initial placement, failing to find a registration. So, prior to starting SA proper, we determine a value for T_0 by perturbing the position of each input volume in turn for several tens of steps by random translation of 10 pixels or less and rotation by 2° or less. For the set of all bad steps ($M_i < M_{i+1}$) we find the average $\Delta M = M_{i+1} - M_i$ and take this to be the expected value. We then start SA with $T_0 = E(\Delta M)/\ln 2$. This way, the expected probability of accepting a step that worsens (increases) the metric is initially 0.5, giving the algorithm a chance to settle down around the correct registration as T decreases.

For some initial volume arrangements, SA will fail to settle down around the true registration. We have observed this to happen when the initial placement was not close to the true registration. Since SA is a stochastic process, we cannot guarantee convergence. However, we have observed that initial registration of OCT data sets cut into overlapping subsets and given an initial registration within 10 pixels of displacement tend to succeed. Rotation is much more sensitive; initial misrotation of 3° or more tends to cause SA to fail, while correct rotation of less than 1° tends to lead to success.

To allow the use of an easy-to-work-with value of $dT = 0.95$, we repeat line 1, 2 and 3 several (hundred) times for each input volume in turn before line 4, rather than decreasing T by an infinitesimal amount each time through the loop. Finally, we set $freeze = T_0 \cdot 10^{-5}$. This results in an annealing schedule which in our experience results in successful registration after an acceptable computation time.

Finally, we save the lowest M_i and its θ_i, and apply that arrangement to the input volumes at the end of the SA process.

3.4 Effect of View Point

The ray casting algorithm can be implemented with either perspective or orthogonal projection of the scene to a viewing plane. Orthogonal projection casts parallel rays normal to the viewing plane through the volume data sets; perspective projection casts rays that diverge from the eye point, pass through the viewing plane and volume data. We implemented perspective projection in this project. Since the rays diverge, ray casting sample points p used for metric calculation are dense close to the eye point and become sparser in regions farther away. Thus, a change in eye point will move the region where p is most densely distributed, likely resulting in a different metric value even with no change in input volume position. This means that only metric values computed from the same eye point are comparable. Practically speaking, the simulated annealing registration must restart when a user moves the global point of view. We have found that repeated experiments with the same initial placement of the input volumes but differing eye points converge to the same correct final registration.

Because our system calculates the registration metric by ray casting through volume overlap regions, we require these regions to be within view, or they will not contribute to the metric. Likewise, "island" volumes that do not overlap others cannot affect the metric, and must by registered by means other than our system.

The common technique of early ray termination cannot be applied to our combined metric and visualization method. Even though a ray may accumulate enough opacity that its color will not change over further sample steps, metric calculation must still be done at those sample steps; so all rays must run to volume exit. However, our algorithm skips all empty space, since it casts rays through BSP leaf nodes, which by construction always intersect at least one input volume.

4 Examples and Results

We have applied our registration system on several classes of input. To demonstrate basic functionality, we show samples from synthetic data sets. Overlapping subsets taken from the *aneurism* data set (courtesy Viatronix, Inc., USA; available at http://www.volvis.org) show successful registration of real-world data. For each example, we show the evolution of the registration metric throughout the optimization run. For test data sets, a metric value of zero indicates perfect registration. A metric value of zero is not possible with real OCT data sets because distinct images of the same area contain random noise. Finally, we show two examples of successfully stitching OCT data sets. Lacking a ground truth for the in vivo OCT data sets, we show several cross-sectional slices through the combined volume to demonstrate the improvement brought to the combined image by our method.

4.1 Test Data Sets

The first test case is two volumes sampled from a data set containing three spherical solids, with centers at $(0.2, 0.4, 0.1)$, $(1.1, 0.1, 0.1)$, and $(0.6, 0.7, 0.6)$ and radius $0.15, 0.40, 0.25$ (arbitrary units), as shown by Fig. 4. Starting from a close but inexact manual registration, the system was able to produce an accurately-registered result. The second test case is two volumes sampled from a data set containing three

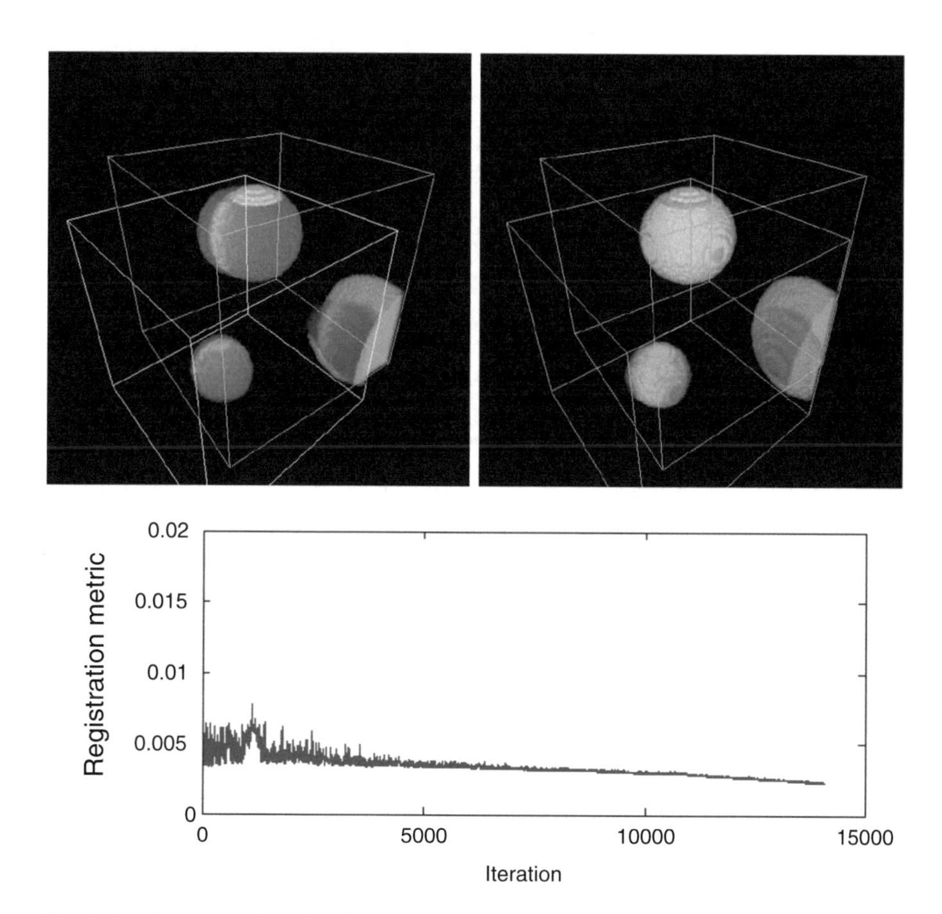

Fig. 4 Synthetic example with three blobs. *Upper-left*: initial placement, before registration. *Upper-right*: result of registration. *Bottom*: registration metric over duration of simulated annealing. The registration metric value depends on the overlapping volume images and the ray casting eye point. A metric value is only significant relative to other values in the same optimization run

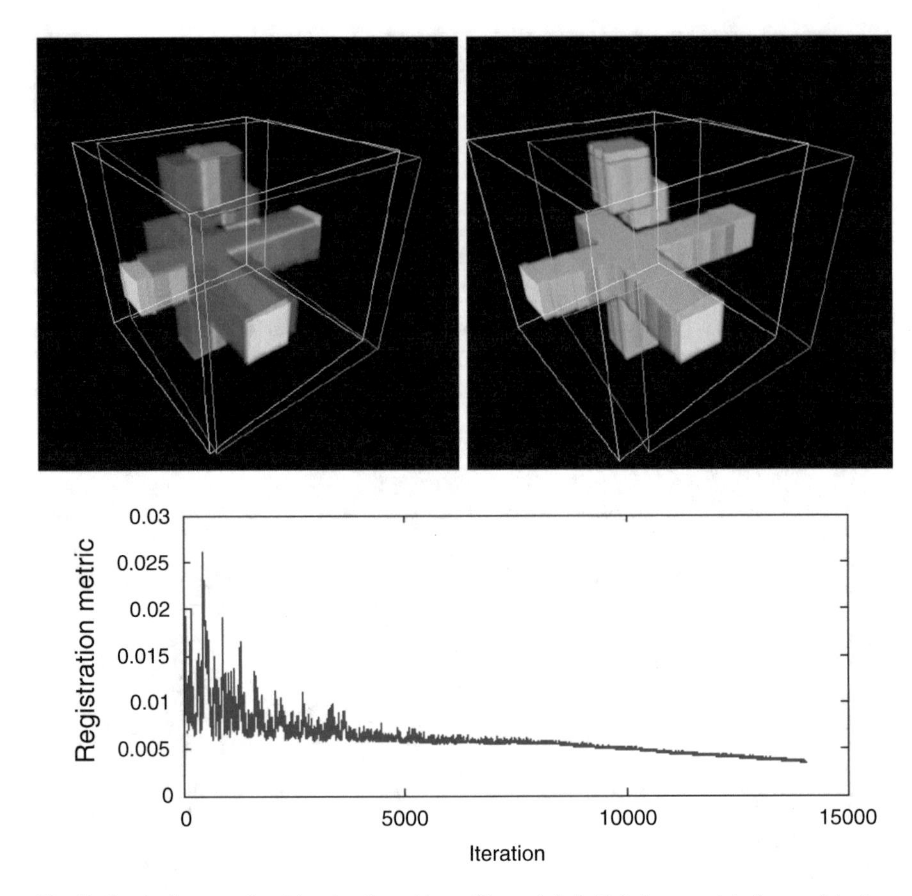

Fig. 5 Synthetic example with axis-aligned bars. *Upper-left*: initial placement, before registration. *Upper-right*: result of registration. *Bottom*: registration metric over duration of simulated annealing

bars, parallel to the $X-$, $Y-$, and $Z-$axes, with a square cross-section 0.2 units on a side and several other features to help quick visual verification of correct orientation. See Fig. 5. The second data set was sampled at a rotation of 0.122 radians about the $Z-$axis. The registration was successful.

The third example shows the ability of our system to simultaneously register several data sets. From the *aneurism* data set, we extracted four overlapping subimages. We manually positioned the subimages in rough alignment, then ran our system to refine the initial placement. See Fig. 6 for overall images before and after registration as well as details of vessels interpenetrating several input volumes.

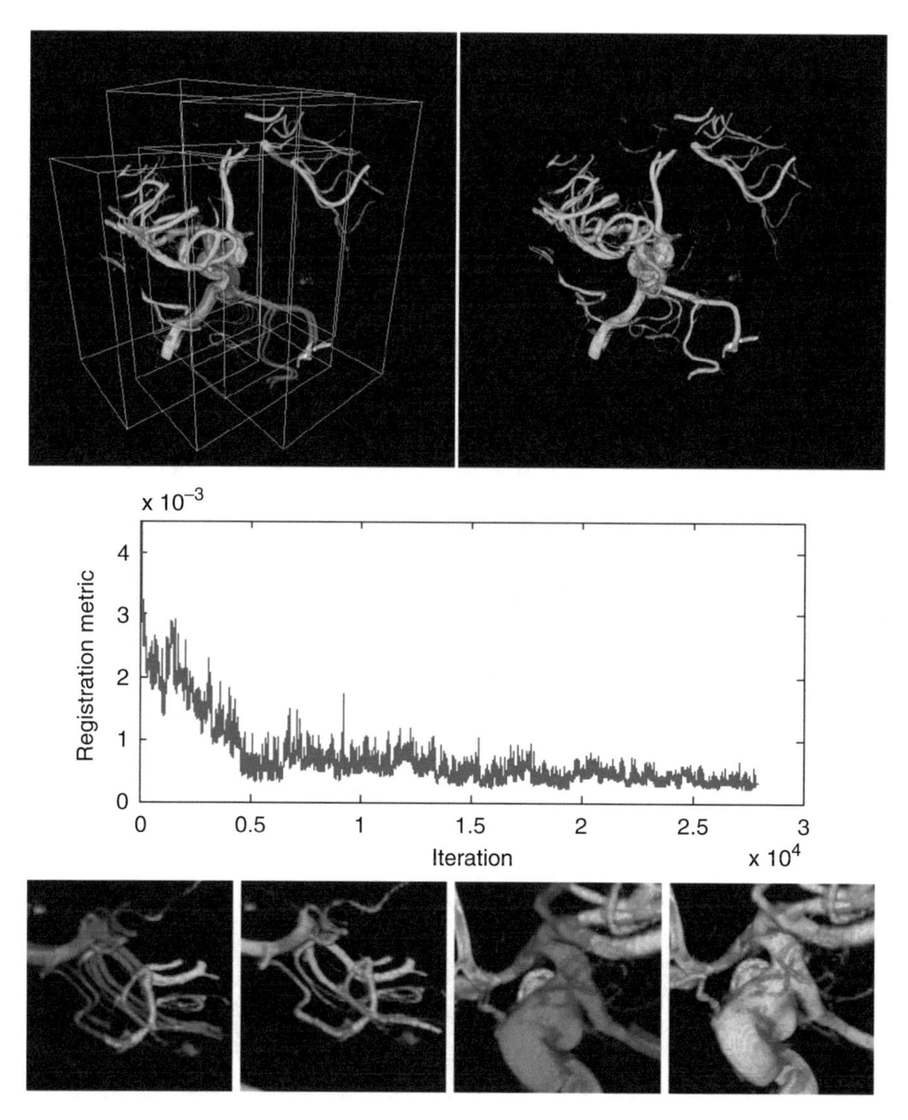

Fig. 6 The *aneurism* volume image. Original image was divided into four overlapping subvolumes, which were then re-registered. *Upper-left*: initial placement, before registration. *Upper-right*: result of registration. *Middle*: registration metric over duration of simulated annealing. *Lower row*: before-and-after detail views

4.2 Application to OCT Volumes

We used our technique to stitch OCT volumes of overlapping retinal regions. A retinal region's location is described by visual angle off the line of sight, in degrees, and direction, in terms of temporal or nasal (toward the temple or the nose,

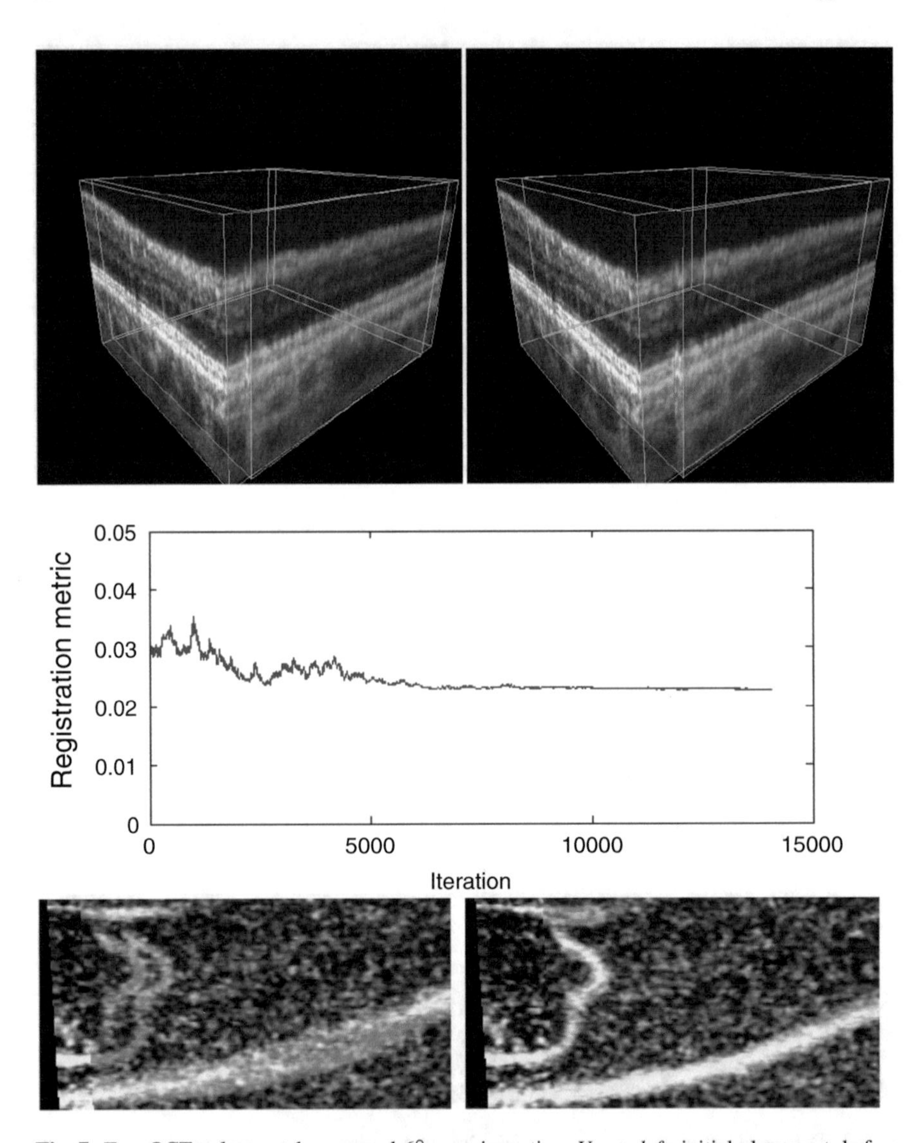

Fig. 7 Two OCT volumes taken around 6° superior retina. *Upper-left*: initial placement, before registration. *Upper-right*: result of registration. *Middle*: registration metric over duration of simulated annealing. *Lower row*: before-and-after detail views showing improvement in vasculature registration

respectively) and superior or inferior (physically above or below the fovea, or center of gaze). Figure 7 shows two overlapping volumes acquired about 6° superior retina, before and after registration. The entire volume is shown, as well as views of various layers within the retina, to illustrate the effect of registration. Likewise, Fig. 8 shows the registration of two data sets taken at 6° temporal, 6° superior retina. In the detail

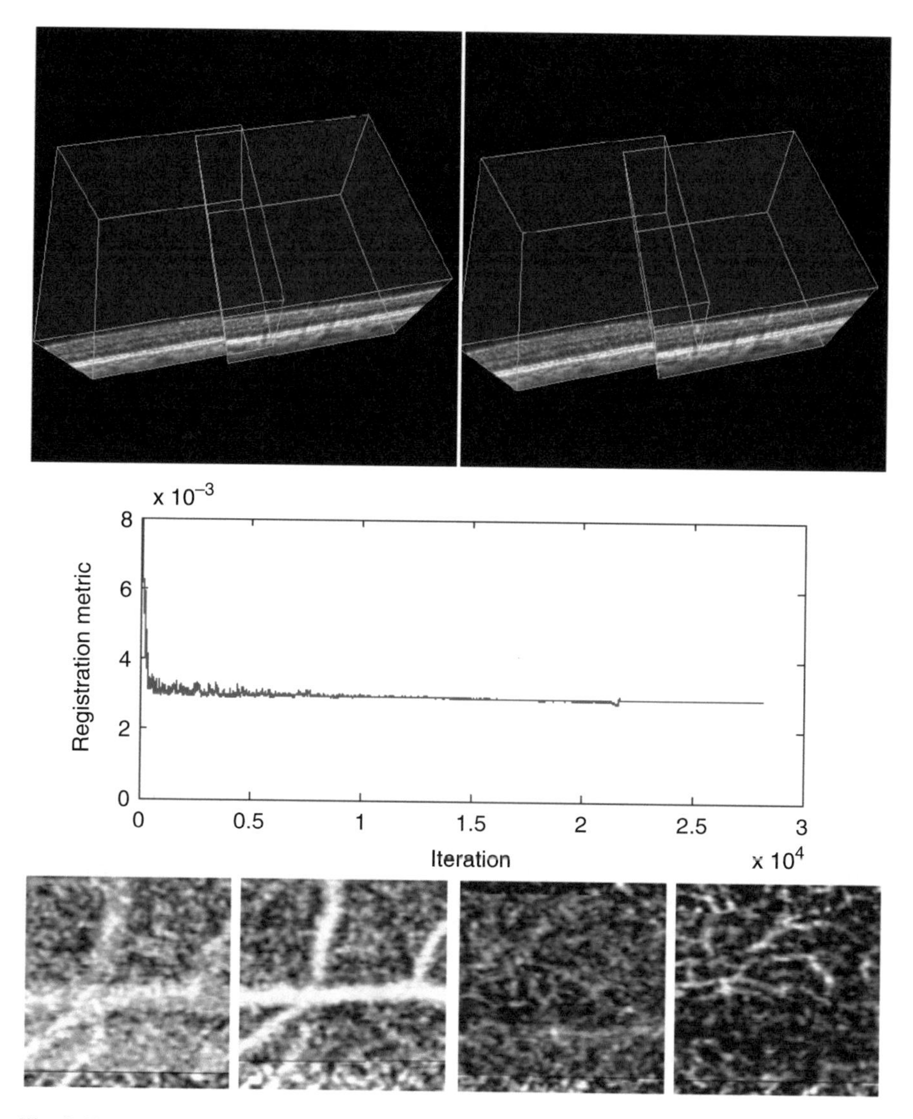

Fig. 8 Two OCT volumes taken around 6° superior, 6° temporal retina. *Upper-left*: initial placement, before registration. *Upper-right*: result of registration. *Middle*: registration metric over duration of simulated annealing. *Lower row*: before-and-after detail views

views of both data sets the effects on vessels are clearly seen. While it is possible to register OCT data sets "by hand," features such as vasculature that guide registration generally occur in the middle of the dense OCT data set and are accessible to the user only with some effort. As shown by the detail views of Figs. 7 and 8, the optimizer gives a major improvement in registration and lets us see and measure detail that would otherwise be lost in noise.

The main difficulty of registering OCT data sets is pervasive eye motion in imaging subjects. Any eye motion will result in distortion in the volume image, and since different motion is present over the course of successive OCT scans the images will not match and will be unable to register without correction for this motion.

5 Conclusions and Further Work

We have presented a system for combined visualization and registration of volume data sets. We perform ray casting through multiple volume data sets. In overlapping regions, we compute the sum of squared difference between volumes at each ray casting sample site. Thus, a metric for the quality of the registration within overlap regions is calculated in real time for each frame displayed by the ray casting engine. We added an optimizer to incrementally perturb the position of each displayed volume, using the simulated annealing algorithm to find an optimal registration. After showing examples using synthetic and example data sets, we successfully applied our system to two pairs of OCT data sets of the human retina. This registration produced a volume image of high resolution and wide coverage.

We have found, however, that our technique is impractical if one or more of the input volumes contain motion distortion in the overlap region. Motion distortion is present to varying degrees in most OCT volumes. Techniques to mitigate motion distortion will increase the applicability of our system to OCT and other scanning image capture modalities.

Acknowledgements The authors gratefully acknowledge the help of VSRI laboratory members Susan Garcia and Raju Poddar. Support was provided by the National Eye Institute (R01 EY014743) and Research to Prevent Blindness (NY). This work was performed under the auspices of the U.S. Department of Energy by Lawrence Livermore National Laboratory under Contract DE-AC52-07NA27344. LLNL-BOOK-644901.

References

1. Bria, A., Silvestri, L., Sacconi, L., Pavone, F., Iannello, G.: Stitching terabyte-sized 3D images acquired in confocal ultramicroscopy. In: 2012 9th IEEE International Symposium on Biomedical Imaging (ISBI), pp. 1659–1662 (2012). doi:10.1109/ISBI.2012.6235896
2. Cai, W., Sakas, G.: Data intermixing and multi-volume rendering. Comput. Graph. Forum **18**(3), 359–368 (1999). doi:10.1111/1467-8659.00356
3. Dalvi, R., Hacihaliloglu, I., Abugharbieh, R.: 3D ultrasound volume stitching using phase symmetry and Harris corner detection for orthopaedic applications. Proc. SPIE **7623**, 762,330.1–762,330.8 (2010). doi:10.1117/12.844608
4. Hadwiger, M., Beyer, J., Jeong, W.K., Pfister, H.: Interactive volume exploration of petascale microscopy data streams using a visualization-driven virtual memory approach. IEEE Trans. Vis. Comput. Graph. **18**(12), 2285–2294 (2012). doi:10.1109/TVCG.2012.240
5. Jacq, J.J., Roux, C.: A direct multi-volume rendering method aiming at comparisons of 3-D images and models. IEEE Trans. Inf. Technol. Biomed. **1**(1), 30–43 (1997)

6. Kainz, B., Grabner, M., Bornik, A., Hauswiesner, S., Muehl, J., Schmalstieg, D.: Ray casting of multiple volumetric datasets with polyhedral boundaries on manycore GPUs. ACM Trans. Graph. **28**(5), 152:1–152:9 (2009). doi:10.1145/1618452.1618498
7. Kirkpatrick, S., Gelatt Jr., C.D., Vecchi, M.P.: Optimization by simulated annealing. Science **220**(4598), 671–680 (1983)
8. Kirmizibayrak, C., Yim, Y., Wakid, M., Hahn, J.: Interactive visualization and analysis of multimodal datasets for surgical applications. J. Digit. Imaging **25**, 792–801 (2012). doi:10.1007/s10278-012-9461-y
9. Lindholm, S., Ljung, P., Hadwiger, M., Ynnerman, A.: Fused multi-volume DVR using binary space partitioning. Comput. Graph. Forum **28**(3), 847–854 (2009). doi:10.1111/j.1467-8659.2009.01465.x
10. Lux, C., Fröhlich, B.: GPU-based ray casting of multiple multi-resolution volume datasets. In: Advances in Visual Computing. Lecture Notes in Computer Science, vol. 5876, pp. 104–116. Springer, Berlin/Heidelberg (2009). doi:10.1007/978-3-642-10520-3_10
11. Nocedal, J., Wright, S.J.: Numerical Optimization. Springer Series in Optimization Research. Springer, Berlin (2000)
12. Poddar, R., Cortés, D.E., Werner, J.S., Mannis, M.J., Zawadzki, R.J.: Three-dimensional anterior segment imaging in patients with type I Boston keratoprosthesis with switchable full depth range swept source optical coherence tomography. J. Biomed. Opt. **18**(086002), 1–7 (2013). Although this citation describes imaging of the anterior segment, we use retinal images produced by the system
13. Rößler, F., Botchen, R., Ertl, T.: Dynamic shader generation for GPU-based multi-volume ray casting. IEEE Comput. Graph. Appl. **28**(5), 66–77 (2008). doi:10.1109/MCG.2008.96
14. Wojtkowski, M., Leitgeb, R., Kowalczyk, A., Bajraszewski, T., Fercher, A.F.: In vivo human retinal imaging by Fourier domain optical coherence tomography. J. Biomed. Opt. **7**, 457–463 (2002). doi:10.1117/1.1482379
15. Zawadzki, R.J., Fuller, A.R., Choi, S.S., Wiley, D.F., Hamann, B., Werner, J.S.: Improved representation of retinal data acquired with volumetric Fd-OCT: co-registration, visualization, and reconstruction of a large field of view. In: Proceedings of Society of Photo-Optical Instrumentation Engineers (SPIE), pp. 68,440C–1. The International Society for Optical Engineering (2008)
16. Zawadzki, R.J., Choi, S.S., Fuller, A.R., Evans, J.W., Hamann, B., Werner, J.S.: Cellular resolution volumetric in vivo retinal imaging with adaptive optics–optical coherence tomography. Opt. Express **17**(5), 4084–4094 (2009). doi:10.1364/OE.17.004084

Feature Lines for Illustrating Medical Surface Models: Mathematical Background and Survey

Kai Lawonn and Bernhard Preim

Abstract This paper provides a tutorial and survey for a specific kind of illustrative visualization technique: feature lines. We examine different feature line methods. For this, we provide the differential geometry behind these concepts and adapt this mathematical field to the discrete differential geometry. All discrete differential geometry terms are explained for triangulated surface meshes. These utilities serve as basis for the feature line methods. We provide the reader with all knowledge to re-implement every feature line method. Furthermore, we summarize the methods and suggest a guideline for which kind of surface which feature line algorithm is best suited. Our work is motivated by, but not restricted to, medical and biological surface models.

1 Introduction

The application of illustrative visualization has increased in recent years. The principle goal behind the concept of illustrative visualization is a meaningful, expressive, and simplified depiction of a problem, a scene or a situation. As an example, running people are represented running stickmans, which can be seen in the Olympic games, and other objects become simplified line drawings, see Fig. 1. More complex examples can be found in medical atlases. Most anatomical structures are painted and illustrated with pencils and pens. Gray's anatomy is one of the famous textbooks for medical teaching. Most other textbooks in this area orient to depict anatomy with art drawing, too.

Other than simplified representation, illustrative visualization is not restricted to these fields. Illustrative techniques are essential for *focus-and-context* visualizations. Consider a scene with anatomical structures and one specific (important)

K. Lawonn (✉)
Institute of Computational Visualistics, University of Koblenz - Landau, Koblenz, Germany
e-mail: lawonn@uni-koblenz.de

B. Preim
Faculty of Computer Science, Department of Simulation and Graphics, Otto-von-Guericke, University Magdeburg, Magdeburg, Germany
e-mail: preim@isg.cs.uni-magdeburg.de

L. Linsen et al. (eds.), *Visualization in Medicine and Life Sciences III*, Mathematics and Visualization, DOI 10.1007/978-3-319-24523-2_5

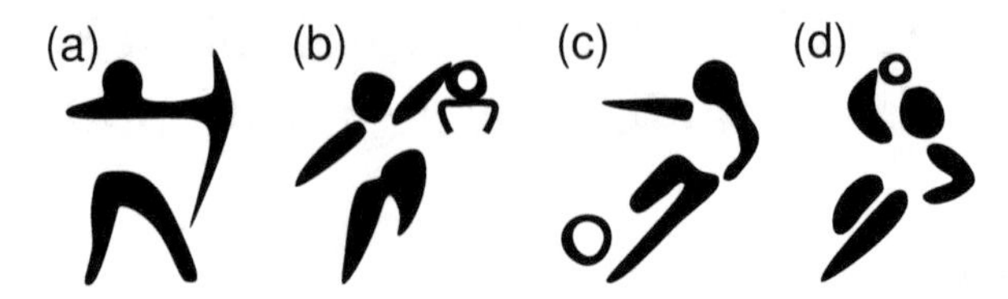

Fig. 1 Visual abstraction of the four Olympic disciplines: (**a**) archery, (**b**) basketball, (**c**) football and (**d**) handball in the style of the pictograms of the Olympic Games 2012 in London

structure. The specific structure may be strongly related to the surrounding objects. Therefore, hiding the other objects is not a viable option. In contrast, depicting all structures leads to visual clutter and optical distraction of the most important structures. Focus-and-context visualization is characterized by a few local regions that are displayed in detail and with emphasis techniques, such as a saturated color. Surrounding contextual objects are displayed in a less prominent manner to avoid distraction from focal regions. Medical examples are vessels with interior blood flow, livers with inner structures including vascular trees and possible tumors, proteins with surface representation and interior ribbon visualization. Focus-and-context visualization is not restricted to medical data. An example is the vehicle body and the interior devices. The user or engineer needs the opportunity to illustrate all devices in the same context.

There are numerous methods for different illustration techniques. This survey is focused on a specific illustrative visualization category: feature lines. Feature lines are a special group of line drawing techniques. Another class of line drawing methods is hatching. Hatching tries to convey the shape by drawing a bunch of lines. Here, the spatial impression of the surface is even more improved. Several methods exist to hatch the surface mesh, see [15, 20, 30, 39, 53, 56]. In contrast, feature lines try to generate lines at salient regions only. Not only for illustrative visualization, feature lines can also be used for rigid registrations of anatomical surfaces [46] or for image and data analysis in medical applications [12]. The goal of this survey is to convey the reader to the different feature line methods and offer a tutorial with all the knowledge to be able to implement each of the methods.

1.1 Organization

We first give an overview of the mathematical background. In Sect. 2, we introduce the necessary fundamentals of differential geometry. Afterwards, we adapt these fundamentals to triangulated surface meshes in Sect. 3. Section 4 discusses general aspects and requirements for feature lines. Next, we present different feature line methods in Sect. 5 and compare them in Sect. 6. Finally, Sect. 7 holds the conclusion of this survey.

2 Differential Geometry Background

This section presents the fundamentals of differential geometry for feature line generation, which will be crucial for the further sections. We present the basic terms and properties. This section is inspired by differential geometry books [10, 11, 28].

2.1 *Basic Prerequisites*

A surface $f: I \subset \mathbb{R}^2 \rightarrow \mathbb{R}^3$ is called a parametric surface if f is an *immersion*. An *immersion* means that all partial derivatives $\frac{\partial f}{\partial x_i}$ are injective at each point. The further calculations are mostly based on the tangent space of a surface. The tangent space $\mathcal{T}_p f$ of f is defined as the linear combination of the partial derivatives of f:

$$\mathcal{T}_p f := \text{span}\left\{ \frac{\partial f}{\partial x_1}\bigg|_{\mathbf{x}=\mathbf{u}}, \frac{\partial f}{\partial x_2}\bigg|_{\mathbf{x}=\mathbf{u}} \right\}.$$

Here, span is the space of all linear combinations. Formally: $\text{span}\{\mathbf{v}_1, \mathbf{v}_2\} := \{\alpha \mathbf{v}_1 + \beta \mathbf{v}_2 \mid \alpha, \beta \in \mathbb{R}\}$. With the tangent space, we can define a normalized normal vector $\mathbf{n}$. The (normalized) normal vector $\mathbf{n}(\mathbf{u})$ at $\mathbf{p} = f(\mathbf{u})$ is defined such that for all elements $\mathbf{v} \in \mathcal{T}_p f$ the equation $\langle \mathbf{v}, \mathbf{n}(\mathbf{u}) \rangle = 0$ holds, where $\langle .,. \rangle$ denotes the canonical Euclidean dot product. Therefore, $\mathbf{n}(\mathbf{u})$ is defined as:

$$\mathbf{n}(\mathbf{u}) = \frac{\frac{\partial f}{\partial x_1}\Big|_{\mathbf{x}=\mathbf{u}} \times \frac{\partial f}{\partial x_2}\Big|_{\mathbf{x}=\mathbf{u}}}{\left\| \frac{\partial f}{\partial x_1}\Big|_{\mathbf{x}=\mathbf{u}} \times \frac{\partial f}{\partial x_2}\Big|_{\mathbf{x}=\mathbf{u}} \right\|}.$$

This map is also called the Gauss map. Figure 2 depicts the domain of a parametric surface as well as the tangent space $\mathcal{T}_p f$ and the normal $\mathbf{n}$.

2.2 *Curvature*

The curvature is a fundamental property to identify salient regions of a surface that should be conveyed by feature lines. Colloquially spoken, it is a measure of how far the surface bends at a certain point. If we consider ourselves to stand on a sphere at a specific point, it does not matter in which direction we go, the bending will be the same. If we imagine we stand on a plane at a specific point, we can go in any direction, there will be no bending. Without knowing any measure of the curvature, we can state that a plane has zero curvature and that a sphere with a small radius has a higher curvature than a sphere with a higher radius. This is due to the fact that a sphere with an increasing radius becomes locally more a plane. Intuitively, the

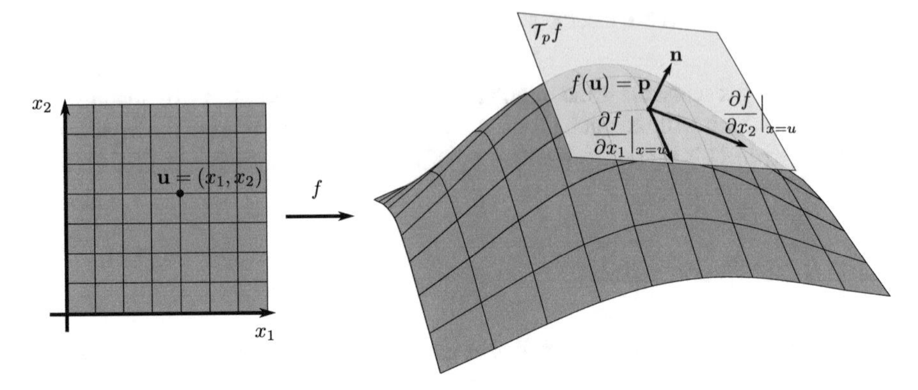

Fig. 2 The basic elements for differential geometry. A parametric surface is given and the partial derivatives create the tangent space

curvature depends also on the direction in which we decide to go. On a cylinder, we have a bending in one direction but not in the other. Painting the trace of a walk on the surface and view it in 3D space, we could treat this as a 3D curve. The definition of the curvature of a curve may be adapted to the curvature of a surface. The adaption of this concepts is explained in the following. Let $c: I \subset \mathbb{R} \to \mathbb{R}^3$ be a 3D parametric curve with $\|\frac{dc}{dt}\| = 1$. The property of constant length of the derivative is called arc length or natural parametrization. One can show that such a parametrization exists for each continuous, differentiable curve that is an immersion. So, if we want to measure the size of bending, we can use the norm of the second derivative of the curve. Therefore, the (absolute) curvature $\kappa(t)$ at a time point t is defined as:

$$\kappa(t) = \left\|c''(t)\right\|.$$

To determine the curvature on a certain point of the surface in a specific direction, we can employ a curve and calculate its curvature. This approach is imperfect because curves that lie in a plane can have non-vanishing curvature, e.g., a circle, whereas we claimed to have zero curvature on a planar surface. Therefore, we have to distinguish which part of the second derivative of the curve contributes to the tangent space and which contributes to the normal part of the surface. Decomposing the second derivative of the curve into tangential and normal part of the surface yields:

$$c''(t) = \underbrace{\mathrm{proj}_{\mathcal{T}_p f}\, c''(t)}_{\text{tangential part}} + \underbrace{\langle c''(t), \mathbf{n}\rangle \mathbf{n}}_{\text{normal part}},$$

where $c(t) = \mathbf{p}$ and $\mathrm{proj}_E\, \mathbf{x}$ means the projection of the point $\mathbf{x}$ onto the space E, see Fig. 3. The curvature $\kappa_c(\mathbf{p})$ of the surface at $\mathbf{p}$ along the curve c is defined as the coefficient of the normal part:

$$\kappa_c(p) = \langle c''(t), \mathbf{n}\rangle. \tag{1}$$

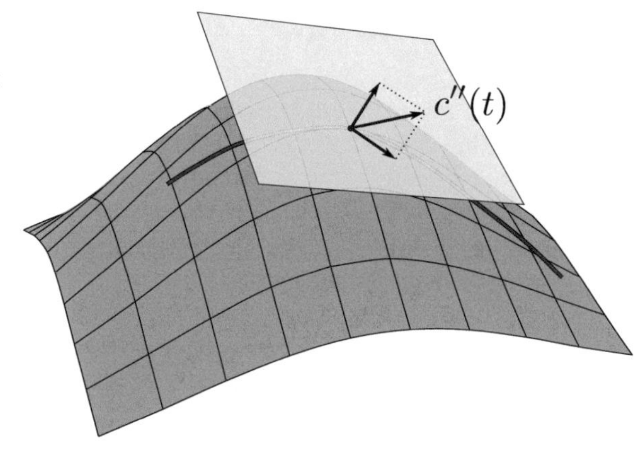

Fig. 3 The curve's second derivative is decomposed into the tangential and normal part

Hence, we know that $c'(t) \in \mathcal{T}_p f$ and $\langle c'(t), \mathbf{n}\rangle = 0$. Deriving the last equation yields:

$$\frac{d}{dt}\langle c'(t), \mathbf{n}\rangle = 0$$
$$\frac{d}{dt}\langle c'(t), \mathbf{n}\rangle = \langle c''(t), \mathbf{n}\rangle + \langle c'(t), \frac{\partial \mathbf{n}}{\partial t}\rangle.$$

We obtain

$$\langle c''(t), \mathbf{n}\rangle = -\langle c'(t), \frac{\partial \mathbf{n}}{\partial t}\rangle.$$

Combining this equation with Eq. (1) yields

$$\kappa_{c'(t)}(\mathbf{p}) = -\langle c'(t), \frac{\partial \mathbf{n}}{\partial t}\rangle. \tag{2}$$

Thus, the curvature of a surface at a specific point in a certain direction can be calculated by a theorem by Meusnier. We call the vectors $\mathbf{v}, \mathbf{w}$ at $\mathbf{p}$ the maximal/minimal principle curvature directions of the maximal and minimal curvature, if $\kappa_{\mathbf{v}}(\mathbf{p}) \geq \kappa_{\mathbf{v}'}(\mathbf{p}), \kappa_{\mathbf{w}}(\mathbf{p}) \leq \kappa_{\mathbf{v}'}(\mathbf{p})$ for all directions $\mathbf{v}' \in \mathcal{T}_p f$. If such a minimum and maximum exists, then $\mathbf{v}$ and $\mathbf{w}$ are perpendicular, see Sect. 2.5 for a proof. If we want to determine the curvature in direction $\mathbf{u}$, we first need to normalize $\mathbf{u}, \mathbf{v}, \mathbf{w}$ and can then determine $\kappa_{\mathbf{u}}(\mathbf{p})$ by:

$$\kappa_{\mathbf{u}}(\mathbf{p}) = \langle \mathbf{u}, \mathbf{v}\rangle^2 \kappa_{\mathbf{v}}(\mathbf{p}) + \langle \mathbf{u}, \mathbf{w}\rangle^2 \kappa_{\mathbf{w}}(\mathbf{p}). \tag{3}$$

The coefficients of the curvature are the decomposition of the principle curvature directions with the vector $\mathbf{u}$.

2.3 Covariant Derivative

The essence of the feature line generation is the analysis of local variations in a specific direction, i.e., the covariant derivative. Therefore, the covariant derivative is a crucial concept for feature line methods. We consider a scalar field on a parametric surface $\varphi: f(I) \to \mathbb{R}$. One can imagine this scalar field as a heat or pressure (as well as a curvature) distribution. The directional derivative of φ in direction $\mathbf{v}$ can be written as $D_{\mathbf{v}}\varphi$ and is defined by:

$$D_{\mathbf{v}}\varphi(\mathbf{x}) = \lim_{h\to 0} \frac{\varphi(\mathbf{x} + h\mathbf{v}) - \varphi(\mathbf{x})}{h}.$$

If φ is differentiable at $\mathbf{x}$, the directional derivative can be simplified:

$$D_{\mathbf{v}}\varphi(\mathbf{x}) = \langle \nabla\varphi(\mathbf{x}), \mathbf{v} \rangle,$$

where ∇ denotes the gradient. The gradient is an operator applied to a scalar field and results in a vector field. When we want to extend the definition of the derivative to an arbitrary surface, we first need to define the gradient of surfaces. In the following, we make use of the *covariant derivative*. The standard directional derivative results in a vector which lies somewhere in the 3D space, whereas the covariant derivative is restricted to stay in the tangent space of the surface. The gradient is a two-dimensional vector. Actually, we need a three-dimensional vector in the tangent space of the surface. Here, we employ the gradient and use it as coefficients of the tangential basis. Unfortunately, this leads to wrong results because of the distortions of the basis of the tangent space, see Fig. 4. The basis is not necessarily an orthogonal normalized basis as in the domain space and can therefore lead to distortions of the gradient on the surface.

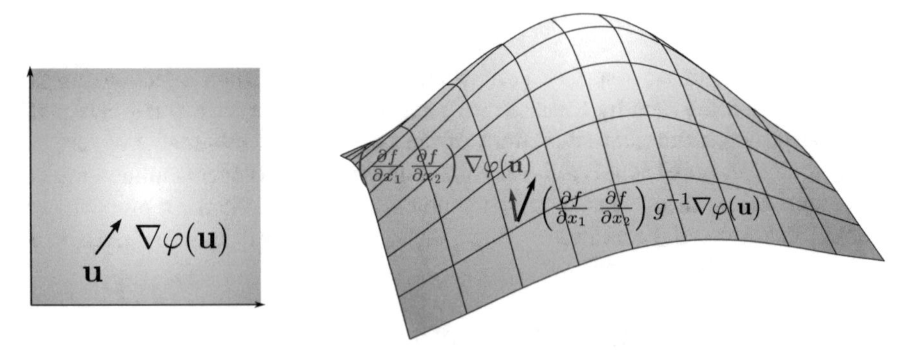

Fig. 4 Given: a scalar field in the domain. Determining the gradient and using it as coefficient for the basis tangent vectors leads to the wrong result (*grey*). Balancing the distortion with the inverse of the metric tensor yields the correct gradient on the surface (*black*)

One way to calculate this vector is to use the plain scalar field $\varphi\colon \mathbb{R}^3 \to \mathbb{R}$. Afterwards, we are able to attain the gradient in three-dimensional space and project it on the tangent space. However, we want to use the gradient of $\varphi\colon \mathbb{R}^2 \to \mathbb{R}$ in the domain of a parametric surface and compensate the length distortion such that we can use it as coordinates with the basis in the tangent space. One important fact is when multiplying the gradient with the i-th basis vector, one obtains the partial derivative of φ with x_i. Hence, we know that the three-dimensional gradient $\nabla\varphi$ lies in the tangent space. Therefore, it can be represented as a linear combination of $\frac{\partial f}{\partial x_1}, \frac{\partial f}{\partial x_2}$ with coefficients α, β:

$$\nabla\varphi = \alpha \cdot \frac{\partial f}{\partial x_1} + \beta \cdot \frac{\partial f}{\partial x_2}.$$

Multiplying both sides with the basis vectors and using the relation $\frac{\partial \varphi}{\partial x_i} = \langle \nabla\varphi, \frac{\partial f}{\partial x_i}\rangle$, we obtain an equation system:

$$\begin{pmatrix} \frac{\partial \varphi}{\partial x_1} \\ \frac{\partial \varphi}{\partial x_2} \end{pmatrix} = \begin{pmatrix} \alpha \cdot \langle \frac{\partial f}{\partial x_1}, \frac{\partial f}{\partial x_1}\rangle + \beta \cdot \langle \frac{\partial f}{\partial x_1}, \frac{\partial f}{\partial x_2}\rangle \\ \alpha \cdot \langle \frac{\partial f}{\partial x_1}, \frac{\partial f}{\partial x_2}\rangle + \beta \cdot \langle \frac{\partial f}{\partial x_2}, \frac{\partial f}{\partial x_2}\rangle \end{pmatrix} = \underbrace{\begin{pmatrix} \langle \frac{\partial f}{\partial x_1}, \frac{\partial f}{\partial x_1}\rangle & \langle \frac{\partial f}{\partial x_1}, \frac{\partial f}{\partial x_2}\rangle \\ \langle \frac{\partial f}{\partial x_1}, \frac{\partial f}{\partial x_2}\rangle & \langle \frac{\partial f}{\partial x_2}, \frac{\partial f}{\partial x_2}\rangle \end{pmatrix}}_{g:=} \begin{pmatrix} \alpha \\ \beta \end{pmatrix}.$$

The matrix g is called the *metric tensor*. This tensor describes the length and area distortion from $\mathbb{R}^2$ to the surface. The last equation yields the coefficients α, β when multiplied with the inverse of g:

$$\begin{pmatrix} \alpha \\ \beta \end{pmatrix} = g^{-1} \begin{pmatrix} \frac{\partial \varphi}{\partial x_1} \\ \frac{\partial \varphi}{\partial x_2} \end{pmatrix}.$$

This leads to a general expression of the gradient for a scalar field $\varphi\colon \mathbb{R}^n \to \mathbb{R}$:

$$\nabla\varphi = \sum_{i,j=1}^{n} \left(g^{ij} \frac{\partial \varphi}{\partial x_j} \right) \frac{\partial}{\partial x_i}, \tag{4}$$

where g^{ij} is the i,j-th matrix entry from the inverse of g and $\frac{\partial}{\partial x_i}$ means the basis. Now, we are able to determine the covariant derivative of a scalar field by first determining its gradient and afterwards using the dot product:

$$D_w\varphi = \langle \nabla\varphi, w\rangle.$$

2.4 Laplace-Beltrami Operator

The Laplace-Beltrami operator is needed for a specific feature line method and will therefore be introduced. The Laplace operator is defined as a composition of the *gradient* and the *divergence*. When interpreting the vector field as a flow field, the divergence is a measure of how much more flow leaves a specific region than flow enters. In the Euclidean space, the divergence div Φ of a vector field $\Phi\colon \mathbb{R}^n \to \mathbb{R}^n$ is the sum of the partial derivatives of the components Φ_i:

$$\operatorname{div} \Phi = \sum_{i=1}^{n} \frac{\partial}{\partial x_i} \Phi_i.$$

The computation of the divergence for a vector field $\Phi\colon \mathbb{R}^n \to \mathbb{R}^n$ in Euclidean space is straightforward. However, for computing the divergence to an arbitrary surface we have to be aware of the length and area distortions. Without giving a derivation of the divergence, the components Φ_i of the vector field have to be weighted by the square root of the determinant $\sqrt{|g|}$ of the metric tensor g before taking the derivative. The square root of the determinant of g describes the distortion change from the Euclidean space to the surface. Formally, the divergence of a vector field $\Phi\colon \mathbb{R}^n \to \mathbb{R}^n$ with a given metric tensor g is given by:

$$\operatorname{div} \Phi = \frac{1}{\sqrt{|g|}} \sum_{i=1}^{n} \frac{\partial}{\partial x_i} \left(\sqrt{|g|}\, \Phi_i \right). \tag{5}$$

Given the definition of the gradient and the divergence, we can compose both operators to obtain the Laplace-Beltrami operator $\Delta\varphi$ of a scalar field $\varphi\colon \mathbb{R}^n \to \mathbb{R}$ on surfaces:

$$\Delta\varphi = \operatorname{div} \nabla\varphi = \frac{1}{\sqrt{|g|}} \sum_{i,j=1}^{n} \frac{\partial}{\partial x_i} \left(\sqrt{|g|}\, g^{ij} \frac{\partial \varphi}{\partial x_j} \right). \tag{6}$$

2.5 Shape Operator

In Sect. 2.2, we noticed that the curvature of a parametric surface at a specific point $\mathbf{p}$ in a certain direction can be determined by Eq. (2):

$$\kappa_{c'(t)}(\mathbf{p}) = -\langle c'(t), \frac{\partial \mathbf{n}}{\partial t} \rangle.$$

Actually, this means that the curvature in the direction $c'(t)$ is a measure of how much the normal changes in this direction, too. Given is $\mathbf{v} \in \mathcal{T}_{\mathbf{p}}f$ with $\mathbf{p} = f(\mathbf{u})$ and

$|\mathbf{v}| = 1$. Then, we determine the coefficients α, β of $\mathbf{v}$ with the basis $\frac{\partial f}{\partial x_1}, \frac{\partial f}{\partial x_2}$:

$$\begin{pmatrix}\alpha \\ \beta\end{pmatrix} = g^{-1}\begin{pmatrix}\langle \mathbf{v}, \frac{\partial f}{\partial x_1}\rangle \\ \langle \mathbf{v}, \frac{\partial f}{\partial x_2}\rangle\end{pmatrix}.$$

We use (α, β) to determine the derivative of $\mathbf{n}$ along $\mathbf{v}$ by using the two-dimensional curve $\tilde{c}(t) = \mathbf{u} + t\binom{\alpha}{\beta}$ and calculate:

$$D_{\mathbf{v}}\mathbf{n} := \frac{d}{dt}\mathbf{n}(\tilde{c}(t)).$$

We define $S(\mathbf{v}) := -D_{\mathbf{v}}\mathbf{n}$. This linear operator is called *Shape Operator* (also *Weingarten Map or Second Fundamental Tensor*). One can see that $S(\frac{\partial f}{\partial x_i}) = \frac{\partial \mathbf{n}}{\partial x_i}$ holds. Note that this operator can directly operate on the 3D space with a three-dimensional vector in the tangent space, as well as the 2D space with the coefficients of the basis. Therefore, it can be represented by a matrix S. Recall Eq. (2), we substitute c' with $\mathbf{v}$ and $\frac{\partial \mathbf{n}}{\partial t}$ by $S\mathbf{v}$:

$$\kappa_{\mathbf{v}}(\mathbf{p}) = \langle \mathbf{v}, S\mathbf{v}\rangle.$$

We want to show that the principle curvature directions are the eigenvectors of S. Assuming $\mathbf{v}_1, \mathbf{v}_2 \in \mathbb{R}^2$ are the normalized eigenvectors with the eigenvalues $\lambda_1 \geq \lambda_2$. Every normalized vector $\mathbf{w}$ can be written as a linear combination of $\mathbf{v}_1, \mathbf{v}_2$: $\mathbf{w} = \alpha\mathbf{v}_1 + \beta\mathbf{v}_2$ with $\|w\| = \|\alpha\mathbf{v}_1 + \beta\mathbf{v}_2\| = \alpha^2 + \beta^2 + 2\alpha\beta\langle \mathbf{v}_1, \mathbf{v}_2\rangle = 1$. Therefore, we obtain:

$$\kappa_{\mathbf{w}}(\mathbf{p}) = \langle \mathbf{w}, S\mathbf{w}\rangle = \frac{1}{2}[(\alpha^2 - \beta^2)(\lambda_1 - \lambda_2) + \lambda_1 + \lambda_2]. \quad (7)$$

One can see from Eq. (7) that $\kappa_{\mathbf{w}}(\mathbf{p})$ reaches a maximum if $\beta = 0, \alpha = 1$, and a minimum is reached if $\alpha = 0, \beta = 1$. If the eigenvalues (curvatures) are not equal, we can show that the principle curvature directions are perpendicular. For this, we need to show that S is a self-adjoint operator. Thus, the equation $\langle S\mathbf{v}, \mathbf{w}\rangle = \langle \mathbf{v}, S\mathbf{w}\rangle$ holds. We show this by using the property $\langle \mathbf{n}, \frac{\partial f}{\partial x_i}\rangle = 0$ and derive this with x_j:

$$\langle \frac{\partial \mathbf{n}}{\partial x_j}, \frac{\partial f}{\partial x_i}\rangle + \langle \mathbf{n}, \frac{\partial^2 f}{\partial x_i \partial x_j}\rangle = 0.$$

We demonstrate that S is a self-adjoint operator with the basis $\frac{\partial f}{\partial x_i}$:

$$\langle S(\frac{\partial f}{\partial x_i}), \frac{\partial f}{\partial x_j}\rangle = \langle -\frac{\partial \mathbf{n}}{\partial x_i}, \frac{\partial f}{\partial x_j}\rangle = \langle \mathbf{n}, \frac{\partial^2 f}{\partial x_i \partial x_j}\rangle = \langle -\frac{\partial \mathbf{n}}{\partial x_j}, \frac{\partial f}{\partial x_i}\rangle = \langle S(\frac{\partial f}{\partial x_j}), \frac{\partial f}{\partial x_i}\rangle.$$

Now, we show that the eigenvectors (principle curvature directions) are perpendicular if the eigenvalues (curvatures) are different:

$$\lambda_1\langle \mathbf{v}_1, \mathbf{v}_2\rangle = \langle S\mathbf{v}_1, \mathbf{v}_2\rangle = \langle \mathbf{v}_1, S\mathbf{v}_2\rangle = \lambda_2\langle \mathbf{v}_1, \mathbf{v}_2\rangle.$$

The equation is only true if $\mathbf{v}_1, \mathbf{v}_2$ are perpendicular (and $\lambda_1 \neq \lambda_2$ holds).

3 Discrete Differential Geometry

This section adapts the continuous differential geometry to discrete differential geometry, the area of polygonal meshes that approximate continuous geometries. The following notation is used in the remainder of this paper. Let $M \subset \mathbb{R}^3$ be a triangulated surface mesh. The mesh consists of vertices $i \in V$ with associated positions $\mathbf{p}_i \in \mathbb{R}^3$, edges $E = \{(i,j) \mid i,j \in V\}$, and triangles $T = \{(i,j,k) \mid (i,j), (j,k), (k,i) \in E\}$. We write $\mathbf{n}_i$ as the normalized normal vector at vertex i. If nothing else is mentioned, we refer to normal vectors at vertices. Furthermore, $\mathcal{N}(i)$ denotes the neighbors of i. So, for every $j \in \mathcal{N}(i), (i,j) \in E$ holds. Furthermore, if we use a triangle for calculation, we always use this notation: given a triangle $\Delta = (i,j,k)$ with vertices $\mathbf{p}_i, \mathbf{p}_j, \mathbf{p}_k$, and the edges are defined as $\mathbf{e}_1 = \mathbf{p}_i - \mathbf{p}_j, \mathbf{e}_2 = \mathbf{p}_j - \mathbf{p}_k, \mathbf{e}_3 = \mathbf{p}_k - \mathbf{p}_i$.

3.1 Voronoi Area

We need to introduce the term Voronoi area, as it is important for the determination of the curvature. So, given are points in a 2D space. Every point is spread out in equal speed. If two fronts collide, they stop to spread out further at this region. After all fronts stopped, every point lies in a region that is surrounded by a front. This region is called a *Voronoi region*. Formally, given distinct points $\mathbf{x}_i \in \mathbb{R}^2$ in the plane, the Voronoi region for the point $\mathbf{x}_k$ is defined as the set of points $V(\mathbf{x}_k)$ with

$$V(\mathbf{x}_k) = \{\mathbf{x} \in \mathbb{R}^2 \, : \, \|\mathbf{x} - \mathbf{x}_k\| \leq \|\mathbf{x} - \mathbf{x}_j\|, j \neq k\}.$$

See Fig. 5a for an example of a Voronoi diagram. To obtain the Voronoi area of a vertex on a surface mesh, the Voronoi area of each incident triangle is accumulated. The Voronoi area calculation is based on the method by Meyer et al. [33]. In case of a non-obtuse triangle, the Voronoi area at $\mathbf{p}_i$ is determined by the perpendicular bisector of the edges incident to $\mathbf{p}_i$. The point of intersection, the midpoint of the incident edges and the point itself define the endpoints of the Voronoi area. The

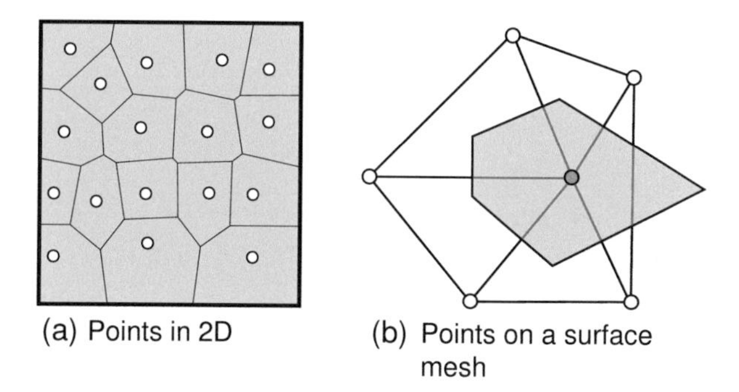

Fig. 5 The Voronoi diagram of different settings. In (**a**) Voronoi diagram of a set of points is determined. In (**b**) the Voronoi area is calculated. If one of the triangles is obtuse, the area leaves the triangle

triangle area of the Voronoi region equals:

$$\mathcal{A}_{\triangle}(\mathbf{p}_i) = \frac{1}{8}\Big(\|\mathbf{e}_1\|^2 \cdot \cot(\mathbf{e}_2, \mathbf{e}_3) + \|\mathbf{e}_3\|^2 \cdot \cot(\mathbf{e}_1, \mathbf{e}_2)\Big).$$

In case of an obtuse triangle, the Voronoi area is equal half of the triangle area if the angle at $\mathbf{p}_i$ is obtuse. Otherwise it is a quarter of the triangle area, see Fig. 5b.

3.2 *Discrete Curvature*

The calculation of the curvatures as well as the principle curvature directions are important for a number of feature line techniques. Several approaches exist to approximate the curvatures. Some methods try to fit an analytic surface (higher order polynomials) to the mesh and determine the curvatures analytically [5, 17]. Another approach estimates the normal curvature along edges first and then estimates the shape operator [6, 18, 33, 38, 48]. Other approaches are based on the calculation of the shape operator S [1, 8, 42]. We use the curvature estimation according to [42]. After S is determined on a triangle basis, it is adapted to vertices. We already defined that $S\mathbf{v}$ yields the change of the normal in the direction of $\mathbf{v}$:

$$S\mathbf{v} = D_{\mathbf{v}}\mathbf{n}.$$

This property is used to assess S for each triangle. When applying S to the edge $\mathbf{e}_1$, it should result in $\mathbf{n}_i - \mathbf{n}_j$ because of the change of the normals along the edge. We need a basis of the tangent space of the triangle:

$$\tilde{\mathbf{e}}_1 = \frac{\mathbf{e}_1}{|\mathbf{e}_1|}, \quad \tilde{\mathbf{e}}_2 = \frac{\mathbf{e}_2}{|\mathbf{e}_2|}.$$

Afterwards, we build the orthogonal normalized basis vectors $\mathbf{x}_\triangle, \mathbf{y}_\triangle$ by:

$$\mathbf{x}_\triangle := \tilde{\mathbf{e}}_1, \qquad \mathbf{y}_\triangle := \frac{\mathbf{x}_\triangle \times (\tilde{\mathbf{e}}_2 \times \mathbf{x}_\triangle)}{\|\mathbf{x}_\triangle \times (\tilde{\mathbf{e}}_2 \times \mathbf{x}_\triangle)\|}. \tag{8}$$

Applying the aforementioned property of the shape operator to all edges according to the basis leads to the following equation system:

$$\begin{aligned}
S\begin{pmatrix}\langle \mathbf{e}_1, \mathbf{x}_\triangle\rangle \\ \langle \mathbf{e}_1, \mathbf{y}_\triangle\rangle\end{pmatrix} &= \begin{pmatrix}\langle \mathbf{n}_i - \mathbf{n}_j, \mathbf{x}_\triangle\rangle \\ \langle \mathbf{n}_i - \mathbf{n}_j, \mathbf{y}_\triangle\rangle\end{pmatrix} \\
S\begin{pmatrix}\langle \mathbf{e}_2, \mathbf{x}_\triangle\rangle \\ \langle \mathbf{e}_2, \mathbf{y}_\triangle\rangle\end{pmatrix} &= \begin{pmatrix}\langle \mathbf{n}_j - \mathbf{n}_k, \mathbf{x}_\triangle\rangle \\ \langle \mathbf{n}_j - \mathbf{n}_k, \mathbf{y}_\triangle\rangle\end{pmatrix} \\
S\begin{pmatrix}\langle \mathbf{e}_3, \mathbf{x}_\triangle\rangle \\ \langle \mathbf{e}_3, \mathbf{y}_\triangle\rangle\end{pmatrix} &= \begin{pmatrix}\langle \mathbf{n}_k - \mathbf{n}_i, \mathbf{x}_\triangle\rangle \\ \langle \mathbf{n}_k - \mathbf{n}_i, \mathbf{y}_\triangle\rangle\end{pmatrix},
\end{aligned} \tag{9}$$

see Fig. 6 for an illustration. Here, we have three unknowns (the matrix entries of the symmetric matrix $S = \left(\begin{smallmatrix} e & f \\ f & g\end{smallmatrix}\right)$) and six linear equations.Thus, a least square method can be applied to fit the shape operator to approximate curvature for each triangle. Next, we need to calculate S for each vertex of the mesh. As the triangle basis normally differs from each vertex tangent space basis, we need to transform the shape operator according to the new coordinate system. First, we assume that the normal $\mathbf{n}_\triangle$ of the face is equal to the incident vertex normal $\mathbf{n}_i$. Hence, the basis $(\mathbf{x}_\triangle, \mathbf{y}_\triangle)$ of the triangle is coplanar to the basis $(\mathbf{x}_i, \mathbf{y}_i)$ of the incident vertex i. Assuming we have the shape operator given in the vertex basis, then the entries can

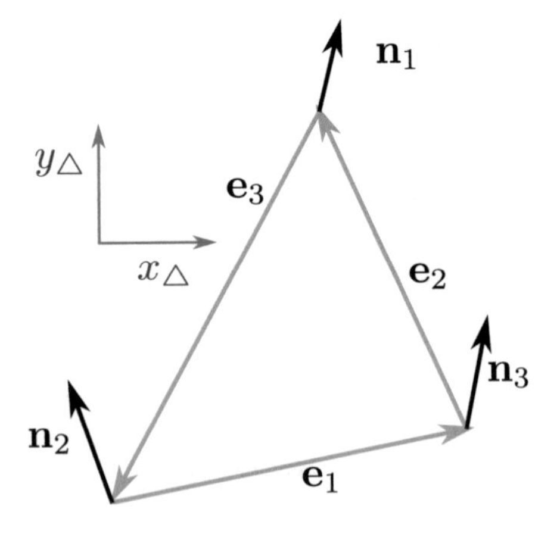

Fig. 6 The shape operator estimation is based on a local coordinate system, the edges and the normals

be determined by:

$$e_p = (1\ 0)\begin{pmatrix} e_p & f_p \\ f_p & g_p \end{pmatrix}\begin{pmatrix} 1 \\ 0 \end{pmatrix} = \mathbf{x}_i^T S \mathbf{x}_i$$

$$f_p = (1\ 0)\begin{pmatrix} e_p & f_p \\ f_p & g_p \end{pmatrix}\begin{pmatrix} 0 \\ 1 \end{pmatrix} = \mathbf{x}_i^T S \mathbf{y}_i$$

$$g_p = (0\ 1)\begin{pmatrix} e_p & f_p \\ f_p & g_p \end{pmatrix}\begin{pmatrix} 0 \\ 1 \end{pmatrix} = \mathbf{y}_i^T S \mathbf{y}_i.$$

As we have determined the shape operator in the basis $(\mathbf{x}_\triangle, \mathbf{y}_\triangle)$, we can express the basis of the vertex by expressing the new coordinate system with the old basis $\mathbf{x}_i = \alpha\mathbf{x}_\triangle + \beta\mathbf{y}_\triangle$:

$$\alpha = \langle \mathbf{x}_i, \mathbf{x}_\triangle \rangle$$
$$\beta = \langle \mathbf{x}_i, \mathbf{y}_\triangle \rangle.$$

The entry e_p can be determined by:

$$e_p = (\alpha\ \beta)\, S \begin{pmatrix} \alpha \\ \beta \end{pmatrix}. \tag{10}$$

The other entries can be calculated by analogous calculations. For the second case, we rotate the coordinate system of the triangle around the cross product of the normal such that the basis of the vertex and the triangle are coplanar. Finally, we use this to determine the shape operator of the vertices. We determine the shape operators for all incident triangles of a vertex. Afterwards, we rotate the coordinate systems of the triangles to be coplanar with the basis of the vertex. Next, we re-express the shape operator in terms of the basis of the vertex. Then, we weight the shape operator according to the Voronoi area of the triangle and accumulate this tensor. Finally, we divide the accumulated shape operator by the sum of the weights. The eigenvalues provide the principle curvatures, and the eigenvectors give the principle curvature directions according to the basis. The pseudo-code 1 summarizes the algorithm.

Please note that this algorithm can be generalized to obtain higher-order derivatives. It can be used to determine the derivative of the curvature as it is important for a specific feature line method. Formally, the derivative of the shape operator has the form:

$$C = (D_{\mathbf{v}} S\ D_{\mathbf{w}} S) = \left(\begin{pmatrix} a & b \\ b & c \end{pmatrix} \begin{pmatrix} b & c \\ c & d \end{pmatrix} \right). \tag{11}$$

Algorithm 1 Pseudo-code for the curvature estimation.

```
for each triangle:
    Build basis accord. to Eq. (8)
    Determine S accord. to Eq. (9)
    for each vertex incident to the triangle:
        Rotate the triangle basis to the vertex basis
        Determine S in the new basis accord. to Eq. (10)
        Add this tensor weighted by the voronoi area
    end
end
for each vertex:
    Divide S by the sum of the weights
    Determine the eigenvalues and eigenvectors
end
```

For the determination of the change of the curvature in direction $\mathbf{u}$, the tensor C has to be multiplied multiple times:

$$D_{\mathbf{u}}\kappa = \langle \mathbf{u}, \big(D_{\mathbf{v}}S \cdot \mathbf{u}\ D_{\mathbf{w}}S \cdot \mathbf{u}\big) \cdot \mathbf{u} \rangle. \tag{12}$$

3.3 Discrete Covariant Derivative

First, we consider a linear 2D scalar field $\varphi(\mathbf{x}) = \alpha \cdot x_1 + \beta \cdot x_2 + \gamma$ and its gradient:

$$\nabla\varphi = \begin{pmatrix} \frac{\partial}{\partial x_1}\varphi \\ \frac{\partial}{\partial x_2}\varphi \end{pmatrix} = \begin{pmatrix} \alpha \\ \beta \end{pmatrix}. \tag{13}$$

To determine the gradient of a triangle $\triangle = (i, j, k)$ with scalar values $\varphi_i := \varphi(\mathbf{p}_i), \varphi_j := \varphi(\mathbf{p}_j)$, and $\varphi_k := \varphi(\mathbf{p}_k)$, we build a basis according to Eq. (8) and transform the points $\mathbf{p}_i, \mathbf{p}_j, \mathbf{p}_k \in \mathbb{R}^3$ to $\mathbf{p}'_i, \mathbf{p}'_j, \mathbf{p}'_k \in \mathbb{R}^2$ by:

$$\mathbf{p}'_i = \begin{pmatrix} 0 \\ 0 \end{pmatrix} \qquad \mathbf{p}'_j = \begin{pmatrix} \langle \mathbf{p}_j - \mathbf{p}_i, \mathbf{x}_\triangle \rangle \\ \langle \mathbf{p}_j - \mathbf{p}_i, \mathbf{y}_\triangle \rangle \end{pmatrix} \qquad \mathbf{p}'_k = \begin{pmatrix} \langle \mathbf{p}_k - \mathbf{p}_i, \mathbf{x}_\triangle \rangle \\ \langle \mathbf{p}_k - \mathbf{p}_i, \mathbf{y}_\triangle \rangle \end{pmatrix}.$$

This transformation describes an isometric and conformal map. The next step is a linearization of the scalar values $\varphi_i, \varphi_j, \varphi_k$. We want to determine a scalar field $\varphi'(\mathbf{x}') = \alpha \cdot x'_1 + \beta \cdot x'_2 + \gamma$ such that

$$\varphi'(\mathbf{p}'_i) = \varphi_i \qquad \varphi'(\mathbf{p}'_j) = \varphi_j \qquad \varphi'(\mathbf{p}'_k) = \varphi_k$$

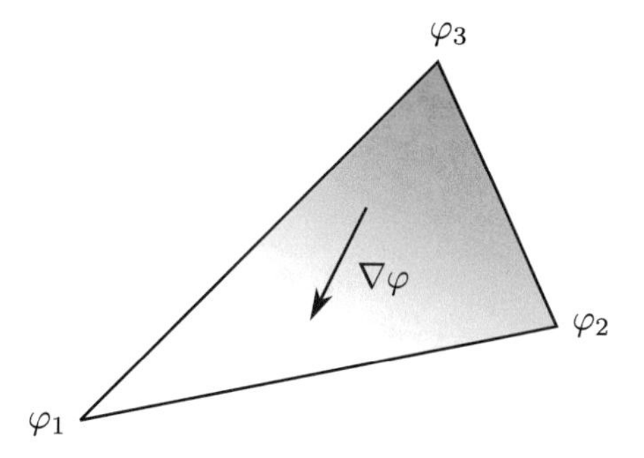

Fig. 7 A triangle with different scalar values

holds. These conditions yield the following equation system:

$$\left(\alpha\ \beta\right)\left(\mathbf{p}'_i\ \mathbf{p}'_j\ \mathbf{p}'_k\right) + \left(\gamma\ \gamma\ \gamma\right) = \left(\varphi_i\ \varphi_j\ \varphi_k\right).$$

With $\mathbf{p}'_i = \binom{0}{0}$ we obtain the following solution:

$$\gamma = \varphi_i,$$
$$\left(\alpha\ \beta\right) = \left(\varphi_j - \varphi_i\ \varphi_k - \varphi_j\right)\left(\mathbf{p}'_j\ \mathbf{p}'_k\right)^{-1}.$$

According to Eq. (13), the gradient of φ' is determined by $\binom{\alpha}{\beta}$. The basis $\mathbf{x}_\triangle, \mathbf{y}_\triangle$ yields the gradient in 3D:

$$\nabla\varphi = \alpha \cdot \mathbf{x}_\triangle + \beta \cdot \mathbf{y}_\triangle.$$

Figure 7 illustrates the gradient of a triangle. To determine the gradient per vertex, we use the same procedure as for the shape operator estimation. We transform the basis and weight the triangle gradient according to its proportion of the Voronoi area.

3.4 Discrete Laplace-Beltrami Operator

Several methods exist to discretize the Laplace-Beltrami operator on surface meshes. For an overview, we recommend the state of the art report by Sorkine [45]. The operator can be presented by the generalized formula:

$$\Delta\varphi(\mathbf{p}_i) = \sum_j w_{ij}\left(\varphi(\mathbf{p}_j) - \varphi(\mathbf{p}_i)\right).$$

Different weights w_j give different discrete Laplace-Beltrami operators. For presenting different versions of this operator it is preferable that it fulfills some properties motivated by the smooth Laplace-Beltrami operator:

(Sym) The weights should be symmetric $w_{ij} = w_{ji}$.
(Loc) If $(i,j) \notin E$ then $w_{ij} = 0$.
(Pos) All weights should be non-negative.
(Lin) If $\mathbf{p}_i$ is contained in a plane and φ is linear, then $\Delta\varphi(\mathbf{p}_i) = 0$ should hold.

In the following, we introduce different discrete Laplace-Beltrami operators.

Combinatorial For the combinatorial Laplace-Beltrami operator we have:

$$w_{ij} = \begin{cases} 1, & \text{if } (i,j) \in E \\ 0, & \text{otherwise.} \end{cases}$$

This version may result in non-zero values on planar surfaces for linear scalar fields. Therefore, it violates (Lin).

Uniform Taubin [49] suggested the uniform Laplace-Beltrami operator. The weights are determined by the number of neighbors of $\mathbf{p}_i$:

$$w_{ij} = \begin{cases} \frac{1}{\mathrm{N}(i)}, & \text{if } (i,j) \in E \\ 0, & \text{otherwise.} \end{cases}$$

These weights also violate (Lin).

Floater's Mean Value Floater [14] proposed the mean value weights by the tangent of the corresponding angles:

$$w_{ij} = \begin{cases} \frac{\tan(\delta_{ij}/2)+\tan(\gamma_{ij}/2)}{\|\mathbf{p}_i-\mathbf{p}_j\|}, & \text{if } (i,j) \in E \\ 0, & \text{otherwise.} \end{cases}$$

See Fig. 8 for the angles. These weights violate (Sym).

Cotangent Weights MacNeal [31] suggested the cotangent weights:

$$w_{ij} = \begin{cases} \frac{\cot(\alpha_{ij})+\cot(\beta_{ji})}{2}, & \text{if } (i,j) \in E \\ 0, & \text{otherwise.} \end{cases}$$

See Fig. 8 for the angles. On general meshes the weights can violate (Pos).

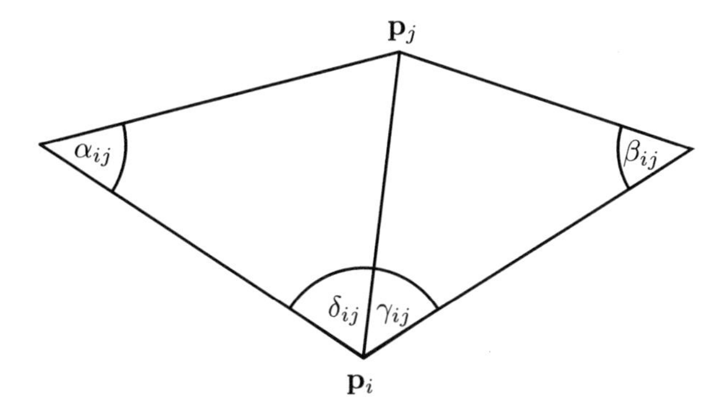

Fig. 8 This figure illustrates the triangles with the angles for the weight calculation

Belkin Weights Belkin [2] suggested to determine weights over the whole surface:

$$\Delta\varphi(\mathbf{p}_i) = \frac{1}{4\pi h^2(\mathbf{p}_i)} \sum_{\Delta_k} \frac{A(\Delta_k)}{3} \sum_{j\in\Delta_k} e^{-\frac{\|\mathbf{p}_i-\mathbf{p}_j\|^2}{4h(\mathbf{p}_i)}} \Big(\varphi(\mathbf{p}_j) - \varphi(\mathbf{p}_i)\Big),$$

where $A(\Delta_k)$ denotes the area of the triangle Δ_k and h corresponds intuitively to the size of the neighborhood. This violates the (Loc) property.

Results The discussion leads to the question if there is any discrete Laplace-Beltrami operator which fulfills all required properties for an arbitrary surface mesh and, furthermore, if this operator converges pointwise to the Laplace-Beltrami operator on smooth surfaces. Wardetzky et al. [52] showed that there is no such operator. The proof is based on a Schönhardt polytope which demonstrate that there is no Laplace-Beltrami operator, which does not violate any condition. One example of a discrete Laplace operator can be obtained using linear finite elements over the polyhedral surface. Hildebrandt et al. [21] proved that if a sequences of surface meshes converges to a smooth surface in Hausdorff distance and the normal fields converge, then the Laplace operators converge in the operator norm. Despite of convergence, this operator is not perfect in the sense of [52], because for meshes with obtuse triangles, their weights may be non-positive weights. As a consequence, a maximum principle for the solutions of the discrete Dirichlet problem cannot be guaranteed.

3.5 *Isolines on Discrete Surfaces*

For feature line methods, it is essential not to restrict the lines to the edges, as it is not desirable to perceive the mesh edges. Given is a surface mesh and a scalar field, we want to depict the zero crossing of the scalar field. Therefore, we linearize the scalar values for each triangle according to the values of the incident points.

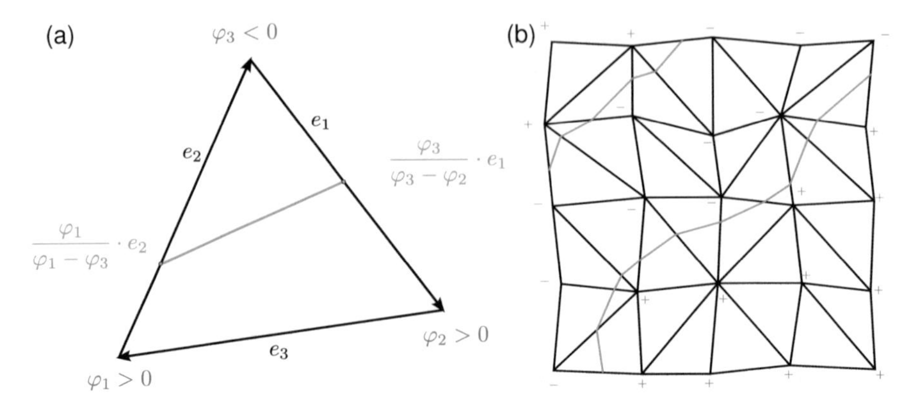

Fig. 9 In (**a**) the position of the zero crossing is determined and the points are connected. In (**b**) the isoline through a mesh is depicted

Afterwards, we look for points on an edge such that the linearized values of the scalar values of the connecting points are equal to zero. Having two points on two edges of a triangle, we connect them. Suppose we have a triangle with scalar values $\varphi_i > 0, \varphi_j > 0$ and $\varphi_k < 0$. Thus, we know that somewhere on edge $\mathbf{e}_2$ and $\mathbf{e}_3$ there is a zero crossing. We determine $t = \frac{\varphi_k}{\varphi_k - \varphi_j}$ and multiply t with edge $\mathbf{e}_2$. This yields the position of the zero crossing on the edge. The position on the edge $\mathbf{e}_3$ is determined as well. Afterwards, both points will be connected, see Fig. 9.

4 General Requirements of Feature Lines

The generation of feature lines leads to several requirements, which have to be considered for acquiring appropriate results.

Smoothing Most of the feature line methods use higher order derivatives. Therefore, the methods assume sufficiently smooth input data. For data acquired with laser scanners or industrial measurement process, smoothness cannot be expected. Discontinuities represent high frequencies in the surface mesh and lead to the generation of distracting (and erroneous) lines. Several algorithms exist, which smooth the surface by keeping relevant features. Depending on the feature line method, different smoothing algorithms can be applied. If the algorithm only uses the surface normals and the view direction, it is sufficient to simply smooth the surface normals. Geometry-based approaches, however, require to smooth the mesh completely. Operating only on scalar values, an algorithm which smoothes the scalar field around a certain region may be applied, too.

Frame Coherence The application of feature line approaches or in general for non-photorealistic rendering makes it crucial to provide methods that are frame-coherent. This means, during the interaction the user should not be distracted by features that pop out or disappear suddenly. A consistent and continuous depiction of features should be provided in consecutive frames of animation.

Filtering Feature line algorithms may generate lines on salient regions as well as lines that result from small local irregularities, which may not be necessary to convey the surface shape or even annoying and distracting. Filtering of feature lines to set apart relevant lines from distracting ones is a crucial part of a feature line generation. User-defined thresholds may control the rate of tolerance for line generation. Some algorithms use an underlying scalar field for thresholding. Lines are only drawn if the corresponding scalar value exceeds the user-defined threshold. Other methods integrate along a feature line, determine the value, and decide to draw the whole line instead of filtering some parts. We will also mention the filtering method of each presented feature line generation method.

5 Feature Lines

Line drawings were used extensively for medical visualization tasks, such as displaying tissue boundaries in volume data [4, 51], vascular structures [41], neck anatomy [27] and brain data [24, 47]. Furthermore, some higher order feature lines were qualitatively evaluated on medical surface data [29]. The importance of feature lines in medical visualization is discussed in [40]. Feature line methods can be divided into *image-based* and *object-based* methods. Image-based approaches are not in the focus of this survey. These methods are based on an image as input. All calculations are performed on the image with the pixels containing, for instance, an RGB or grey value. Usually, the image is convolved with different kernels to obtain the feature lines. The resulting feature lines are represented by pixels in the image space. These lines are mostly not frame-coherent. Comprehensive overviews of different feature line methods in image space are given by Muthukrishnan and Radha [34], Nadernejad et al. [35], and Senthilkumaran and Rajesh [44]. This section presents selected object-based feature line methods. We will explain the methods and limitations. Further information on line drawings can be found in [40, 43].

5.1 Contours

We refer to a silhouette as a depiction of the outline of an object as this is the original definition by Étienne de Silhouette. The contour is defined as the loci of points

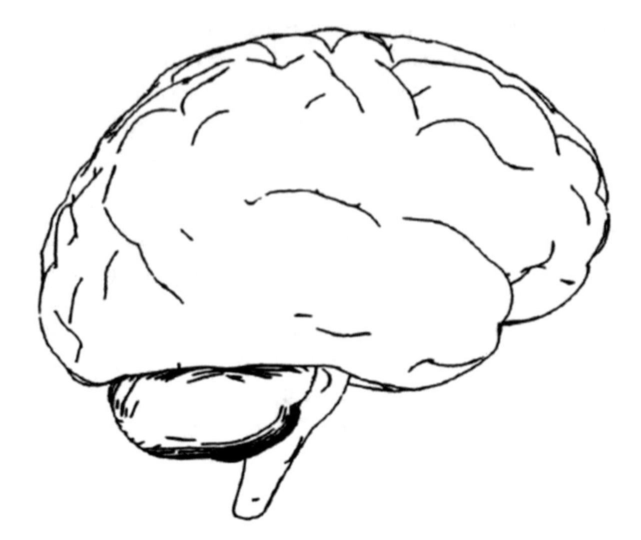

Fig. 10 The brain model with contours

where the normal vector and the view vector are mutually perpendicular (Fig. 10):

$$\boxed{\langle \mathbf{n}, \mathbf{v} \rangle = 0,}$$

where $\mathbf{n}$ is the normal vector and $\mathbf{v}$ is the view vector which points towards the camera. For the discrete case, we highlight edges as a contour whenever the sign of the dot product of the view vector with the normals of the incident triangle normals changes. The contour yields a first impression of the surface mesh. On the other hand, it is not sufficient to depict the surface well. The contour is not appropriate to gain a spatial impression of the object. Furthermore, it cannot depict salient regions, for instance strong edges.

Summary In the first place, the contour is necessary for gaining a first impression on the shape of the object. Unfortunately, spatial cues, as for instance strong edges, are not depicted.

5.2 *Crease Lines*

Crease lines are a set of edges where incident triangles change strongly. The dihedral angle, i.e., the angle of the normals of the corresponding incident triangles, along the edges is calculated. The edge belongs to a crease line if the dihedral angle exceeds a user-defined threshold τ. As the change of the normals is an indicator of the magnitude of the curvature, one can state that all points contribute to a feature

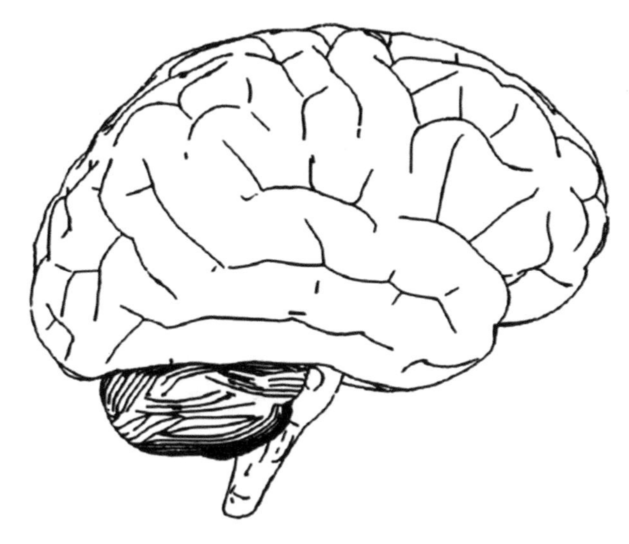

Fig. 11 The brain model with crease lines and contours

line if the underlying absolute value of the maximum curvature exceeds a threshold (Fig. 11):

$$\boxed{\kappa_i \geq \tau \text{ or } \langle \mathbf{n}_i, \mathbf{n}_j \rangle \geq \tau',}$$

for adjacent triangles with corresponding normals $\mathbf{n}_i, \mathbf{n}_j$. Afterwards, all adjacent vertices which fulfill the property are connected. These feature lines need to be computed only once, since they are not view-dependent. Furthermore, these lines are only drawn along edges.

Summary Crease lines display edges where the dihedral angle is large. Strong edges are appropriately depicted, but if the object has small features, this method is not able to depict only important edges. This is caused by the local determination of the dihedral angle without concerning a neighborhood. Even smoothing the surface mesh would not deliver proper line drawings. Furthermore, this method is only able to detect features on edges.

5.3 Ridges and Valleys

Ridges and valleys were proposed by Interrante et al. [23] and adapted to triangulated surface meshes by Ohtake et al. [37]. These feature lines are curvature-based and not view-dependent. The computation is based on the principle curvature κ_1 as well as the associated principle curvature direction $\mathbf{k}_1$ with $|\kappa_1| \geq |\kappa_2|$. Formally,

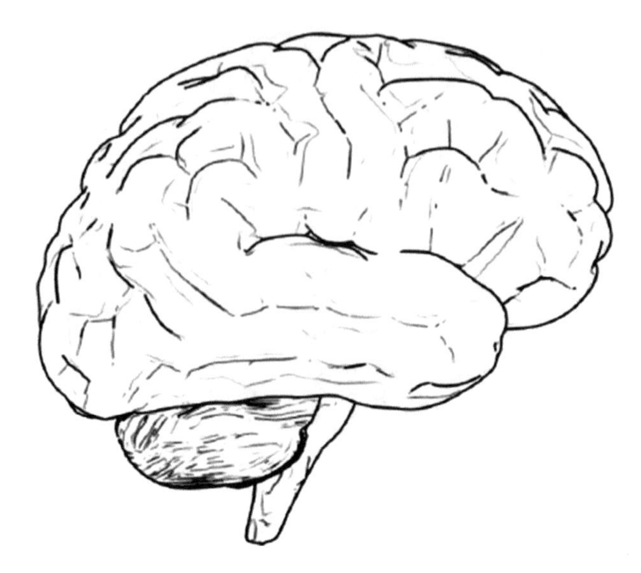

Fig. 12 The brain model with ridges and valleys, and contours

ridges and valleys are defined as the loci of points at which the principle curvatures assume an extremum in the principle direction:

$$\boxed{D_{\mathbf{k}_1}\kappa_1 = 0.}$$

According to two constraints, the sets of points are called (Fig. 12)

$$D_{\mathbf{k}_1}D_{\mathbf{k}_1}\kappa_1 \begin{cases} < 0, & \text{and } \kappa_1 > 0\text{: ridges} \\ > 0, & \text{and } \kappa_1 < 0\text{: valleys.} \end{cases} \tag{14}$$

To determine the ridge and valley lines, we first need to compute the principle curvatures and their associated principle curvature directions, recall Sect. 3.2. Afterwards, we determine the gradient of κ_1 for each vertex, see Sect. 3.3. Finally, we compute the dot product of the gradient and the associated principle curvature direction $\mathbf{k}_1$. This yields the scalar value of $D_{\mathbf{k}_1}\kappa_1$ for each vertex. Next, we distinguish between ridges and valleys and determine $D_{\mathbf{k}_1}D_{\mathbf{k}_1}\kappa_1$ for each vertex. Here, we need again the gradient of each vertex with the value $D_{\mathbf{k}_1}\kappa_1$ and determine the dot product of the result with $\mathbf{k}_1$. Hence, we gain two scalar values per vertex: $D_{\mathbf{k}_1}\kappa_1$ and $D_{\mathbf{k}_1}D_{\mathbf{k}_1}\kappa_1$. Afterwards, we assess the zero-crossing of the first scalar value, recall Sect. 3.5. We connect the zero crossings in every triangle for which one condition of Eq. (14) holds. The filtering of the lines is again performed by employing an user-defined threshold. The integral along each ridge and valley line is determined according to the underlying curvature. If the magnitude of the integral exceeds the threshold for ridges or valleys, the line is drawn.

Summary The calculation is solely based on the curvature and therefore view-independent. This method is able to detect small features. The filtering depends on the underlying curvature and the length of the curve. Therefore, a long line with small curvature has also the chance to be drawn as a small line with high curvature. This strategy emphasizes also long feature lines. Ridges and valley lines are very susceptible to noise, since this method is of third order. Therefore, small discontinuities on the surface mesh lead to erroneous derivatives and this error propagates for each further derivative. A crucial task for this method is to guarantee a smoothed mesh to obtain reasonable results. Other approaches use surface fitting [37, 55] or a combination of discrete curvature approximation and smoothing techniques [22] to obtain a smoothed surface. From an artist's point of view, some features may be more highlighted than others from different points of view. This is caused by the different perception of an object and by various light positions. For this, the ridge and valley lines are not appropriate due to the restriction of view-independent results.

5.4 *Suggestive Contours*

Suggestive contours are view-dependent feature lines introduced by DeCarlo et al. [9]. They extend the definition of the contour. These lines are defined as the set of minima of $\langle \mathbf{n}, \mathbf{v} \rangle$ in the direction of $\mathbf{w}$, where $\mathbf{n}$ is the surface normal, $\mathbf{v}$ is the view vector which points towards the camera, and $\mathbf{w} = (\mathbf{Id} - \mathbf{nn}^T)\mathbf{v}$ is the projection of the view vector on the tangent plane. Formally:

$$\boxed{D_{\mathbf{w}} \langle \mathbf{n}, \mathbf{v} \rangle = 0 \text{ and } D_{\mathbf{w}} D_{\mathbf{w}} \langle \mathbf{n}, \mathbf{v} \rangle > 0.}$$

Another equivalent definition of the suggestive contours is given by the radial curvature κ_r. It is defined as the curvature in direction of $\mathbf{w}$. As seen in Eq. (3), this curvature can be determined by knowing the principle curvature directions as well as the corresponding curvatures. Therefore, the definition of the suggestive contours is equivalent to the set of points at which the radial curvature κ_r is equal 0 and the directional derivative of κ_r in direction $\mathbf{w}$ is positive (Fig. 13):

$$\boxed{\kappa_r = 0 \text{ and } D_{\mathbf{w}} \kappa_r > 0.}$$

The filtering strategy is to apply a small threshold to eliminate suggestive contour points where the radial curvature in direction of the projected view vector is very low. Additionally, a hysteresis threshold is applied to increase granularity.

Summary Suggestive contours extend the normal definition of the contour. This method depicts zero crossing of the diffuse light in view direction. This can be seen as inflection points on the surface. This method is of second order only and thus less susceptible to noise. Unfortunately, suggestive contours are not able to depict some sorts of sharp edges, which are in fact noticeable features. For instance, a rounded cube has no suggestive contours.

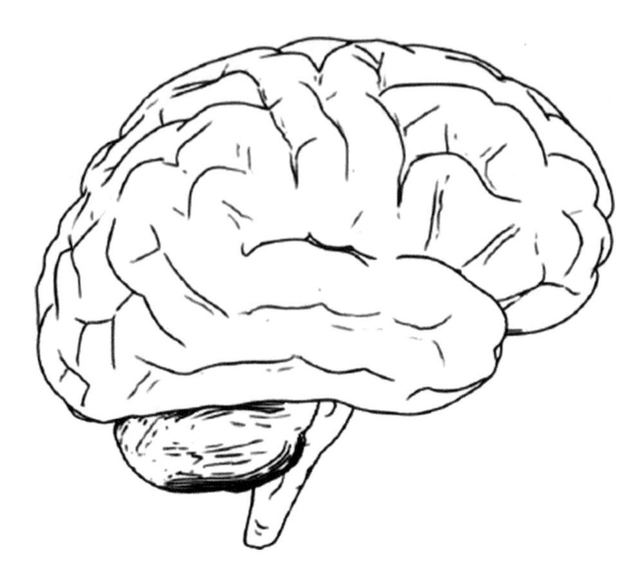

Fig. 13 The brain model with suggestive contours and contours

5.5 Apparent Ridges

Apparent ridges were proposed by Judd et al. [25]. These feature lines extend the definition of ridges by a view-dependent curvature term. Therefore, a projection operator P is used to map the vertices on a screen plane V. The orthonormal basis of the screen plane is given by $(\mathbf{v}_1, \mathbf{v}_2)$. Assume we have a parametrized surface $f: I \subset \mathbb{R}^2 \to \mathbb{R}^3$. Then the projection of f onto V is given by:

$$P(\mathbf{x}) = \begin{pmatrix} \langle \mathbf{v}_1, f(\mathbf{x}) \rangle \\ \langle \mathbf{v}_2, f(\mathbf{x}) \rangle \end{pmatrix}.$$

The Jacobian J_P of P can be expressed as:

$$J_P = \begin{pmatrix} \langle \mathbf{v}_1, \frac{\partial f}{\partial x_1} \rangle & \langle \mathbf{v}_1, \frac{\partial f}{\partial x_2} \rangle \\ \langle \mathbf{v}_2, \frac{\partial f}{\partial x_1} \rangle & \langle \mathbf{v}_2, \frac{\partial f}{\partial x_2} \rangle \end{pmatrix}.$$

In the discrete case with surface meshes, the Jacobian can be expressed by a basis for the tangent plane $(\mathbf{e}_1, \mathbf{e}_2)$:

$$J_P = \begin{pmatrix} \langle \mathbf{v}_1, \mathbf{e}_1 \rangle & \langle \mathbf{v}_1, \mathbf{e}_2 \rangle \\ \langle \mathbf{v}_2, \mathbf{e}_1 \rangle & \langle \mathbf{v}_2, \mathbf{e}_2 \rangle \end{pmatrix}.$$

If a point $\mathbf{p}'$ on the screen plane is not a contour point, there exists a small neighborhood where the inverse of P exists. Normal vectors $\mathbf{n}'$ at a point $\mathbf{p}'$ on the screen plane are defined as $\mathbf{n}'(\mathbf{p}') := \mathbf{n}(p^{-1}(\mathbf{p}'))$. The main idea is to build a view-dependent shape operator S' at a point $\mathbf{p}'$ on the screen as

$$S'(\mathbf{w}') = D_{\mathbf{w}'}\mathbf{n}'$$

where $\mathbf{w}'$ is a vector in the screen plane. The view-dependent shape operator is therefore defined as:

$$S' = SJ_P^{-1}.$$

Here, the basis of the tangent space expressing S and J_P must be the same. In contrast to the shape operator, the view-dependent shape operator is not a self-adjoint operator, recall Sect. 2.5. Therefore, it is not guaranteed that S' has two eigenvalues, but it has a maximum singular value κ_1' (Fig. 14):

$$\kappa_1' = \max_{\|\mathbf{w}\|=1} \|S'(\mathbf{w}')\|.$$

This is equivalent to find the maximum eigenvalue of $S'^T S'$ and to take the square root. The corresponding singular eigenvector $\mathbf{t}'$ is called the maximum view-

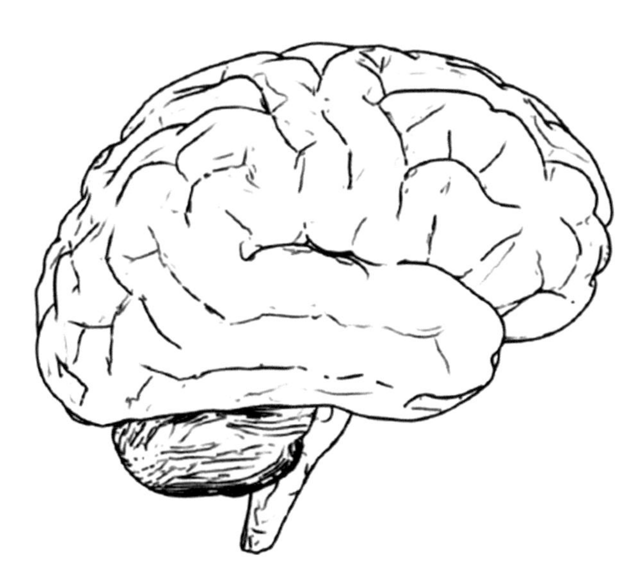

Fig. 14 The brain model with apparent ridges

dependent principle direction. The rest of the method is similar to the ridge and valley methods. Formally, apparent ridges are defined as the loci of points at which the view-dependent principle curvature assumes an extremum in the view-dependent principle direction:

$$\boxed{D_{\mathbf{t}'}\kappa_1' = 0 \text{ and } D_{\mathbf{t}'}D_{\mathbf{t}'}\kappa_1' < 0.}$$

The sign of κ' is always positive. To distinguish between ridge lines and valley lines, we may compare the sign of the object-space curvature:

$$\kappa_1 \begin{cases} < 0, & \text{ridges} \\ > 0, & \text{valleys.} \end{cases}$$

The calculation of the directional derivative is different from the other methods. This calculation is performed with finite differences. Therefore, we transform the singular eigenvector $\mathbf{t}'$ to object space $\mathbf{t}$ using the corresponding basis of the associated vertex i. Furthermore, we need the opposite edges of the vertex and determine two points $\mathbf{w}_1, \mathbf{w}_2$ on the edges such that $\mathbf{t}$ and the edges are orthogonal and $\mathbf{w}_1, \mathbf{w}_2$ are the dropped perpendiculars of $\mathbf{t}$ to the corresponding edges. The directional derivatives are determined by averaging the finite differences of the curvatures between $\mathbf{p}_i$ and $\mathbf{w}_1, \mathbf{w}_2$. The curvature of $\mathbf{w}_1, \mathbf{w}_2$ is assessed by linear interpolation of the endpoints of the associated edge. Having the principle view-dependent curvature direction $\mathbf{t}'$, we need to make it consistent over the mesh because it is not well-defined. Therefore, $\mathbf{t}'$ is flipped in opposite direction whenever it does not point to the direction where the view-dependent curvature is increasing. The zero-crossings are determined by checking if the principle view-dependent curvature directions of the vertices along an edge point are in the same direction. Only in this case there is no zero-crossing. Pointing in different directions means that the enclosing angle is greater than 90 degrees. The zero crossing is determined by interpolating the values of the derivatives. To locate only maxima, a perpendicular is dropped from each vertex to the zero crossing line. If the perpendiculars of the vertices of an edge make an acute angle with their principle view-dependent curvature directions, the zero crossing is a maximum. Otherwise, the zero crossing is a minimum. To eliminate unimportant lines, a threshold based on the view-dependent curvature is used.

Summary Apparent ridges incorporate the advantages of the ridges and valley lines as well as the view dependency. They extend the ridge and valley definition by introducing view-dependent curvatures. This method is able to depict salient regions as sharp edges. Unfortunately, the third order computation leads to low frame rates and to visual clutter if the surface mesh is not sufficiently smoothed.

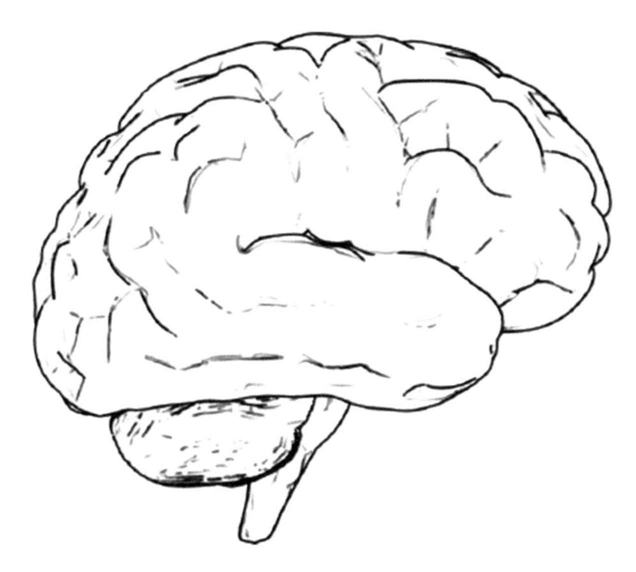

Fig. 15 The brain model with photic extremum lines

5.6 Photic Extremum Lines

Photic extremum lines (PELs) were introduced by Xi et al. [54]. These feature lines depict regions of the surface mesh with significant variations of illuminations. This method is based on the magnitude of the light gradient. Formally, these lines are defined as the set of points where the variation of illumination along its gradient direction reaches a local maximum (Fig. 15):

$$\boxed{D_{\mathbf{w}}\|\nabla f\| = 0 \text{ and } D_{\mathbf{w}}D_{\mathbf{w}}\|\nabla f\| < 0,}$$

with $\mathbf{w} = \frac{\nabla f}{\|\nabla f\|}$. Normally, f is used as the headlight illumination: $f := \langle \mathbf{n}, \mathbf{v} \rangle$ with $\mathbf{n}$ as the normal vector and $\mathbf{v}$ as the view-vector. PELs have more degrees of freedom to influence the result by adding more light sources. Thus, the scalar value of f changes by adding the light values of the vertices by other lights. Noisy photic extremum lines are filtered by a threshold which is based on the integral of single connected lines. The strength T of a line with points $\mathbf{x}_0, \ldots, \mathbf{x}_n$ is determined by:

$$T = \int \|\nabla f\| = \sum_{i=0}^{n-1} \frac{\|\nabla f(\mathbf{x}_i)\| + \|\nabla f(\mathbf{x}_{i+1})\|}{2} \|\mathbf{x}_i - \mathbf{x}_{i+1}\|.$$

If T is less than a user-defined threshold, the line is canceled out.

Summary Photic extremum lines are strongly inspired by edge detection in image processing and by human perception of a change in luminance. It uses the variation of illumination. The result may be improved by adding lights. Beside the filtering strategy to integrate over the lines and accumulate the magnitude of the gradient, the noise can also be reduced by adding a spotlight that directs to certain regions. Nevertheless, smoothing is necessary to gain reasonable results. Here, the smoothing of the normal is sufficient as the computation is mainly based on the normals. However, the computation has high performance costs. The original work was improved by Zhang et al. [57] to significantly increase the runtime.

5.7 Demarcating Curves

Demarcating curves were proposed by Kolomenkin et al. [26]. These feature lines are defined as the transition of a ridge to a valley line. To determine these lines, the derivative of the shape operator has to be calculated, recall Eq. (11):

$$C = \left(D_{\mathbf{v}}S \; D_{\mathbf{w}}S \right).$$

The demarcating curves are defined as the set of points where the curvature derivative is maximal (Fig. 16):

$$\boxed{\langle \mathbf{w}, S\mathbf{w} \rangle = 0 \text{ with } \mathbf{w} = \arg \max_{\|\mathbf{v}\|=1} D_{\mathbf{v}}\kappa.}$$

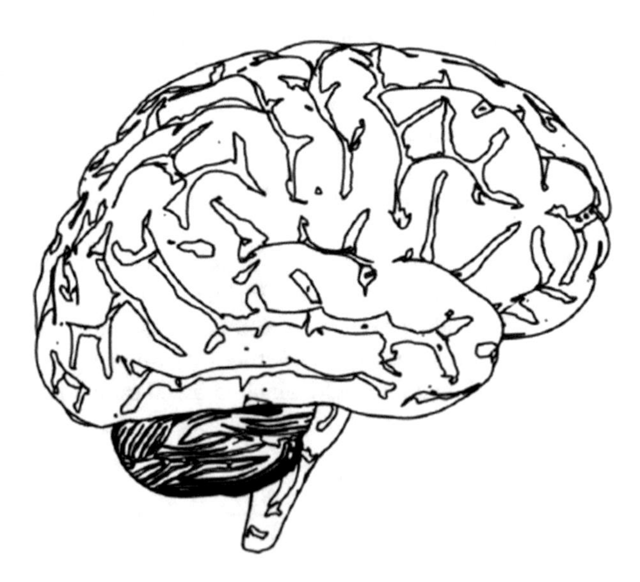

Fig. 16 The brain model with demarcating curves and contours

The values for $\mathbf{w}$ can be analytically found as the roots of a third order polynom. This is obtained by setting $\mathbf{v} = \binom{\sin(\theta)}{\cos(\theta)}$ and combining this with Eq. (12). A user-defined threshold eliminates demarcating curves, if it exceeds the value of $D_{\mathbf{w}}\kappa$.

Summary Demarcating curves are view-independent feature lines displaying regions where the change of the curvature is maximal. Therefore, higher-order derivatives are used. A $2 \times 2 \times 2$ rank-3 tensor is determined. This method can be used to illustrate bumps by surrounding curves. The advantage of the method is to enhance small features. Especially when combined with shading, this approach has its strength in illustrating archaeology objects where specific details are important, e.g., old scripts. For this application, view-dependent illustration techniques are not recommended because details need to be displayed for every camera position. Contrary, due to higher-order derivatives, the method is sensitive to noise and is not well suited for illustrative visualization.

5.8 Laplacian Lines

Laplacian lines were proposed by Zhang et al. [58]. The introduction of these lines was inspired by the Laplacian-of-Gaussian (LoG) edge detector in image processing and aims at a similar effect for surface meshes. The idea of the LoG method is to determine the Laplacian of the Gaussian function and to use this kernel as a convolution kernel for the image. Laplacian lines calculate the Laplacian of an illumination function f and determine the zero crossing as feature lines. To remove noisy lines, the lines are only drawn if the magnitude of the illumination gradient exceeds a user-defined threshold τ:

$$\boxed{\Delta f = 0 \text{ and } \|\nabla f\| \geq \tau,}$$

where Δ is the discrete Laplace-Beltrami operator on the surface mesh and f is the illumination with $f := \langle \mathbf{n}, \mathbf{v} \rangle$. Here, the discrete Laplace-Beltrami operator with the Belkin weights is used, as introduced in Sect. 3.4. The advantage of this method is the simplified representation of the Laplacian of the illumination:

$$\begin{aligned} \Delta f(\mathbf{p}) &= \Delta \langle \mathbf{n}, \mathbf{v} \rangle \\ &= \langle \Delta \mathbf{n}, \mathbf{v} \rangle. \end{aligned}$$

Here, $\Delta \mathbf{n}$ is the vector Laplace operator in the Euclidean space (Fig. 17).

This is just a composite of the Laplacian of the different components. Thus, the algorithm consists of a preprocessing step to calculate the Laplace-Beltrami operator with the Belkin weights of the components of the normal $\Delta \mathbf{n}$. During runtime, the algorithm detects the zero crossings of $\langle \Delta \mathbf{n}, \mathbf{v} \rangle$ and checks if the magnitude of $\|\nabla f\|$ exceeds the user-defined threshold.

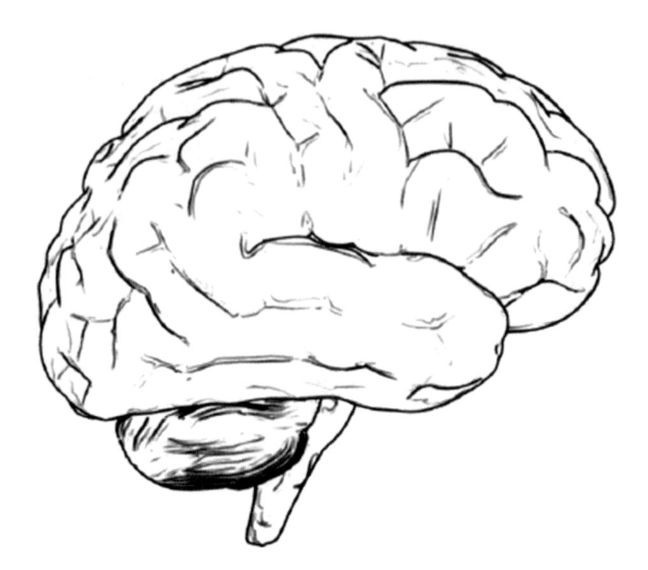

Fig. 17 The brain model with Laplacian lines

Summary The Laplacian lines are strongly inspired by edge detection algorithms in image processing. This method is based on the Laplacian-of-Gaussian. Basically, the method searches for zero crossings in the Laplacian of the illumination. The computational effort can be simplified by a preprocessing step. Thus, interactive frame rates for geometric models of moderate size are possible during the interaction. Similar to other higher order methods, this approach also assumes well smoothed surface normals. The Belkin weights for the Laplace-Beltrami operator have a smoothing effect for the Laplacian line generation. This method illustrates sharp edges well, but is not suitable for round corners.

6 Discussion and Comparison

This section deals with general properties of the different feature line methods. We discuss the different approaches to derive first recommendations which method may be used for which kind of geometry. First, we list all feature line methods in Table 1 and name different properties and the order of the corresponding method. Furthermore, in Fig. 18 the higher-order feature lines are illustrated on an analytic function.

The benefit of feature lines is motivated by the visual perception. In [32] it is stated that the first stage of the assessment of the shape is done by extracting features, such as contours. These characteristics help to understand the shape. The illustration of shapes with feature lines cannot be seen as an alternative to shading. It is rather an additional concept. Kolomenkin et al. [26] showed that their

Table 1 List of different feature line methods with derivative order and view dependency

Name	Order	View-dep.
Contours	1	Yes
Crease lines	1	No
Ridges & Valleys	3	No
Suggestive contours	2	Yes
Apparent ridges	3	Yes
Photic extremum lines	3	Yes
Demarcating curves	3	No
Laplacian lines	3	Yes

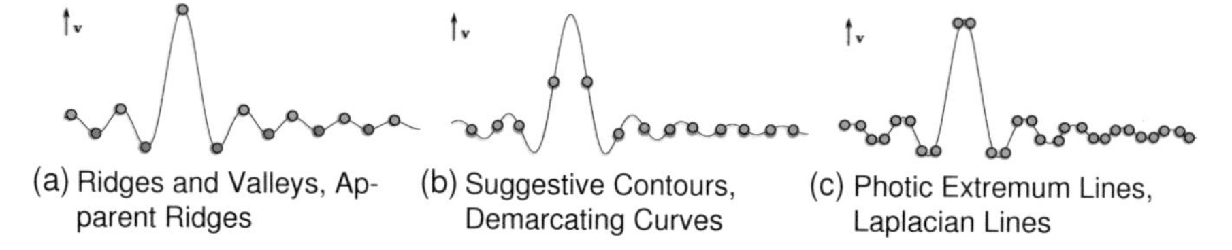

Fig. 18 Drawing of an analytic function with illustrated feature line positions. In (**a**) the ridges are denoted in orange and the valleys are illustrated in cyan. For this function with fixed view direction, the apparent ridges coincide with ridge and valley lines. In (**b**) the suggestive contours and the demarcating curves are the same. In (**c**) the photic extremum lines and the Laplacian lines coincide

demarcating lines support the shading and can extract text from archaeology objects. However, for examining structures where the whole object inherits important information, feature lines should not be used solely. For data where the scene can be divided into focus and context objects, feature lines can be applied to the context objects. Furthermore, feature lines can also be used to enhance focus with additional shading.

Depending on the underlying model, we may recommend different techniques. Most of the feature lines are able to depict the contour, but this depends strongly on the bending of the surface at the contour. Especially apparent ridges and photic extremum lines are able to draw contour lines, but in our experiments we noticed that activating the contour enhances the visual impression because some parts of the contour were missing. If the surface model is an assembly with sharp edges we recommend to use ridge and valley lines or apparent ridges. These features often appear in medical models like implants or prostheses. For simple models with only a few sharp edges, crease lines may be appropriate as well. If the models have a lot of round edges the answer for the right feature line method is a matter of taste. These features appear in models like vascular surfaces or organs. For scenarios where it is important to illustrate details for every camera position, we recommend ridges and valleys as well as demarcating curves. From an artistic point of view, suggestive contours, photic extremum lines, and Laplacian lines should be chosen. The reason for this suggestion is that especially for a rounded cube the photic extremum lines and Laplacian lines generate double lines around the feature to denote the rounded edge. If the user wants to visualize the line along the edge, the ridge and valley lines or the apparent ridges should be used. For this case, the crease line approach is not useful because it depicts only edges with specific greater value of the dihedral angle. Therefore, too many lines may be generated. If the surface has many crevices, we recommend the suggestive contours. They illustrate the inflection points of valleys. Round corners are often represented in many organic structures like livers or bones, see Fig. 20 for a femur model or a skull model.

Table 2 lists all possible features and Fig. 19 shows the different features. Please note that the assessment of the suitability of a method—marked in the table—necessarily is a subjective assessment by the authors and two artists. For instance, regarding the property whether the methods are able to detect round edges, we mean if it detects the specific round feature. As already mentioned, it does not reflect the ability to enhance the round edge from an illustrative or artist point of view. This concerns the ability to depict bumps. In agreement with artist, the bump shown from a sideway (s.w.) perspective would be illustrated such that it depicts the smooth transition from the ground to the dent. The drawing of the surrounding circle of the bump is not desirable as it conveys a sharp transition from the bump to the ground. For bumps shown from the top perspective it is sufficient if a round circle is drawn. Bumps can occur as polyps or blebs on a cerebral aneurysm. Especially blebs are important anatomical features to be detected because they are an indicator for rupture. Blebs can also occur as polyps in CT colonography.

We also listed the property *deformation* in the table. This characteristic means if the corresponding method is able to illustrate the features of deformable surfaces, e.g., animated objects, in real-time. As an example, Oeltze et al. [36] analyzed myocardial perfusion data. The focus lies on the examination of the infarction scarf on the left ventricle. In this paper, the left ventricle is illustrated as context information. Using the time-dependent data, it would also be possible to illustrate the context information with some feature line methods during the animation.

For example suggestive contours have two definitions of how to assess the feature lines. One is curvature-based and the other is light-based. With the second definition, no preprocessing is needed to assess the curvature and the principle curvature directions. This is in contrast to ridges and valleys and apparent ridges. Therefore, these algorithms are not able to compute the feature lines during the deformation. Photic extremum lines are also able to compute the feature lines during runtime because of the light and view dependency. The Laplacian lines need to precompute the Laplacian of the normals. Hence, this method is not suited for deformations.

Furthermore, Fig. 20 shows some exemplary models illustrated with higher order feature lines. Three typical models in the discrete differential geometry field (cow, Buddha, Max Planck) as well as three models from the medical image data (brain, femur, skull) are presented.

In summary, current feature lines are not suitable for the depiction of anatomical structures directly derived from medical image data because the underlying surfaces are too noisy. Advanced smoothing algorithms are necessary to reduce artifacts, but preserve important anatomical structures. For the depiction of a sparse representation of the model in a context-aware manner, the feature line methods can be used.

Table 2 List of supported feature by the methods

Name	Sharp edges	Round edges	Bumps (s.w.)	Bumps (top)	Contour	Deformation
Contour	☒	☒	☒	☒	☑	☑
Crease lines	☑	☒	☒	☒	☒	☑
Ridges & Valleys	☑	☑	☒	☑	☒	☒
Suggestive contours	☒	☒	☑	☑	☒	☑
Apparent ridges	☑	☑	☑	☑	☑	☒
Photic extremum lines	☑	☒	☒	☑	☑	☑
Demarcating curves	☒	☒	☒	☑	☒	☒
Laplacian lines	☑	☒	☒	☑	☑	☒

The different features are illustrated in Fig. 19

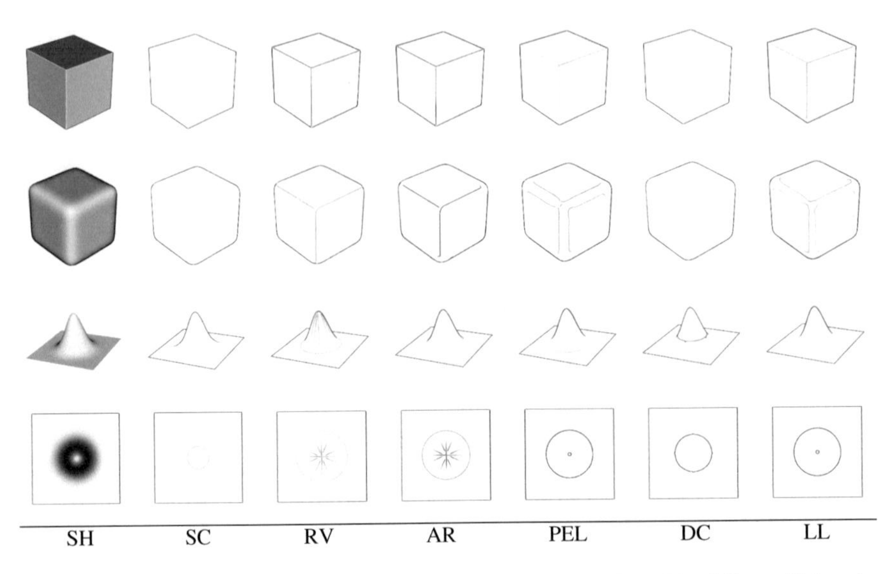

Fig. 19 Different surface features are illustrated with shading (SH) and in different high-order feature line methods: suggestive contours (SC), ridge and valley (RV), apparent ridges (AR), photic extremum lines (PEL), demarcating curves (DC), and Laplacian lines (LL)

6.1 Medical Application

As stated, feature line methods can be seen as an illustrative visualization method that can enhance shading or as an alternative in the focus-and-context visualization. In this section, we list different application fields where illustrative visualization is useful for effectively depicting medical data. At the end, we list possible fields where feature lines can be used to encode context information.

Fischer et al. [13] proposed to use illustrative visualizing tools to depict structures of hidden surfaces. The rendering style is tailored for understanding spatial relationships and for visualizing hidden objects. Born et al. [3] used illustrative techniques to depict stream surfaces. Their techniques are very useful for the visualizing of complex flow structures. In the area of brain data, Jainek et al. [24] suggested to use a hybrid visualization method to illustrate mesh and volume rendering. Their approach is efficient for the exploration in clinical research. Chu et al. [7] proposed a guideline of various rendering techniques. They combined, e.g., isophote-based line hatching and silhouette drawing, for illustrative vascular visualization. Different rendering techniques for medical applications were presented by Tietjen et al. [50]. An example for illustrative visualization for liver surgery can be found in [19].

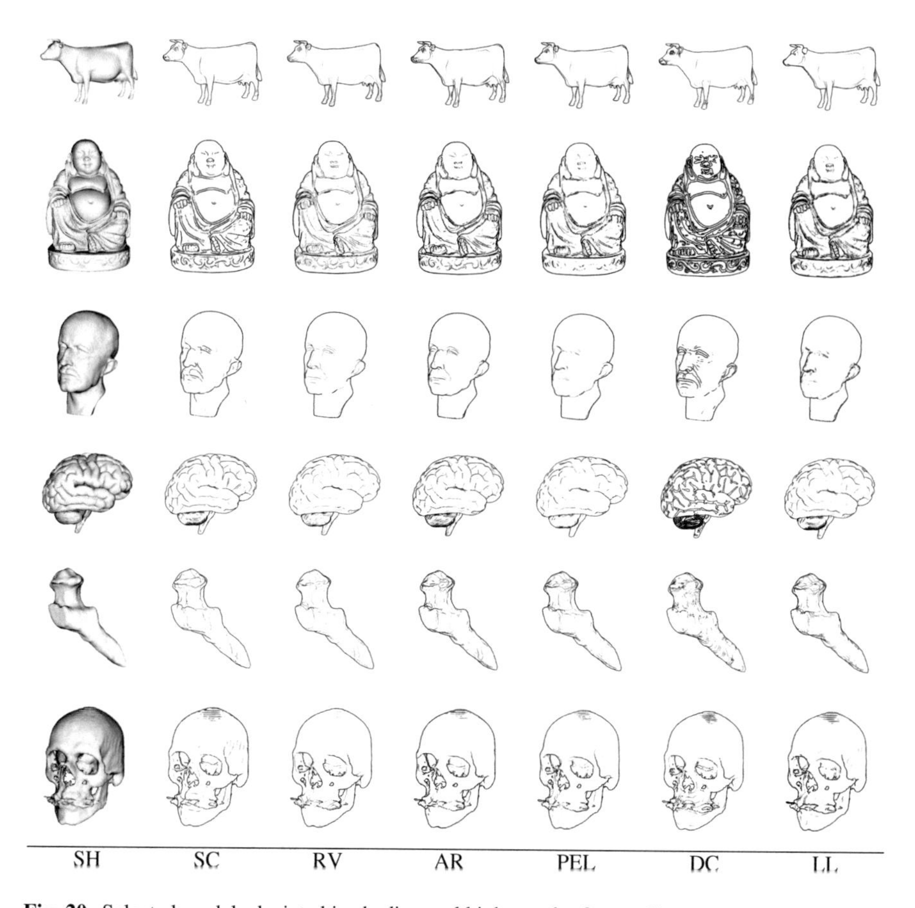

Fig. 20 Selected models depicted in shading and higher-order feature lines

Glaßer et al. [16] presented an approach to visualize 3D cluster results. Here, the medical researcher can analyze the whole 3D scene with different cluster results where he can also select interesting objects. The surrounding objects become context information. Thus, we propose to illustrate them with feature lines. In this case, we used the contour because the objects does not inherit much features. Figure 21a illustrates the main object with unselected objects illustrated with feature lines.

In the field of endoscopic views, the identification of polyps is necessary. Once the polyps are detected, they can be illustrated in such a way that the endoscopic views are used for context information. In Fig. 21b, we used suggestive contours for the vessel and diffuse shading for the polyps.

In Fig. 21c, we visualized the portal vein and three liver segments. The portal vein is illustrated in diffuse shading in red. The liver segments are visualized in diffuse shading with transparency and photic extremum lines.

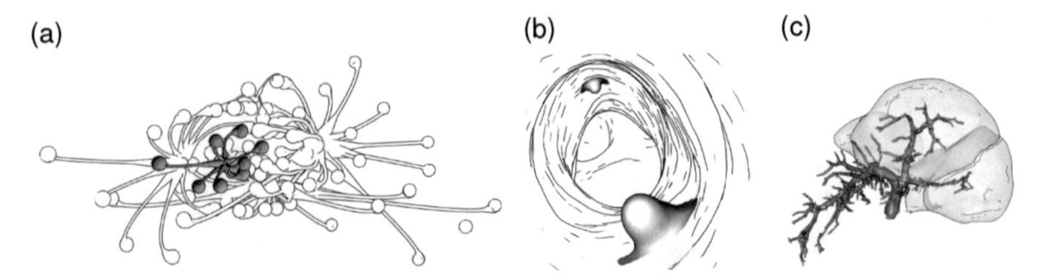

Fig. 21 Different medical application fields where feature lines can be used to illustrate surrounding objects. In (**a**) a 3D cluster results is depicted. One cluster is shaded and the others are illustrated with contour lines. In (**b**) the vessel is illustrated with suggestive contours and shading is used for the polyps. In (**c**), we visualized the portal vein with diffuse shading and three liver segments visualized in diffuse shading with transparency and photic extremum lines.

7 Conclusion

We have summarized the most common feature line methods for object space-based presentations of 3D meshes as they are frequently used in medicine and molecular biology. The presentation of the different methods was also covered by two basic sections. We did not only list the most common feature line methods and their calculation in the discrete space, but also provided the mathematical background to explain the calculation from the differential geometry point of view. Our goal was to present an extensive list of feature lines on the one hand, and to equip the reader with basic knowledge of differential geometry on the other hand. The graduated student may be able to follow the different methods and to implement every feature line algorithm based on our explanations in the field of discrete differential geometry. Therefore, our survey and tutorial may also be used by students who are new in the field of illustrative rendering. Furthermore, this survey may also be used as a starting point for the development of new feature line methods. The potential of advanced and recently introduced feature line techniques is currently not exploited in the display of medical surface models. The careful application of these methods and perception-based evaluations are left open for future work.

References

1. Alliez, P., Cohen-Steiner, D., Devillers, O., Lévy, B., Desbrun, M.: Anisotropic polygonal remeshing. In: Proceedings of ACM SIGGRAPH, vol. 22, pp. 485–493 (2003)
2. Belkin, M., Sun, J., Wang, Y.: Discrete laplace operator on meshed surfaces. In: Proceedings of Symposium on Computational Geometry, pp. 278–287. ACM, New York (2008)
3. Born, S., Wiebel, A., Friedrich, J., Scheuermann, G., Bartz, D.: Illustrative stream surfaces. IEEE Trans. Vis. Comput. Graph. **16**(6), 1329–1338 (2010)
4. Burns, M., Klawe, J., Rusinkiewicz, S., Finkelstein, A., DeCarlo, D.: Line drawings from volume data. In: Proceedings of ACM SIGGRAPH, vol. 24, no. 3, pp. 512–518 (2005)
5. Cazals, F., Pouget, M.: Estimating differential quantities using polynomial fitting of osculating jets. In: Proceedings of ACM SIGGRAPH, pp. 177–187 (2003)

6. Chen, X., Schmitt, F.: Intrinsic surface properties from surface triangulation. In: Proceedings of the European Conference on Computer Vision, pp. 739–743 (1992)
7. Chu, A., Chan, W.Y., Guo, J., Pang, W.M., Heng, P.A.: Perception-aware depth cueing for illustrative vascular visualization. In: International Conference on BioMedical Engineering and Informatics, vol. 1, pp. 341–346 (2008)
8. Cohen-Steiner, D., Morvan, J.M.: Restricted delaunay triangulations and normal cycle. In: Proceedings of Symposium on Computational Geometry, pp. 312–321. ACM (2003)
9. DeCarlo, D., Finkelstein, A., Rusinkiewicz, S., Santella, A.: Suggestive contours for conveying shape. In: Proceedings of ACM SIGGRAPH, vol. 22, pp. 848–855 (2003)
10. do Carmo, M.P.: Differential Geometry of Curves and Surfaces. Prentice-Hall, Englewood Cliffs (1976)
11. do Carmo, M.P.: Riemannian Geometry. Birkhäuser, Boston (1992)
12. Eberly, D.: Ridges in image and data analysis. Computational Imaging and Vision. Springer, Berlin (1996)
13. Fischer, J., Bartz, D., Straßer, W.: Illustrative display of hidden iso-surface structures. In: Proceedings of IEEE Visualization, pp. 663–670 (2005)
14. Floater, M.S.: Mean value coordinates. Comput. Aided Geom. Des. **20**(1), 19–27 (2003)
15. Girshick, A., Interrante, V., Haker, S., Lemoine, T.: Line direction matters: an argument for the use of principal directions in 3d line drawings. In: Proceedings of the Non-Photorealistic Animation and Rendering, pp. 43–52 (2000)
16. Glaßer, S., Lawonn, K., Preim, B.: Visualization of 3D cluster results for medical tomographic image data. In: Proceedings of VISIGRAPP/GRAPP, pp. 169–176 (2014)
17. Goldfeather, J., Interrante, V.: A novel cubic-order algorithm for approximating principal direction vectors. ACM Trans. Graph. **23**(1), 45–63 (2004)
18. Hameiri, E., Shimshoni, I.: Estimating the principal curvatures and the darboux frame from real 3-d range data. Trans. Syst. Man Cybern. B **33**(4), 626–637 (2003)
19. Hansen, C., Wieferich, J., Ritter, F., Rieder, C., Peitgen, H.O.: Illustrative visualization of 3d planning models for augmented reality in liver surgery. Comput. Assist. Radiol. Surg. **5**(2), 133–141 (2010)
20. Hertzmann, A., Zorin, D.: Illustrating smooth surfaces. In: Proceedings of ACM SIGGRAPH, pp. 517–526 (2000)
21. Hildebrandt, K., Polthier, K., Wardetzky, M.: On the convergence of metric and geometric properties of polyhedral surfaces. Geom. Dedicata **123**, 89–112 (2005a)
22. Hildebrandt, K., Polthier, K., Wardetzky, M.: Smooth feature lines on surface meshes. In: Proceedings of the Third Eurographics Symposium on Geometry Processing, Vienna, vol. 255, pp. 85–90 (2005b)
23. Interrante, V., Fuchs, H., Pizer, S.: Enhancing transparent skin surfaces with ridge and valley lines. In: Proceedings of IEEE Visualization, pp. 52–59 (1995)
24. Jainek, W.M., Born, S., Bartz, D., Straer, W., Fischer, J.: Illustrative hybrid visualization and exploration of anatomical and functional brain data. Comput. Graph. Forum **27**(3), 855–862 (2008)
25. Judd, T., Durand, F., Adelson, E.: Apparent ridges for line drawing. In: Proceedings of ACM SIGGRAPH, p. 19 (2007)
26. Kolomenkin, M., Shimshoni, I., Tal, A.: Demarcating curves for shape illustration. In: Proceedings of ACM SIGGRAPH Asia, pp. 157:1–157:9 (2008)
27. Krüger, A., Tietjen, C., Hintze, J., Preim, B., Hertel, I., Strauß, G.: Analysis and exploration of 3d visualization for neck dissection planning. Comput. Assist. Radiol. Surg. **1281**(0), 497–503 (2005)
28. Kühnel, W.: Differential Geometry: Curves - Surfaces - Manifolds. Student Mathematical Library. American Mathematical Society, Providence (2006)
29. Lawonn, K., Gasteiger, R., Preim, B.: Qualitative evaluation of feature lines on anatomical surfaces. In: Bildverarbeitung für die Medizin (BVM), pp. 187–192 (2013)
30. Lawonn, K., Mönch, T., Preim, B.: Streamlines for illustrative real-time rendering. Comput. Graph. Forum **33**(3), 321–330 (2013)

31. MacNeal, R.: The solution of partial differential equations by means of electrical networks. Ph.D Thesis. California Institute of Technology (1949)
32. Marr, D.: Early processing of visual information. Philos. Trans. R. Soc. Lond. Ser. B Biol. Sci. **275**(942), 483–519 (1976)
33. Meyer, M., Desbrun, M., Schröder, P., Barr, A.H.: Discrete differential-geometry operators for triangulated 2-manifolds. In: Proceedings of Visuality & Mathematics, pp. 35–57 (2002)
34. Muthukrishnan, R., Radha, M.: Edge detection techniques for image segmentation. Int. J. Comput. Sci. Inf. Technol. **3**(6) (2011)
35. Nadernejad, E., Sharifzadeh, S., Hassanpour, H.: Edge detection techniques: Evaluations and comparisons. Appl. Math. Sci. **2**(31), 1507–1520 (2008)
36. Oeltze, S., Hennemuth, A., Glaßer, S., Kühnel, C., Preim, B.: Glyph-based visualization of myocardial perfusion data and enhancement with contractility and viability information. In: Visual Computing for Biology and Medicine, pp. 11–20 (2008)
37. Ohtake, Y., Belyaev, A., Seidel, H.-P.: Ridge-valley lines on meshes via implicit surface fitting. ACM SIGGRAPH **23**, 609–612 (2004)
38. Page, D.L., Koschan, A., Sun, Y., Paik, J., Abidi, M.A.: Robust crease detection and curvature estimation of piecewise smooth surfaces from triangle mesh approximations using normal voting. In: Computer Vision and Pattern Recognition, pp. 162–167 (2001)
39. Praun, E., Hoppe, H., Webb, M., Finkelstein, A.: Real-time hatching. In: Proceedings of ACM SIGGRAPH, pp. 579–584 (2001)
40. Preim, B., Botha, C.: Visual Computing for Medicine, 2nd edn. Morgan Kaufmann, Amsterdam (2013)
41. Ritter, F., Hansen, C., Dicken, V., Konrad, O., Preim, B., Peitgen, H.O.: Real-time illustration of vascular structures. IEEE Trans. Vis. Comput. Graph. **12**(5), 877–884 (2006)
42. Rusinkiewicz, S.: Estimating curvatures and their derivatives on triangle meshes. In: Symposium on 3D Data Processing, Visualization, and Transmission (2004)
43. Rusinkiewicz, S., Cole, F., DeCarlo, D., Finkelstein, A.: Line drawings from 3d models. In: Proceedings of ACM SIGGRAPH, pp. 39:1–39:356 (2008)
44. Senthilkumaran, N., Rajesh, R.: Edge detection techniques for image segmentation – a survey of soft computing approaches. Int. J. Recent Trends Eng. **1**(2), (2009)
45. Sorkine, O.: Laplacian Mesh Processing, pp. 53–70. Eurographics Association, Dublin (2005). Eurographics 05 STAR
46. Stylianou, G.: A feature based method for rigid registration of anatomical surfaces. In: Geometric Modeling for Scientific Visualization, Mathematics and Visualization, pp. 139–149. Springer, Berlin/Heidelberg (2004)
47. Svetachov, P., Everts, M.H., Isenberg, T.: DTI in context: illustrating brain fiber tracts in situ. Comput. Graph. Forum **29**(3), 1023–1032 (2010)
48. Taubin, G.: Estimating the tensor of curvature of a surface from a polyhedral approximation. In: Proceedings of International Conference on Computer Vision, pp. 902–907. IEEE Computer Society (1995)
49. Taubin, G.: A signal processing approach to fair surface design. In: Proceedings of ACM SIGGRAPH, pp. 351–358 (1995)
50. Tietjen, C., Isenberg, T., Preim, B.: Combining silhouettes, surface, and volume rendering for surgery education and planning. In: The Eurographics Conference on Visualization, pp. 303–310 (2005)
51. Treavett, S.M.F., Chen, M.: Pen-and-ink rendering in volume visualisation. In: Ertl, T., Hamann, B., Varshney, A. (eds.) Proceedings of IEEE Visualization, pp. 203–210 (2000)
52. Wardetzky, M., Mathur, S., Kälberer, F., Grinspun, E.: Discrete laplace operators: no free lunch. In: Symposium on Geometry Processing, pp. 33–37. Eurographics Association (2007)
53. Webb, M., Praun, E., Finkelstein, A., Hoppe, H.: Fine tone control in hardware hatching. In: Non-Photorealistic Animation and Rendering, pp. 53–58 (2002)
54. Xie, X., He, Y., Tian, F., Seah, H.S., Gu, X., Qin, H.: An effective illustrative visualization framework based on photic extremum lines (PELS). IEEE Trans. Vis. Comput. Graph. **13**, 1328–1335 (2007)

55. Yoshizawa, S., Belyaev, A., Seidel, H.-P.: Fast and robust detection of crest lines on meshes. In: Proceedings of the 2005 ACM Symposium on Solid and Physical Modeling, Cambridge, pp. 227–232 (2005)
56. Zander, J., Isenberg, T., Schlechtweg, S., Strothotte, T.: High quality hatching. Comput. Graph. Forum **23**(3), 421–430 (2004)
57. Zhang, L., He, Y., Seah, H.S.: Real-time computation of photic extremum lines (PELs). Vis. Comput. **26**(6–8), 399–407 (2010)
58. Zhang, L., He, Y., Xia, J., Xie, X., Chen, W.: Real-time shape illustration using laplacian lines. IEEE Trans. Vis. Comput. Graph. **17**, 993–1006 (2011)

Remote Visualization Techniques for Medical Imaging Research and Image-Guided Procedures

Peter Kohlmann, Tobias Boskamp, Alexander Köhn, Christian Rieder, Andrea Schenk, Florian Link, Uwe Siems, Marcus Barann, Jan-Martin Kuhnigk, Daniel Demedts, and Horst K. Hahn

Abstract There has been a tremendous increase in medical image computing research and development over the last decade. This trend continues to gain further speed, driven by the sheer amount of multimodal medical image data but also by the broad spectrum of computer-assisted applications. At the same time, user expectations with respect to diagnostic accuracy, robustness, speed, automation, workflow efficiency, broad availability, as well as intuitive use have reached a high level already. More recently, cloud computing has entered the field of medical imaging, providing means for more flexible workflows including the support of mobile devices and even a medical imaging equivalent of the App Store paradigm. This paper discusses requirements for modern medical software systems with a focus on image analysis and visualization. It provides examples from different areas of application covering collaborative multi-center imaging trials with online reading and advanced analysis as well as an intraoperative augmented-reality scenario for translating liver surgery planning data directly into the operating room through a mobile multi-touch device. A combination of remote rendering and visualization techniques with an efficient modular development framework (MeVisLab) is presented as a basis for fast implementation, early evaluation, and iterative optimization in these applications.

P. Kohlmann • A. Köhn • C. Rieder • A. Schenk • J.-M. Kuhnigk • D. Demedts • H.K. Hahn (✉)
Fraunhofer MEVIS, Bremen, Germany
e-mail: peter.kohlmann@mevis.fraunhofer.de; alexander.koehn@mevis.fraunhofer.de; christian.rieder@mevis.fraunhofer.de; andrea.schenk@mevis.fraunhofer.de; jan-martin.kuhnigk@mevis.fraunhofer.de; daniel.demedts@mevis.fraunhofer.de; horst.hahn@mevis.fraunhofer.de

T. Boskamp • F. Link • U. Siems • M. Barann
MeVis Medical Solutions AG, Bremen, Germany
e-mail: tobias.boskamp@mevis.de; florian.link@mevis.de; uwe.siems@mevis.de; marcus.barann@mevis.de

L. Linsen et al. (eds.), *Visualization in Medicine and Life Sciences III*, Mathematics and Visualization, DOI 10.1007/978-3-319-24523-2_6

1 Introduction

The outsourcing of services and applications to an external IT infrastructure increasingly takes place in various private and professional application areas. In medical image computing, the introduction of hosted applications and cloud technologies bears the potential to significantly transform the handling of medical images in clinics and medical research facilities including the associated workflows: data storage, quality assurance, integration of data from multiple sources, complex modeling methods, knowledge-based decision support, as well as visualization and analysis of medical image data are selected tasks which are no longer required to be available locally or through complex offline processes. Respective developments are driven by the ongoing rapid diversification of medical imaging applications, increase of data volumes, and increased user expectations and requirements regarding both the availability as well as the quality of specialized tools.

A *Medical Image Computing Cloud*, implemented as either a private cloud within the hospital firewall or through secure web technologies, will typically be operable via browser technologies. The information which is stored in the cloud like original and derived data, analysis results, treatment plans, study reports, patient history, and other clinical information can be accessed at any time with multiple (including mobile) devices. In particular, multi-centric research consortia will greatly benefit from centralized collaboration services not only for data storage, but also viewing and automated as well as interactive data analysis. Stored data will often contain highly sensitive information, which requires the implementation of thorough measures to ensure data privacy and security. An article about cloud computing in medical imaging highlights various aspects including technical requirements as well as a discussion about ethical and security issues in a comprehensive manner [13].

Ultimately, an equivalent of the well-known *App Store* paradigm for specialized medical software applications appears as an attractive solution for the rapid diversification mentioned above. Currently, medical visualization and analysis research only rarely find their way into clinical application, because of the high cost and time involved to transform research results into medical software products. Moreover, initial research results often require iterative development cycles including intensive end user feedback in order to achieve the level of optimization needed for routine application. Both issues could be addressed by a functioning, interoperable *Medical Imaging App Store* which provides a pay-per-case infrastructure and a user community. In such a store, users can preview and test new application functionality.

Fraunhofer MEVIS, Institute for Medical Image Computing, has a history in research and development in the fields of image acquisition, image processing, modeling, visualization, user interfaces, and application design. Together with the commercial spin-off MeVis Medical Solutions MeVisLab was developed as a powerful, modular framework for the development of image processing algorithms, visualization, and interaction methods, with a special focus on medical imaging.

In the following, the MeVisLab framework is described with a special focus on its recently introduced remote visualization capabilities. These new features allow the deployment of MeVislab-based applications for client-server environments. Two example applications are presented and discussed in this paper. First, it is demonstrated how an image viewer is embedded into a cloud infrastructure for clinical trials and collaborative imaging-based research and made available as web application via a modern web browser. A second example presents an augmented-reality application to support liver surgery, which is made available as an iPad App.

2 Remote Image Analysis and Visualization Framework

In the past years, several approaches for remote rendering on digital networks were proposed. At the same time, advanced medical image analysis and visualization methods were presented. However, cloud computing has just entered the field of medical imaging. In this section, the underlying remote rendering technology used by the concrete applications described in the following chapters is presented. Firstly, a brief description of MeVisLab, the modular development framework for medical image processing applications in which the proposed approach was implemented, is provided. Then, the decision for implementing an approach based on server-based rendering and asynchronous client-server communication is motivated. After giving an overview of related work in the field of remote applications, the main part of this chapter describes the developed methods and their integration into MeVisLab.

2.1 The MeVisLab Framework

Faced by the need for a common foundation for software development projects in the field of medical image computing, researchers and software engineers at Fraunhofer MEVIS and MeVis Medical Solutions have developed the R&D platform MeVisLab [21]. MeVisLab combines a feature-rich, extensible toolbox of image processing and visualization components with the possibility to add new, self-developed algorithms and to create prototypical applications that allow an evaluation of these methods in clinical settings. Originally designed as an internal development tool, MeVisLab has been made publicly available in 2004 and is used at research and development sites worldwide. A core features is that it allows efficient development of all aspects of a medical image analysis and visualization application:

- Powerful DICOM I/O and composition of native DICOM frames into 3D or 4D volumes.
- Efficient low-level image processing and visualization using C++ modules.
- Graphical programming to combine low-level modules into complex processing networks.

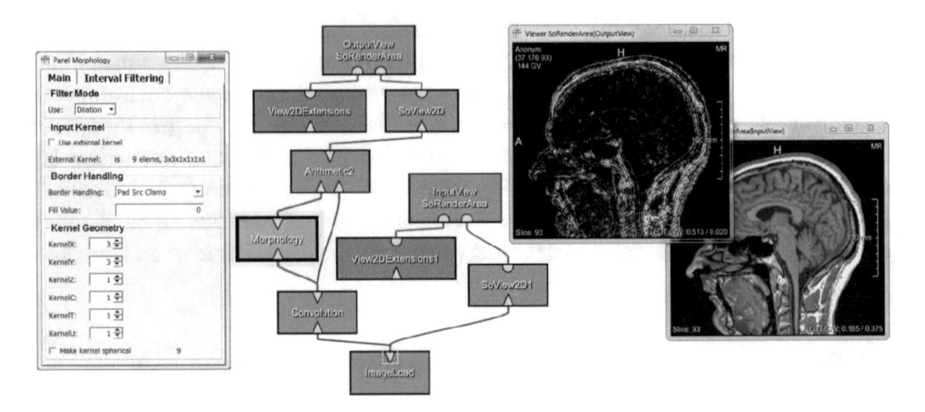

Fig. 1 Image processing example network in MeVisLab, implementing a simple edge filter. Modules are represented as boxes, data flow is visualized by connecting lines. A module's operation can be controlled through its user panel (shown for the Morphology module). Viewer panels can be opened to display the renderings generated by visualization modules

- Application development through python binding and a simple description language for GUI creation.

To make MeVisLab-based applications available to external users, they can be conveniently packaged to be deployed as standalone executables, so that, in a conventional workstation setting, they can be installed on a PC within a clinical site and locally operated by the physician (Fig. 1).

Including automatically wrapped parts of the open source libraries ITK [12], VTK [40], and Open Inventor [36], MeVisLab comes with over 2000 C++ modules plus hundreds of ready-to-use processing networks wrapped in so-called Macro Modules. Most of them can be classified into the categories image enhancement, image segmentation, image registration, quantification, visualization, and computer-aided detection (see [32] for more details).

The visualization of different types of data, including 2D images, 3D image volumes, dynamic image data, geometric surface models and meshes, as well as graphical annotations and text overlays, belongs to the core features of MeVisLab. The visualization engine is based on OpenGL [23] and Open Inventor, complemented by the integration of VTK. Some of its main characteristics include:

- A unified framework for 2D, 3D and 4D rendering.
- Interactive, real-time rendering of image data, surfaces, and contours.
- Utilization of modern GPUs and the OpenGL Shading Language [33].
- An extensible rendering architecture including a modular shader pipeline [31].

The visualization framework includes a GPU-based image volume renderer specifically designed for the visualization of large volumes and to offer a wide range of rendering options [17]. Recent developments in MeVisLab's volume rendering and 3D visualization library include the addition of multi-volume rendering, with applications in medical image visualization [41]. For a more comprehensive overview of MeVisLab, please refer to [3, 9, 15, 21, 32].

2.2 *Concept Discussion*

There are many possible ways to distribute the components of a typical medical image analysis application—image database, image import, image analysis, image visualization, application logic, and GUI—between server and client. Moreover, client-server communication can be either synchronous or asynchronous. Following requirements were considered for the presented approach:

1. According to the cloud paradigm, applications must be ubiquitously accessible on-demand.
2. There should be no limitations with respect to the visualization capabilities, including full support for dynamic and interactive visualizations.
3. The same application shall be deployable in different scenarios, including thin and zero-footprint clients (e.g., web clients), as well as mobile clients for tablet PCs and smartphones, with little additional development effort.
4. The infrastructure shall support environments with limited bandwidth and noticeable latencies (e.g., WIFI, DSL, or LTE connections).
5. The user interface shall remain responsive during time-consuming operations.

In the view of the authors, requirements 1–3 can be best fulfilled by a client-server architecture that does not depend on the specific (and typically uncontrollable) setup of the client platform and leaves almost the entire application including data handling, processing, and even rendering to the server. Requirements 4 and 5 suggest an asynchronous remote process communication protocol. Requirements 2 and 3 are especially relevant for R&D frameworks like MeVisLab which already come with a lot of visualization features and are often used as platforms for research and rapid prototyping. In exploratory situations, neither the final functionality nor the target platform is fixed in advance. Thus, (re-)implementing a feature that is often already available on the MeVisLab-based server for a particular client platform is not desirable. Finally, application development becomes more complex if more *intelligence* is distributed between client and server.

A drawback of server-based rendering is poor interactive speed in high-latency environments, which can be a severe disadvantage in comparison to client-based solutions with sufficient rendering power. Still, it has to be kept in mind that instead of only the current view, the entire dataset or rendering scene has to be downloaded by the client first. Due to this, on top of the extra implementation effort described in the previous paragraph, the increased responsiveness typically comes along with longer startup times. Asynchronous client-server communication makes programming more difficult, because keeping the application state synchronized between client and server is challenging. This is one more reason for keeping the client as thin as possible and making the server master of the application state. The implementation of the chosen architecture within MeVisLab is described in detail in the paragraphs following the related work discussion.

2.3 Related Work

The idea of remote desktops and applications is not new. The well-known virtual network computing (VNC) system introduced by Richardson et al. [30] is a client-server system, which allows the client to view the entire desktop environment of the server from anywhere over the Internet. One of the first applications for remote volume visualization was presented by Lippert et al. [18]. The underlying concept is especially designed for distributed and networked applications, where a remote server maintains large scale volume data sets, for being inspected, browsed-through, and rendered interactively by a local client. Hendin et al. [10] introduced a volume visualization tool based on the Virtual Reality Modeling Language (VRML), which uses three stacks of perpendicular slices. A VRML-plugin is controlled by a JAVA applet which allow the embedding of the tool into HTML. Engel et al. [7] extended this approach with techniques for fast volume clipping, collaborative work, and data size reduction. Ma et al. [19] presented a system that allows to explore remote visualization of time-varying data sets using a web browser. It computes images on a visualization server, which are transferred to the client and inserted into a graph. The parallel visualization tool Visapult [2] combines minimized data transfers and workstation-accelerated rendering. Yoon et al. [42] proposed image-based rendering acceleration and compression methods to increase efficiency of systems using 3D graphics on the web. Stegmaier et al. [37] presented a generic solution for hardware-accelerated remote visualization which works transparently for all OpenGL-based applications and scene graphs. A distributed architecture where a remote rendering server is able to manage complex models and to code a MPEG stream to be sent to the client allowing for interactive visualization on the client was introduced by Lamberti et al. [16].

Silicon Graphics provides a commercial solution called OpenGL VizServer [24], which enables lightweight client workstations to access the hardware-accelerated rendering capabilities of SGI Onyx servers. VirtualGL [39] is an open-source toolkit that gives any Unix/Linux remote display software the ability to run OpenGL applications with full 3D hardware acceleration. The Chromium Renderserver [26] is an open-source software infrastructure which allows one or more users to run and view image output from unmodified, interactive OpenGL and X11 applications on a remote parallel computational platform equipped with graphics hardware accelerators via industry-standard Layer-7 network protocols and client viewers.

In the medical imaging community, several systems allow a combination of visualization and image processing algorithms. Bitter et al. [3] compared four freely available frameworks for image processing and visualization which use ITK [12]. A survey of the most successful open-source libraries and prototyping frameworks for medical application development was presented by Caban et al. [4]. Another

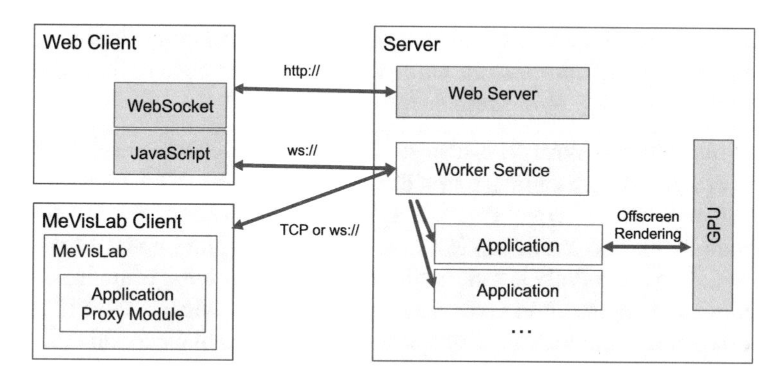

Fig. 2 The MeVisLab client-server architecture. Multiple MeVisLab-based applications may be executed on the server *(right)*, each in its own process. GPU-based off-screen rendering is used for visualization tasks. Communication between client and server process is routed through the Worker Service background process, which also controls the applications' life cycle. Different client platforms are supported, including client applications based on MeVisLab (*bottom left*) and web applications implemented using current web standards (*top left*)

prototyping environment is the eXtensible Imaging Platform (XIP) [27], which has a comparable functionality as MeVisLab.

2.4 MeVisLab for Client-Server Applications

Based on the reasoning provided in Sect. 2.2, a client-server architecture for MeVisLab applications was implemented that combines server-based rendering technology with a proprietary, asynchronous remote process communication protocol (see Fig. 2). This architecture is described in the subsequent paragraphs.

2.4.1 Remote Rendering

As discussed in Sect. 2.2, the ability to utilize the full extent of the server's visualization feature set (in this case MeVisLab's visualization libraries) as part of client-server applications was one of the main reasons for the decision to keep visualization functionality strictly on the server. Therefore, a remote rendering technology using OpenGL's frame buffer objects was implemented in MeVisLab. Using this technique, a visualization scene is rendered off-screen into a buffer in GPU memory on the server. The rendered image is then encapsulated into a message and sent to the client. For image encoding, both the PNG and the JPEG format can be used with different options for lossless and lossy data compression.

A key factor for the usability of server-based rendering applications is transmission speed. In order to minimize the amount of data that needs to be transferred from server to client, two key optimizations were implemented:

- **Dynamic compression:** The idea is that while the entire scene is changing rapidly (e.g., during user interaction like slicing through a 3D dataset), the user is not able to notice the small details lost by using a lossy compression algorithm. As soon as no new rendering is requested for a certain amount of time (e.g., 500 ms), a lossless PNG is sent to allow detailed inspection of unmodified data as required for application in the medical field. For a typical $512 \times 512 \times 24$ bits per pixel view of a native CT image with colored annotations, this means that the compressed size of each transferred image shrinks from 31.3 % (i.e. 240 kB) for a lossless PNG to 3.8 % (i.e. 29 kB) for a lossy JPEG at the default setting of 75 % quality. Moreover, the JPEG compression is almost three times faster than JPEG compression (MeVisLab uses libPNG and TurboJPEG, respectively) and thus better suited for setups where server speed is the limit.
- **Tiled rendering:** In certain frequent use cases such as placing a marker, or continuously displaying the value under the mouse cursor, only a small are of the view changes. These use cases can be optimized substantially by subdividing the rendered image into N tiles (a typical size is $N = 64$ for a 512×512 view) transmitting only those tiles that differ from their predecessors. While this will have no effect for global image changes, the above-mentioned use cases will benefit by a transmission data reduction of a factor N. It can be combined with the dynamic compression technique, in which compression is performed separately on each changed tile.

There are further ideas for data reduction. For instance, as a generalization of the tiled rendering approach, the difference of the rendered image to its predecessor could be transmitted, assuming that redundancy will result in better compression ratios. However, as there is currently no native browser support for such delta encoding, an implementation would have to be implemented in JavaScript on a per-pixel basis, making the performance of low-end clients the bottleneck. A more promising improvement might be the utilization of movie codecs which make use of such heuristics and are already supported by browsers. However, codecs currently available in browsers are not optimized for providing interactive updates. Further, they typically require more than just the previous image for efficient encoding which results in increased latency. Still, these are issues that many remote applications currently struggle with, so perhaps codec development will address this in the future. Currently, movie encoding could be used for optimizing certain use-cases where the scene changes without constant user interaction, such as cine modes, or certain fly-through visualizations.

2.4.2 Communication Protocol

MeVisLab's proprietary remote rendering protocol supports dynamic and interactive visualizations. A typical message sequence that occurs in an interactive client-server application can be outlined as follows (see Fig. 3):

1. A user event in the client viewing area is recorded and encoded in an event message that is sent to the server.
2. The server receives the event and feeds it into the application's visualization scene.
3. If the user event results in a visualization update, the application notifies the client about the pending update.
4. The client responds to the update message by sending a frame request message to the server (additional update notifications are combined and delayed until the requested frame was received).
5. The server renders the updated scene and sends the new frame to the client.
6. The client displays the received frame to the user.

Steps 3 and 4 ensure that no frames are sent that the client has not requested and let the client directly control the update speed. A severe drawback of this strictly request-driven approach is the fact that two client-server round trips occur between the original user event and the display of the updated scene at the client. This can lead to noticeable latency and limitation of the effective frame rate. In order to overcome both effects, the MeVisLab remote rendering protocol offers a streaming mode in which update and frame request messages are omitted and the server responds to the initial event message with a frame message which contains the updated scene rendering. As an effect, the latency is reduced by 50 % which allows higher frame rates. However, a fast sequence of user events may now lead to a frame rate which exceeds the channel's bandwidth and may yield undesired overload effects. This is avoided by introducing a special acknowledge message issued by the

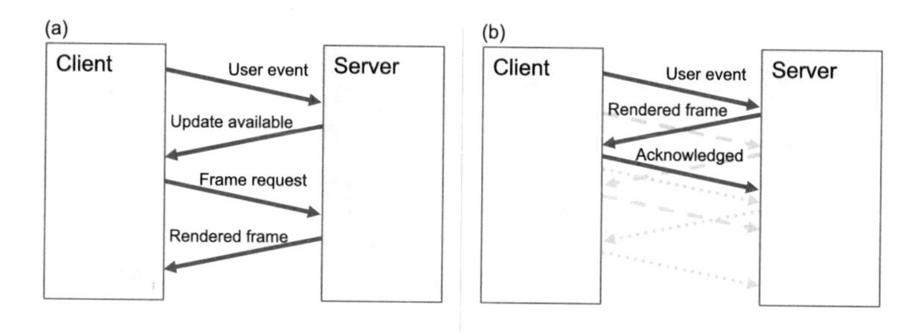

Fig. 3 Remote rendering protocol in an interactive visualization scenario. Request driven mode (**a**): two round-trip cycles between a user event and the resulting viewer refresh because an updated frame is only rendered on explicit client request. Streaming mode (**b**): the server renders and sends updated frames immediately which reduces the latency to one round-trip cycle. Acknowledge messages are gathered in order to limit the number of frames that are "on the way"

client to signal the receipt of a rendered frame message. The server checks incoming acknowledge messages and delays the rendering of new frames as soon as more than a specified number (typically 2–5) of previous frame messages have not been acknowledged (see Fig. 3). Thus, the effective frame rate is automatically adjusted depending on the channel's bandwidth and latency.

Using streaming mode and dynamic compression at the default of 75 % quality, the required bandwidth to slice through a CT volume with a 1024 × 1024 client viewer at 5, 15, or 30 frames per second (fps) is 0.5, 1.5, or 3.0 MB/s, respectively, so that today's LAN or WLAN bandwidth is more than sufficient. In this test setup, the speed at which the server compressed images was the limiting factor, resulting in a frame rate of 33 fps for both LAN and WLAN connection. Using an LTE connection from a smartphone, a frame rate of 11 fps was achieved. On average, 289 ms elapsed between an image request and receiving the mobile client's acknowledgment (of which JPEG compression took 24 ms). The tiling optimization was ineffective for these experiments since the test operation (slicing) always affects the entire image region.

2.4.3 Remote Module Interfaces

The remote rendering technology and the remote module interface represent the middleware for server-client communication. For a complete client-server application, a user interface has to be added to the client side, to enable user access to the application's features and output data. MeVisLab currently offers two options (see Fig. 4). While the deployment of a MeVisLab-based client is supported, this is

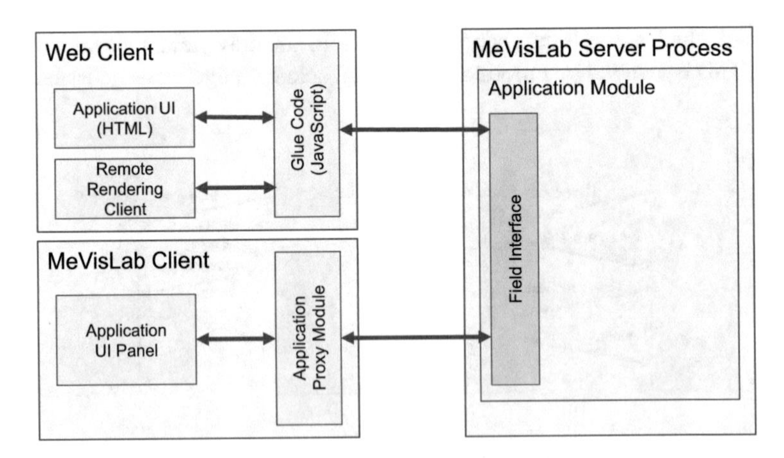

Fig. 4 Client implementation in different deployment scenarios. The server application (*right*) communicates with the client through its field interface. In the client implementation (*bottom left*) this field interface is replicated through an automatically generated proxy module. The application UI is generated using standard MeVisLab mechanisms. In web-based clients, the UI is implemented in HTML and linked to the application's field interface through JavaScript glue code. Visualizations are displayed by means of a JavaScript remote rendering client

mostly used for prototyping. For most end-user applications such as the ones described in Sects. 3 and 4, a platform-independent deployment via zero-footprint web clients is preferable.

In a typical zero-footprint deployment scenario, the client is executed in a web browser. This scenario is supported by the MeVisLab Web Toolkit (MWT), an extension which is not available in the public MeVisLab distribution. MWT includes JavaScript implementations of the remote module interface, a remote rendering client, as well as utility code to connect to the server process and to dynamically add UI elements in the browser page that are automatically generated from native MeVisLab GUI descriptions. MWT is based on modern web standards, such as HTML5 and Websockets [8], supports secure SSL sessions and is compatible with most current web browsers.

3 Cloud Application for Multi-Center Image-Based Clinical Trials

A good example to demonstrate the benefits of a cloud application are image-based clinical studies. They are often performed by research teams at (university) hospitals, the pharmaceutical industry, or contract research organizations. Different steps of the studies are increasingly performed in multi-center networks which might be spread over different countries. Those steps include image acquisition, quality assurance (QA), reading. Figure 5 illustrates how an image viewer can be embedded in a cloud infrastructure to interact with a study portal and data storage entities. Applications and services are made available as web application.

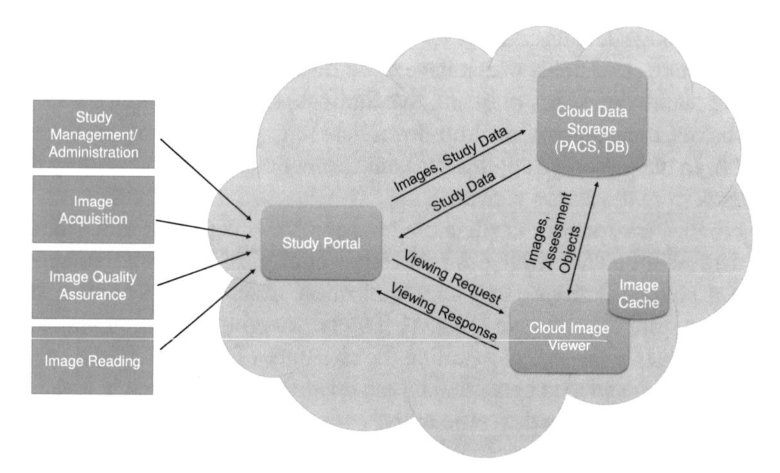

Fig. 5 Embedding of an image viewer into a cloud infrastructure for clinical trials. Different user groups can access cloud services/applications through a study portal via a modern web browser

3.1 User Groups

User groups which perform roles with different data access rights in a clinical study can connect to a study portal via web browser from any location and with different devices (e.g., PC, notebook, tablet). Roles which are essential in a typical image-based clinical trial are:

- **Study management/administration:** study setup which includes the definition of trial documents such as an eCRF (electronic case report form) and the control of access rights for the study portal.
- **Image acquisition:** acquisition of images according to the study protocol. Acquisition sites upload images via the study portal.
- **Image quality assurance:** images have to fulfill defined quality criteria. An image QA manager can access all uploaded images to perform a control step and to approve or revoke the images for reading. This is typically done by inspecting the images with the image viewer.
- **Image reading:** image reading is performed for images which are approved during the QA step. Those images are loaded with the image viewer to perform qualitative and quantitative assessments to fill the eCRF with case-specific data.

3.2 Core Elements

The study portal is started via web browser as entry point of the cloud application. Depending on the user group, different feature sets are available. Study management/administration needs access to configuration tools to grant access rights, to assign roles, and to setup eCRFs. An eCRF is a questionnaire to collect the data from participating sites. Its complexity depends very much on the specific study but typically includes text fields which have to be filled out manually (e.g., description of patient's health condition, or qualitative findings) and further input controls, such as checkboxes, radiobuttons, or drop-down lists (e.g. patient's sex, smoking/drinking habits). In the presented application some input fields are automatically filled by assessments from the image viewer.

The cloud infrastructure needs storage space for image data and other study-related data with backup strategies and access control mechanisms. Image data is stored in a picture archiving and communication system (PACS) which supports operations to store and retrieve DICOM (Digital Imaging and Communications in Medicine) images. Non-image data like the structure of the eCRF, trial meta data, user management, and data generated by the image viewer is stored in an additional database to submit/retrieve data efficiently.

For simple image viewing of DICOM frames, the image display capabilities of a cloud PACS which uses remote rendering to generate images on the server and sends those images to the remote client or endpoint devices (see [13] for more details) might be sufficient. If advanced image processing/preprocessing, complex interaction with the images, or 3D visualization is required, the capabilities of a cloud PACS are usually not sufficient. A flexible and powerful MeVisLab-based cloud image viewer is presented in Sect. 3.4.

3.3 Data Flow

Figure 5 illustrates the data streams between core elements. As those applications/services might be installed on different virtual machines or different physical servers, high bandwidth for data transfer has to be available (especially for image transfer) to ensure high performance of the overall system.

Image data is uploaded from the study portal to the PACS and the image viewer imports a selection of DICOM images from the PACS. If the image viewer is used to modify images or image meta data (DICOM tags), e.g., during image QA, or generates new DICOM images, it stores the images in the PACS. For storage of non-image data a database management system (DBMS) is available. A database server to provide database services to other applications is often part of the DBMS. The portal utilizes the DBMS to store and retrieve all accumulated trial data. Moreover, the image viewer might need access to the DBMS to store/retrieve information such as performed image annotations or a current state to suspend/resume an image viewing session. Finally, an information exchange channel between portal and viewer has to be established. When the image viewer is started from the portal it needs information about the images which should be processed and information about the assessments which have to be performed. The viewer sends back the results of performed assessments such as measurement results.

3.4 Cloud Image Viewer

This section describes a cloud image viewer implemented with MeVisLab. The viewer is a zero-footprint web application presented as an interactive website which can be accessed by any modern web browser. The client's hardware requirements do not depend on the complexity of the web application since image processing and rendering is done on the server (see Sect. 2.4). Versions of this viewer are part of T-System's *Trial Connect* platform [38] and integrated into the e-health platform SATMED which aims to improve public health in emerging and developing countries [34].

The viewer is started from a study portal. It has to be set-up for data exchange with the cloud storage components and requires an interface to exchange information with the study portal. A suitable DBMS is the open-source system MySQL [22] which implements a client-server model for database access. For database access, the viewer configuration has to contain information about the DBMS (hostname and port), and information about a specific database which is used for data storage (database name, username, and password). A suitable open-source PACS is dcm4che [6]. To store and retrieve images, the viewer configuration has to contain information about the PACS (hostname, port, and application entity name) and calling information which are registered at the PACS (calling application entity name and calling port). In addition, a workspace path has to be defined which points to disk space where the viewer can store retrieved and processed images temporarily.

For a specific viewing session, an interface is implemented in JSON (JavaScript Object Notation) [11] which allows the transfer of information between study portal and viewer. The image viewer can be configured by the study portal with so-called viewing requests. Apart from image loading details, a viewing request contains information about assessments which have to be performed. An assessment request consists of a description (e.g., aorta cross section), a list of requested values (e.g., area, perimeter), and the corresponding list of preferred units (e.g., mm^2, mm). This allows a dynamic configuration of the image viewer for a specific trial. The viewing response contains results which are used to fill the eCRF with quantitative values automatically which is highly beneficial compared to error-prone manual transfer of values.

To move images between viewer and PACS over a network, the DICOM toolkit DCMTK [5] is used. During viewer initialization, all DICOM frames for the specific viewing request are retrieved from the PACS and stored in the viewer workspace. Subsequently, a DICOM importer provided by MeVisLab composes volume cache files which allow efficient viewer access to DICOM frames which are now logically sorted into 3D or 4D volumes. Often, such volumes are composed from frames which belong to a series according to DICOM tag information, but the DICOM importer allows the definition of custom sorting rules which can be configured for the requirements of a specific trial.

A screenshot of the cloud image viewer is shown in Fig. 6. Thumbnails represent volumes which can be loaded per drag-and-drop. A list is presented which shows

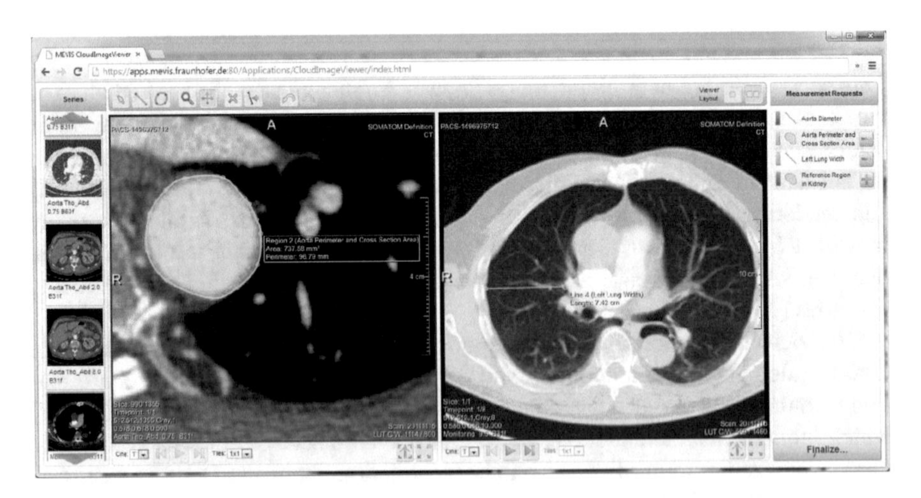

Fig. 6 The MeVisLab-based cloud image viewer has access to a rich set of image processing and visualization components. Image data courtesy of German Heart Institute Berlin

the assessment requests from the study portal. The viewer offers convenient tools to navigate to spatial or temporal slices, to zoom and pan images, and to control the contrast of the images. To perform measurements, line and contour drawing tools are implemented. The UI offers convenient controls to establish connections between requests and drawings. After all assessments are performed the session is finalized and the result values are sent to the study portal for further processing. Due to the viewer's suspend/resume and finalize capabilities with respect to the current viewing session, an audit trail is available which supports an uninterrupted tracing of user interactions which is a requirement for most clinical studies.

4 Augmented-Reality Application of Mobile Devices in the Operating Room

Augmented reality (AR) can be seen as a bridge between the virtual and the real world. It allows to enhance or augment the real world with information from virtual worlds. It is a widely used technique in various disciplines. In medicine, there has been also a lot of research in the last decades. Head-mounted displays have been used, e.g., to provide anesthetist with virtual information about the patient's condition during an operation [25]. But these devices are heavy and complicated to set-up. With the rise of tablets and wearable glasses AR actually becomes more feasible to be integrated into the clinical workflow. Reitinger et al. [29] used virtual reality glasses, a tracked helmet, and instruments to interact with liver planning data to prepare a surgery. Rassweiler et al. [28] demonstrated the use of a tablet with AR during a percutaneous kidney intervention.

4.1 Medical Background

AR in the operating room (OR) requires virtual models or planning data derived from pre-existing images of the patient. Typically, radiological data such as CT or MRI data is used and processed before surgery. After segmentation and further image analysis steps, 3D models of the individual anatomy showing organs, pathologies, risk structures, and resection lines can be overlaid with real patient anatomy.

Planning and risk analysis for liver interventions represents an established research topic and routine application at Fraunhofer MEVIS and MeVis Medical Solutions [35]. A service offered by MeVis Distant Services allows liver surgeons worldwide to send CT or MRI data for analysis and preparation of a patient-individual resection plan by means of the HepaVision software. The extracted objects include the liver, tumors, intrahepatic vascular systems, perfusion territories, and different resection proposals with dedicated risk analysis. 2D and 3D results can be downloaded by the physicians via a secure Internet connection and displayed on a workstation using specialized software (e.g., Liver Viewer or Liver Explorer) or with a 3D-PDF. With these anatomical models and resection proposals, surgeons can more easily understand the patient's anatomy, can compare different resection proposals, choose an optimal surgical strategy, and finally can perform the surgery more safely.

However, the transfer of results (e.g., resection plans and risk volumes) into the OR has been a challenge. Often surgeons assess planning data in the OR by paper print-outs or transfer them only mentally. The wish to have the planning data available in the OR becomes apparent if one understands the difficulty of liver operations. These interventions usually last many hours because the organ is difficult to resect and the surgeon has to be very careful to avoid a cut into larger vessels and into tumors in oncological interventions. The liver hosts branching vessel structures where a cut in an inappropriate place can cause severe blood loss. In addition, doctors must ensure that the patient retains enough sufficiently perfused organ volume for survival. To accomplish this, doctors need to know the location of blood vessels accurately and the depending liver territories both before and during an operation.

Fraunhofer MEVIS has developed the Mobile Liver Explorer, which runs on a tablet and allows to interactively access all virtual, patient-individual planning data. It provides an AR mode, which allows overlaying the virtual planning data onto the real liver and thus helps to locate critical structures. This app has been developed during a scientist's research stay in Japan in close cooperation with surgeons from Yokohama City University Hospital. Clinical requirements and expert feedback have been incorporated throughout the development process. Since the stay was limited to 3 months, the implementation was focused on major clinical demands and their solutions:

1. **Access to all planning data:** The requirement with highest priority was the availability of all planning data in the OR and the possibility to examine it like on a workstation. This includes 2D slice renderings with orthogonal and multiplanar reformatting, as well as standard viewing tools like changing contrast/brightness, zoom level, and slicing. The 3D view should be able to display surface models of the segmented objects like liver, tumors and other lesions, hepatic vascular systems, and resection proposals.
2. **Augmented reality:** The second priority was an AR mode that allows to overlay the selected 3D models with the video stream from the inbuilt back-facing camera. Alignment was planned to be done manually because automated alignment has no high priority for the surgeons. Manual interaction is fast (few seconds) and there is no continuous movement of the device during the intervention that would require automatic alignment between the real organ and the 3D model position.
3. **Tools supporting adaptation and visualization in the OR:** Due to the limited time, only three, more complex tools were implemented and further will follow. These are a vessel branch dependent perfusion analysis, a vessel length measurement algorithm, and a vessel eraser method (see [14] for more details).

4.2 System Design

The system is comprised of a server and a mobile client, in this case an Apple iPad. The server runs the host application, a variant of the Liver Explorer, which has been adapted to support the remote rendering capabilities described in Sect. 2.4.1. In addition, it is set-up as a WiFi hotspot. The iPad runs the Mobile Liver Explorer, which connects to the server using WiFi. For the app, a MeVisLab remote rendering protocol for iOS was implemented in Objective C. Due to performance issues, especially when streaming fullscreen images, a native app was implemented instead of browser-based one. Since a local WLAN connection is used, the bandwidth is not the issue but rather the render update of the tablet. While 25 fps are achieved in native mode, the HTML5 canvas-based version delivered only around 10 fps on an iPad3. An iPad Air offers comparable performance between browser-based and native apps (40–50 fps).

After having established the connection to the server, the app uses the remote messaging protocol (see Sect. 2.4.3) to receive the patient list. Thus, without a connection no patient data can be viewed, which is crucial for patient data security. After selection of a patient, the app requests so-called *collections*. A collection is a predefined view on the planning data, including display objects, their visual properties, the camera position and orientation. By tapping a collection it gets loaded by the server and the app switches to the third view which comprises a 2D view, a 3D view, and interaction tools. Communication with the server is required for:

- **Rendering:** The content of each view is rendered remotely and streamed to the client. The image is decoded and directly loaded into a UIImage Cocoa UI element which happens on each open inventor scene change. Remote rendering was chosen since all needed interaction functionality is already implemented in MeVisLab and the expected latency is below 50 ms due to the direct WLAN connection which ensures good response times.
- **Gesture processing:** Touch interactions are sent to the server where the gesture recognition is done and processed by dedicated interaction modules. This approach is more flexible compared to gesture recognition on the device since generic MeVisLab modules do not have to be adapted to support touch interactions. Only high latency which is not an issue in the presented set-up would impact server-side gesture recognition in a negative way.
- **System state changes:** All other commands, such as tool button selection, are sent using the general remote messaging protocol. The asynchronous nature has to be incorporated, so a state change is communicated back to the client. The UI only updates after the server acknowledged a successful state change.

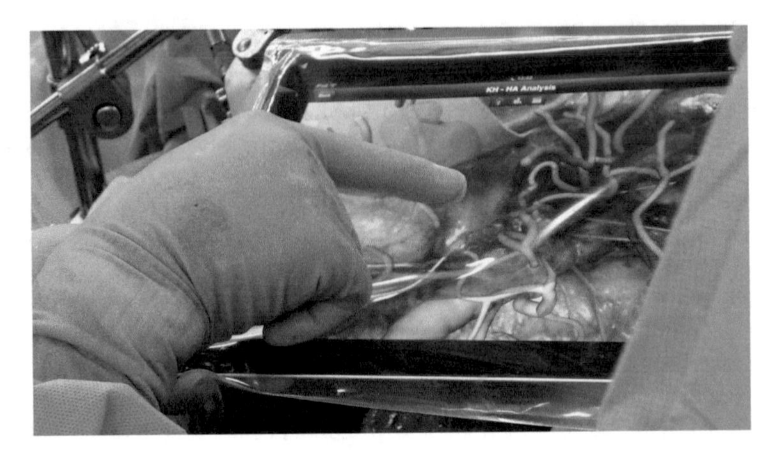

Fig. 7 Augmented reality with the iPad in the OR showing the virtual hepatic arteries

4.3 Augmented Reality for Liver Surgery

For over 15 years, preoperative planning of liver surgeries helps surgeons to decide about treatment, enhance the outcome, and even enables surgeries that would not have been done without 3D planning and risk analysis. The next logical step is to provide the planning data during surgery and to allow a direct comparison of the virtual plan and the real resection by using AR techniques. The 3D view of the app is used to display the live stream of the iPad camera if activated. The virtual camera is adapted to the intrinsic camera parameters. In addition, a remote message is sent to the server to enable the transmission of an alpha channel and RGB values of the remotely rendered image in PNG format. The alpha channel is needed to blend the rendering of the virtual models with the live image stream. The alignment of the virtual camera with the real world is done manually by the user. As discussed with the clinicians this is sufficient for their purpose, since information provided by the AR already helps to coarsely identify vessel locations (see Fig. 7). Further work should speed up the initial alignment by taking into account possible iPad positions in relation to the situs, data from the gyroscope sensor, and image information. A fixation device to mount the iPad over the patient to allow a stable positioning of the tablet is already in use in the hospital.

4.4 Clinical Application

The Mobile Liver Explorer was tested during its development at Yokohama University Hospital and is currently evaluated in a clinical study of 50 tumor resections there. Initial experience from first surgeries showed the practicability and usefulness of the new software and device. The touch functionality through a sterile sleeve works well, and multi-touch gestures allow quick zooming and rotating of the 3D model. The AR overlay of vascular structures enabled the identification of single branches, while the vessel length measurements provided valuable information for vascular anastomoses. Display of hepatic veins over real liver tissue prior to resection helped to identify single risk vessels and to avoid bleeding. After 13 resections, a comparison with similar liver resections already showed a significantly reduced operation time and a lower blood loss [20].

The system has been tested in Germany in the OR as well. Prof. Dr. Karl Oldhafer, Chief of the Department of Surgery at the Asklepios Klinik Barmbek in Hamburg made a statement about the AR feature of the app: "Using this function, we can virtually look into the organ and make the tumor and vessel structures visible. This simplifies comparison to determine whether the intervention has gone according to plan. With this new technology, we are able to better implement computer-supported operation planning for tumor removal."

While it seems that AR on mobile devices will be very helpful in the complex field of liver surgery, this technique may be useful also for other interventions. Examples are thorax surgery, where the location of tumors in relation to arteries, veins, and bronchi may influence the surgical strategy, and pancreas surgery, where the organ is surrounded by a large number of arteries and veins.

5 Discussion

The client-server architecture implemented in the MeVisLab platform has been specifically designed to meet requirements for the development of web-based application prototypes in the field of medical image computing. A main goal was to reduce the complexity for the developer and to allow high development efficiency. This led to a design where application logic and state, as well as visualization engines reside on the server and are clearly separated from the client's UI layer. While this approach may show noticeable latency effects when using interactive applications over slow networks, the remote rendering protocol was optimized in order to reduce such undesired effects. As an inherent consequence of the presented client-server architecture, MeVisLab-based server applications are always stateful. The state of an application is bound to a specific server process, and when the connection is lost or the server process crashes, the state is lost. However, this seems to be a reasonable approach for the type of applications under consideration because a stateless implementation would require continuously persisting of complex application states associated with large amounts of data in a server-side database. The automatic code generation facility of the MWT allows very short development cycles and a high level of reuse between deployment scenarios. The applicability, however, is currently limited by the set of supported UI elements, and current development efforts aim to reduce these limitations.

Several applications were already implemented based on the novel remote rendering capabilities provided by MeVisLab and two of them were discussed in this paper. It can be observed that there is a strong demand for such applications from both industrial and clinical/research partners. The cloud image viewer is already embedded into a cloud infrastructure for multi-center image-based clinical trials. This infrastructure very much fits to the requirements of customers from industry and research which need to manage their scientific data cost- and time-efficiently with access from all participating sites. Further it was recently embedded into the SATMED e-health platform. The current viewer implementation provides a basic set of features for image viewing and measurement tools but has access to the rich set of modules available in MeVisLab and can be easily extended on demand. This allows to provide new plugins for trial-specific assessment functionality. For example, a volume measurement tool including complex segmentation algorithms can be activated for trials dealing with tumor response evaluation. In addition, MeVisLab-based web applications are at the core of the German National Cohort MR imaging sub-study, which proves their reliance and robustness. They are used on a daily basis to perform visual quality assurance and, most importantly, the distributed incidental findings reading of currently more than 30 subjects per day reviewed by trained radiologists at five sites across Germany [1].

For the Mobile Liver Explorer, the new framework enabled fast development of a mobile device app. Existing features and tools in MeVisLab were used without need for reimplementation and the remote rendering allowed visualizing the basic patient data and derived models without need to transfer the data itself to the tablet computer. With this concept, the specific implementation effort for the client, a mobile device in this case, is small compared to the capabilities of the resulting app. It also makes it much easier to offer similar apps for different tablet types.

References

1. Bamberg, F., Kauczor, H.U., Weckbach, S., et al. Whole-body MR imaging in the German National Cohort: rationale, design, and technical background. Radiology **277**(1), 206–220 (2015)
2. Bethel, W.: Visualizaton dot com. Comput. Graph. Appl. **20**(3), 17–20 (2000)
3. Bitter, I., van Uitert, R., Wolf, I., et al. Comparison of four freely available frameworks for image processing and visualization that use ITK. IEEE Trans. Vis. Comput. Graph. **13**(3), 483–493 (2007)
4. Caban, J.J., Joshi, A., Nagy, P.: Rapid development of medical imaging tools with open-source libraries. J. Digit Imag. **20**(Suppl. 1), 83–93 (2007)
5. DCMTK - DICOM Toolkit. http://dcmtk.org/dcmtk.php.en (2015). Accessed 10 Nov 2015
6. dcm4che. http://www.dcm4che.org (2015). Accessed 10 Nov 2015
7. Engel, K., Sommer, O., Ertl, T.: Framework for interactive hardware accelerated remote 3D-visualization. In: de Leeuw, W., van Liere, R.: (eds.) Data Visualization, pp. 167–177. Springer, Vienna (2000)
8. Fette, I., Melnikov, A.: The WebSocket Protocol. http://tools.ietf.org/html/rfc6455 (2011). Accessed 10 Nov 2015
9. Hahn, H.K., Link, F., Peitgen, H.-O.: Concepts for rapid application prototyping in medical image analysis and visualization. In: Proceedings of Simulation and Visualization, pp. 283–298. SCS Publishing House, Ghent (2003)

10. Hendin, O., John, N.W., Shochet, O.: Medical volume rendering over the WWW using VRML and JAVA. In: Westwood, J.D., Hoffman, H.M., Stredney, D., et al. (eds.) Proceedings of Medicine Meets Virtual Reality, vol. 6, pp. 34–40. IOS Press and Ohmsha, Amsterdam (1998)
11. Introducing JSON. http://www.json.org (2015). Accessed 10 Nov 2015
12. Insight Segmentation and Registration Toolkit (2015). www.itk.org. Accessed 10 Nov 2015
13. Kagadis, G.C., Kloukinas, C., Moore, K., et al.: Cloud computing in medical imaging. Med. Phys. **40**(7), 070901 (2013)
14. Köhn, A., Matsuyama, R., Endo, I., et al. Liver surgery data and augmented reality in the operation room: experiences using a tablet device. Int. J. Comput. Assist. Radiol. Surg. **9**(Supplement 1) , 111 (2014)
15. König, M., Spindler, W., Rexilius, J., et al. Embedding VTK and ITK into a visual programming and rapid prototyping platform. In: Cleary, K.R., Galloway Jr, R.L. (eds.) Proceedings of SPIE Medical Imaging, vol. 6141, San Diego, 2006, pp. 796–806 (2006)
16. Lamberti, F., Sanna, A.: A streaming-based solution for remote visualization of 3D graphics on mobile devices. IEEE Tran Vis Comput Graph **13**(2), 247–260 (2007)
17. Link, F., König, M., Peitgen, H.-O.: Multi-resolution volume rendering with per object shading. In: Kobbelt, L., Kuhlen, T., Aach, T., et al. (eds.) Proceedings of Vision, Modeling, and Visualization, pp. 185–191. AKA, Berlin (2006)
18. Lippert, L., Gross, M.H., Kurmann, C.: Compression domain volume rendering for distributed environments. Comput. Graph. Forum **16**(3), C95–C107 (1997)
19. Ma, K.-L., Camp, D.M.: High performance visualization of time-varying volume data over a wide-area network. In: Proceedings of Supercomputing '00. IEEE Computer Society, Los Alamitos (2000)
20. Matsuyama, R., Taniguchi, K., Mori, R, et al.: IPad guided right hemihepatectomy with a new application designed specifically for navigation surgery: initially clinical experience for perihilar cholangiocarcinoma. In: IHPBA World Congress 2014 (2014)
21. MeVisLab. http://www.mevislab.de (2015). Accessed 10 Nov 2015
22. MySQL. http://www.mysql.com (2015). Accessed 10 Nov 2015
23. OpenGL. http://www.opengl.org (2015). Accessed 10 Nov 2015
24. OpenGL Vizserver. http://www.sgi.com/products/software/vizserver (2015). Accessed 10 Nov 2015
25. Ormerod, D.F., Ross, B., Naluai-Cecchini, A.: Use of an augmented reality display of patient monitoring data to enhance anesthesiologists' response to abnormal clinical events. Stud. Health Technol. Inform. **94**, 248–250 (2003)
26. Paul, B., Ahern, S., Bethel, E.W., et al. Chromium renderserver: scalable and open remote rendering infrastructure. IEEE Trans. Vis. Comput. Graph. **14**(3), 627–639 (2008)
27. Prior, F.W., Erickson, B.J., Tarbox, L.: Open source software projects of the caBIG In Vivo Imaging Workspace Software special interest group. J. Digit Imag. **20**(Suppl. 1), 94–100 (2007)
28. Rassweiler, J.J., Müller, M., Fangerau, M., et al. iPad-assisted percutaneous access to the kidney using marker-based navigation: initial clinical experience. Eur Urol **61**(3), 628–631 (2012)
29. Reitinger, B:, Bornik, A., Beichel, R., et al. Tools for augmented-reality-based liver resection planning. In: Galloway Jr, R.L. (ed.) Proceedings of SPIE Medical Imaging, vol. 5367, San Diego, pp. 88–99 (2004)
30. Richardson, T., Stafford-Fraser, Q., Wood, K.R., et al. Virtual network computing. IEEE Internet Comput. **2**(1), 33–38 (1998)
31. Rieder, C., Palmer, S., Link, F., et al.: A shader framework for rapid prototyping of GPU-based volume rendering. Comput Graph Forum **30**(3), 1031–1040 (2011)
32. Ritter, F., Boskamp, T., Homeyer, A., et al.: Medical image analysis: a visual approach. IEEE Pulse **2**(6), 60–70 (2011)
33. Rost, R.J., Licea-Kane, B.M., Ginsburg, D., et al.: OpenGL Shading Language, 3rd ed. Addison-Wesley Professional, Boston (2009)
34. SATMED E-Health Platform. http://satmed.lu (2015). Accessed 10 Nov 2015

35. Schenk, A., Haemmerich, D., Preusser, T.: Planning of image-guided interventions in the liver. IEEE Pulse **2**(5), 48–55 (2011)
36. SGI Open Inventor. http://oss.sgi.com/projects/inventor (2015). Accessed 10 Nov 2015
37. Stegmaier, S., Magallón, E.T.: A generic solution for hardware-accelerated remote visualization. In: Ebert, D., Brunet, P., Navazo, I. (eds.) Proceedings of EG/IEEE TCVG Symposium on Visualization , pp. 87–94. ACM, New York (2002)
38. T-Systems Trial Connect. https://www.telekom-healthcare.com/en/hospitals/healthcare-content-management/trial-connect/data-management-system-for-managing-study-data-32016 (2015). Accessed 10 Nov 2015
39. VirtualGL. http://www.virtualgl.org (2015). Accessed 10 Nov 2015
40. Visualization Toolkit (VTK). http://www.vtk.org (2015). Accessed 10 Nov 2015
41. Weiler, F., Rieder, C., David, C.A., et al.: On the value of multi-volume visualization for preoperative planning of cerebral AVM surgery. In: Ropinski, T., Ynnerman, A., Botha, C.P., et al. (eds.) Proceedings or the Eurographics Workshop on Visual Computing for Biomedicine. Eurographics Association, Norrköping, pp. 49–56 (2012)
42. Yoon, I., Neumann, U.: Web-based remote rendering with IBRAC. Comput. Graph. Forum **19**(3), 321–330 (2000)

Part III
Visualization for Diffusion-Weighted Imaging

Visualization of MRI Diffusion Data by a Multi-Kernel LIC Approach with Anisotropic Glyph Samples

Mark Höller, Uwe Klose, Samuel Gröschel, Kay-M. Otto, and Hans-H. Ehricke

Abstract In diffusion weighted magnetic resonance imaging (DW-MRI), high angular resolution imaging techniques have become available, allowing a voxel's diffusion profile to be measured and represented with high fidelity by a fiber orientation distribution function (FOD), even in situations of crossing and branching white matter fibers. Fiber tractography algorithms, such as streamline tracking, are used for visualizing global relationships between brain regions. However, they are prone to errors, e.g., may miss to visualize relevant fiber branches or provide incorrect connections. Line integral convolution (LIC), when applied to diffusion datasets, yield a more local representation of white matter patterns, and due to the local restriction of its convolution kernel is less susceptible to visualizing erroneous structures. In this paper we propose a multi-kernel LIC approach, which uses anisotropic glyph samples as an input pattern. Derived from FOD functions, multi-cylindrical glyph samples are generated by analysis of a highly-resolved FOD field. This provides a new sampling scheme for the anisotropic packing of samples along integrated fiber lines. Based on this input pattern two- and three-dimensional LIC maps can be constructed, depicting fiber structures with excellent contrast and resolving crossing and branching fiber pathways. We evaluate our approach by simulated DW-MRI data as well as in vivo studies with a healthy volunteer and a brain tumor patient.

M. Höller (✉) • K.-M. Otto • H.-H. Ehricke
Institute for Applied Computer Science (IACS), Stralsund University, Zur Schwedenschanze 15, D-18439 Stralsund, Germany
e-mail: mark.hoeller@fh-stralsund.de; hans.ehricke@fh-stralsund.de; kay.otto@fh-stralsund.de

U. Klose
MR Research Group, Department of Diagnostic and Interventional Neuroradiology, University Hospital Tübingen, Hoppe-Seyler-Straße 3, D-72076 Tübingen, Germany
e-mail: uwe.klose@med.uni-tuebingen.de

S. Gröschel
Department of Pediatric Neurology, University Children's Hospital Tübingen, Hoppe-Seyler-Straße 1, D-72076 Tübingen, Germany
e-mail: samuel.groeschel@med.uni-tuebingen.de

L. Linsen et al. (eds.), *Visualization in Medicine and Life Sciences III*, Mathematics and Visualization, DOI 10.1007/978-3-319-24523-2_7

1 Introduction

Within neural tissue the diffusion of water molecules is not free but rather hindered by cell membranes and cytoskeleton. Diffusion-weighted magnetic resonance imaging (DW-MRI) uses the application of pulsed field gradients to make the image contrast sensitive to motion of water. With Diffusion Tensor Imaging (DTI) the local diffusion profile is modelled by an ellipsoidal tensor with three eigenvectors and corresponding eigenvalues. But the diffusion tensor model is only capable of resolving a single anisotropy direction for each voxel. In more complex fiber architectures, such as crossing, branching and kissing fibers, diffusion tensor imaging is inadequate and may result in misleading and inaccurate interpretations [9, 18]. With high angular resolution diffusion imaging (HARDI), the diffusion profile is acquired with a large number of gradient directions (> 50) providing a better angular resolution. From the diffusion profile the fiber orientation distribution function (FOD) can be derived which is capable of resolving more than one anisotropy direction (Fig. 1). Diffusion tensor and FOD can be used to visualize the underlying brain architecture.

In DW-MRI, only a few white matter visualization approaches have gained clinical relevance. Among these, color-coded FA maps and tractography-based streamlines and streamtubes have had the widest clinical application. Color-coded FA maps combine the fractional anisotropy (FA) of each voxel with the color-coded principal eigenvector of the diffusion tensor [28]. Although from color-coded slice images the major pathways may be mentally reconstructed, they primarily reveal only local anisotropic information. Geometric models using streamlines [26] and streamtubes [42] can be used to visualize results of deterministic and probabilistic tractography algorithms. Furthermore, more complex approaches using hyperstreamlines [6] and tensorlines [39] have also been explored. Whilst these methods reveal global relationships, such as connections between brain regions, they fail to reliably depict uncertainties in the presented fiber anatomy, due to problems of data acquisition and signal processing. Rather, they represent data interpretations and depend on processing parameters, including choice of seed

Fig. 1 Surfaces of fiber orientation distribution function in a crossing region

regions, tracking algorithms and track termination criteria. This is also true for results of feature extraction methods [30], which generate a pathway's complex hull e.g. by segmentation or fiber clustering [8, 12, 24, 25]. In order to visualize the diffusion tensor's characteristics, such as anisotropic diffusion direction and magnitude, different types of glyphs are used.

The first glyph to be applied to diffusion tensor imaging (DTI) data was an ellipsoid [29]. The shape of the diffusion ellipsoid depends on the local diffusion profile and is determined by three eigenvalues and corresponding eigenvectors. Other glyphs, constructed from geometric primitives like cuboids and cylinders, have also been proposed. Kindlmann [19] introduced superquadric glyphs, which combine symmetry properties of ellipsoids with the shape and orientation of cuboids and cylinders. This facilitates the distinction between different shapes of the local diffusion profile. For the visualization of HARDI datasets orientation distribution function (ODF) glyphs may be used. These are constructed by the deformation of a sphere surface according to ODF values distributed over a half sphere. Directional color-coding of ODF glyphs helps to reveal local anisotropy directions [36, 37]. Glyphs can give the full diffusion profile information for each voxel, whereas the geometry of pathways and thus connectivity information are difficult to identify. Merging ellipsoids [5] is a technique used to visualize diffusion tensor imaging data by combining local tensor information with global connectivity features. Ellipsoids are placed along integral curves, generated by tractography, and merged under the control of weighted influence functions, derived from tensor properties. Another method that combines local tensor information with probabilistic tractography is Fiber Stippling [11]. The shape of line stipples are determined by the principal diffusion direction. Those line stipples are then projected into a slice.

In visualization theory there is another class of algorithms, visualization by textures, which may help to bridge the gap between local and global representations of the diffusion field. Triggered by advances in computer graphics hardware, texture-based visualization has gained great attention in the scientific community and various algorithms have been proposed, some of which may be applied to flow fields. Among these, line integral convolution (LIC), which was originally proposed by Cabral and Leedom [4], seems to be applicable to DW-MRI datasets. This technique smoothes a noise input image with a vector field, using a convolution kernel, which is locally adapted by vector field integration. McGraw et al. applied the LIC algorithm to a smoothed field of principal eigenvectors to visualize rat spinal cords [23]. Hsu was one of the first to apply LIC to a diffusion tensor field, using the tensor's principal eigenvector to guide the construction of a fixed-length filter kernel [16]. When using LIC the choice of the input noise texture strongly influences the resulting image. Kiu and Banks [2] applied a multi-frequency input noise scheme to visualize flow magnitude and direction. By modifying different parameters of the input image, Hotz et al. [15] encoded the eigenvalues of a positive-definite metric with the same topological structure as the tensor field. Wünsche et al. [41] proposed a three-dimensional LIC volume to be visualized by direct volume rendering. It is also possible to compute a single LIC-image for each of the three eigenvectors and overlay the resulting images to get a fabric-like texture

[15]. In a previous paper we described an adaptation of the LIC method to high angular resolution diffusion imaging, utilizing orientation distribution functions as representations of the local diffusion profile. Thus we were able to consider more than one anisotropy direction and correctly visualize crossing and kissing fiber pathways [14]. Moreover, we proposed a color-coding scheme for directional encoding of LIC slices, enhancing fiber continuity perception. In contrast to streamline tractography methods the LIC approach is easily parametrized, it does not require the definition of seed or goal regions and it applies tracking only in a very small region around a voxel, thus avoiding error propagation along longer integral curves. Therefore, it is less prone to visualizing non-existent tracts or failing to visualize existing fiber pathways. However, in our previous work, the LIC map lacked contrast, which made visualization of smaller fiber bundles difficult. In this paper we introduce a methodology for generating anisotropic glyph samples as an input pattern for an LIC algorithm. The geometry of glyph samples is derived from fiber orientation distribution functions (FODs) and according to the local diffusion profile can represent multiple anisotropy directions. As a glyph sample we use the three-dimensional geometry of the FOD shape model, rasterized with a super-resolution voxel grid. We also use the more simple shape of multiple cylinders which are constructed in line with anisotropy directions. Placement and packing of samples is controlled by the combination of a stochastic approach with local fiber line integration. This allows LIC maps to be generated with improved contrast, in which the viewer may visually track fibers, even through regions of crossing, branching or kissing fiber pathways.

2 Overview

Originally, the LIC approach was designed for flow visualization by engraving a vector field's structure onto a noise texture. LIC is essentially a filtering technique that blurs an input texture locally along a given vector field, thus providing highly correlated voxel intensities along field lines. Initially, for each voxel of the input texture, a field line is integrated over a fixed number of voxels, using the voxel as a tracking seed. The field line is used as the kernel of a convolution operator, which averages voxel intensities along the line. With 3D diffusion data the contrast of the LIC result is not very high when using white noise as input pattern to the LIC algorithm. Furthermore crossing fibers cannot be resolved adequately (Fig. 2).

In the method described here, a white noise texture is not used as originally proposed by Cabral and Leedom, rather an anisotropic spot pattern, which is continuously sampled along integral lines, is implemented. Furthermore, we propose a multi-kernel approach allowing more than a single anisotropy direction to be visualized for each voxel. We compute LIC smoothing kernels by integration over the FOD's first and (if available) second maximum directions. Then, we average over all sample values of the LIC input volume covered by the kernels.

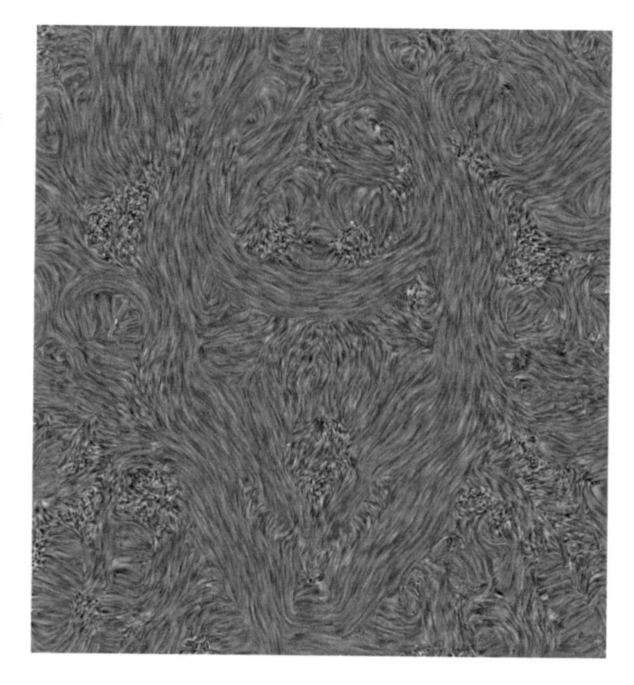

Fig. 2 LIC result of a coronal slice produced with white noise as input pattern. The corpus callosum and the pyramidal tracts are visible

Figure 3 shows an overview of the approach adopted, depicting the most relevant processing steps and data elements. From the acquired HARDI diffusion dataset FODs are computed by spherical deconvolution [35]. The FOD volume is used to create a high-resolution volume of glyph samples, which are used as the input pattern to a multiple-kernel LIC algorithm. The resulting LIC volume is a three-dimensional gray scale texture representing regional anisotropic behavior. Additionally, a direction volume is generated, in which the averaged anisotropy direction within the LIC convolution kernel is stored as a direction vector for each voxel. This is used for directional color encoding of slices through the LIC volume. By the application of volume rendering techniques, the LIC volume can be three-dimensionally visualized as a whole or after definition of a volume of interest. The details of (1) input pattern generation, (2) multi-kernel line integral convolution and (3) visualization are described in the next chapters.

3 Input Pattern Generation

In order to prepare and structure input data for the LIC algorithm, various methods have previously been proposed. The most common approaches utilize high frequency white noise [4] or sparse noise [17, 38] fields. It has been demonstrated that by applying dot-like structures, the contrast of the LIC output can be enhanced. Ellipses are structures well suited to this purpose, since they can represent local

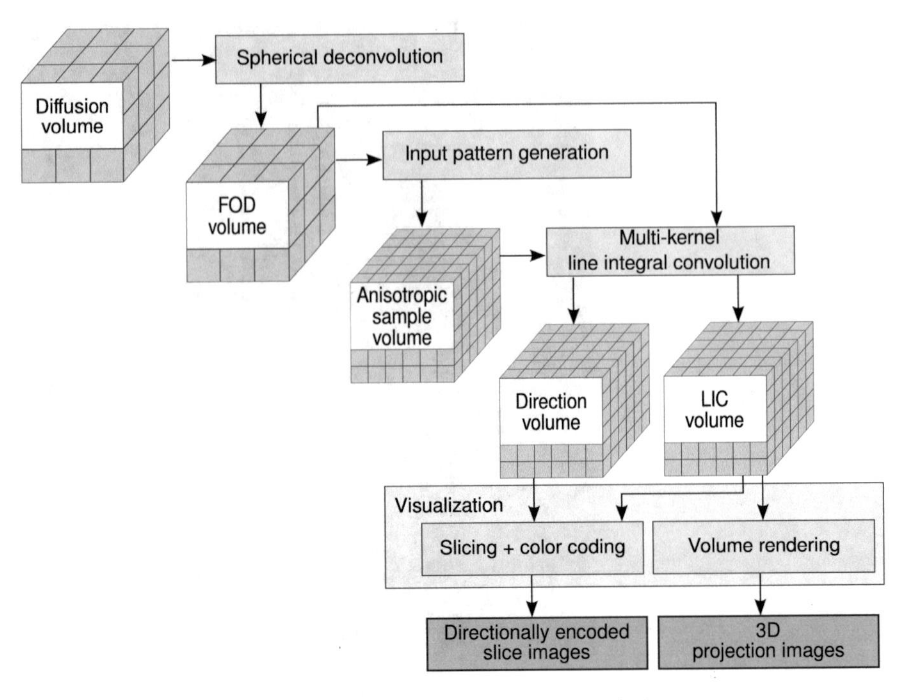

Fig. 3 Pre-processing chain with all data elements used in our method

features of the flow field, such as flow direction and magnitude. As the shape of elliptical spots is defined by the parameters of the diffusion tensor, difficulties arise when trying to visualize multiple local fiber orientations. We have previously demonstrated that by using ODF glyphs, more complex fiber anatomies can be visualized with the LIC method [14]. Several methods provide a higher angular resolution of kissing, branching and crossing fibers in a voxel, including sharpening ODFs by regularization algorithms [7, 27], ODF reconstruction with constant solid angle (CSA) [1], tensor decomposition [31] or spherical deconvolution, leading to the fiber orientation distribution (FOD) function [34, 35]. Instead of elliptical or ODF-based samples, we propose the use of three-dimensional glyph samples derived from the FOD as the input pattern for the LIC algorithm. We decided to use the FOD method as it is highly relevant for clinical applications and FODs can easily be calculated from HARDI signals.

3.1 Glyph Sampling

In order to represent the shape of the FOD in a three-dimensional regular grid, we construct a super-resolution LIC input grid with a spatial resolution in the order of 0.1–0.2 mm. Sampling of the FOD is performed by marking all voxels in the super-

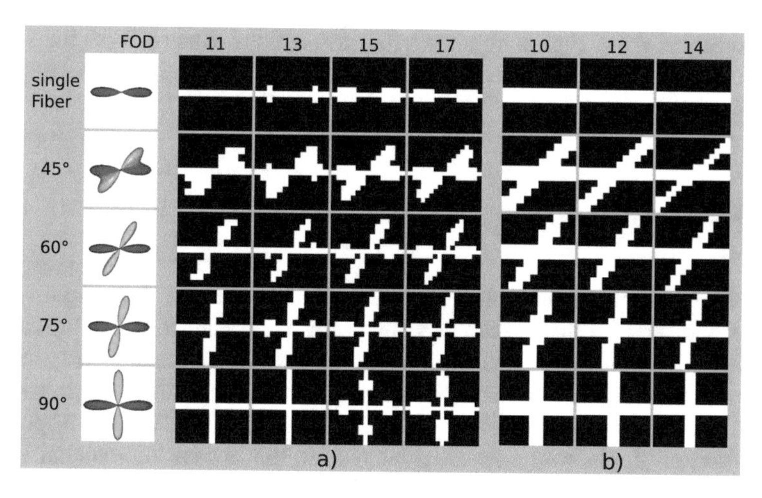

Fig. 4 Glyph sampling with different anisotropy direction angles: (**a**) sampling of FOD glyph with sampling sizes from $11 \times 11 \times 11$ to $17 \times 17 \times 17$ voxels, (**b**) sampling of cylindrical glyphs with sampling sizes from $10 \times 10 \times 10$ to $14 \times 14 \times 14$ voxels and cylinder width 2

resolution grid which are in contact with the FOD shape model, depicted in white and empty voxels in black (Fig. 4a). The figure shows the results of different glyph sizes between $10 \times 10 \times 10$ to $17 \times 17 \times 17$ super-resolution voxels. The figure clearly demonstrates that the accuracy of the representation improves with rising sampling ratio, which is particularly relevant for multidirectional FODs with smaller angles between fiber orientations. However, with an increasing sampling ratio we have to increase the size of the super-resolution grid. This requires more computer memory and processing power. In evaluation experiments, we found sample sizes of $9 \times 9 \times 9$ to $13 \times 13 \times 13$ to be an acceptable compromise between accuracy of representation and computational workload. To place up to eight ($= 2 \times 2 \times 2$) glyph samples within a voxel of the original diffusion dataset, a super-resolution dataset with a spatial resolution of 1/18 to 1/26 of the voxel length of the original dataset is required.

Since the binary input pattern for the LIC algorithm should primarily represent fiber directions, it is sufficient to use the simpler glyph of a cylinder instead of the FOD. The main axis of the cylinder denotes a fiber direction, which is derived from a local FOD maximum (see Sect. 4.1). If an FOD has only one local maximum, a single cylinder is constructed with orientation and length defined by the FOD maximum. Multiple crossing cylinders represent an FOD with two or more global maxima, thus making use of multi-cylindrical samples. For each maximum we place one cylinder into the grid. For better fiber continuity perception after the LIC the length of the cylinder is always set to 100 % of the sampling size. Since the width of the cylinder is not determined by signal parameters or the FOD's shape, it can be freely chosen depending on the grid resolution. As an extreme, the cylinder can be represented by a straight line, which is sufficient, particularly when using low-resolution grids. Figure 4b shows that for sample sizes of 9–13 the sampled cylinders do not differ too much from FODs, and fiber orientation angles of 45° can be resolved.

3.2 Glyph Placement

Distributing samples over a regular grid as an input to the LIC algorithm can produce undesirable effects, such as visual artifacts by line shifts and breaks. Feng et al. constructed a set of non-overlapping ellipses as an input to a generalized anisotropic Lloyd relaxation process, resulting in textures similar to those generated by the reaction diffusion approach [10, 22]. Kindlmann and Westin proposed the distribution of glyphs over a non-regular grid. In this approach, potential energies between neighboring tensors were calculated and glyphs placed into a dense packing throughout the field. Glyph packing may be applied either to a slice or to the volume dataset as a whole [20]. Kindlmann et al. also used the tensor volume to simulate a reaction-diffusion process of two chemicals in an anisotropic medium in order to produce a texture of spots, distributed over a non-regular grid [21].

In the context of fiber visualization and diffusion signal processing, placement of samples for use with LIC should reflect the underlying fiber pattern as represented by the diffusion data. In order to avoid discontinuities in the fiber texture image produced by the LIC algorithm, it is useful to place samples along fiber pathways within the super-resolution grid, so that samples are placed with improved fiber continuity, rather than on the regular grid of the original diffusion dataset. Therefore the following algorithm for sample placement is proposed:

1. Create a super-resolution grid.
2. Randomly place seedpoints distributed over the super-resolution grid by a uniform random technique.
3. To each seedpoint apply deterministic tracking over a short distance.
4. For each tracked streamline: place glyph samples (cylinders/FODs) along the streamline by tight packing.

Firstly, a super-resolution grid is created. The resolution factor $1/F_E$ relates to the voxel length of the original diffusion dataset. Each voxel of the diffusion dataset is divided into $F_E \times F_E \times F_E$ subvoxels. A diffusion dataset with 2.0 mm voxelsize typically results in a super-resolution dataset with a subvoxel size of approximately 0.1 mm. For the random placement of seedpoints, iterations are made until 1 % of the points of the super-resolution grid are filled with seeds. A rate of 1 % was found to be sufficient to fill the grid with tightly packed glyph samples. The deterministic tracking uses interpolated FODs and integrates along local maxima, using a Runge-Kutta integration scheme. Tracking stops after n integration steps, where n depends on the grid's resolution and is chosen so that the tracking distance is in the order of the voxelsize of the original diffusion dataset. This minimizes the cumulative tracking error ensuring that we do not follow non-existent pathways. A

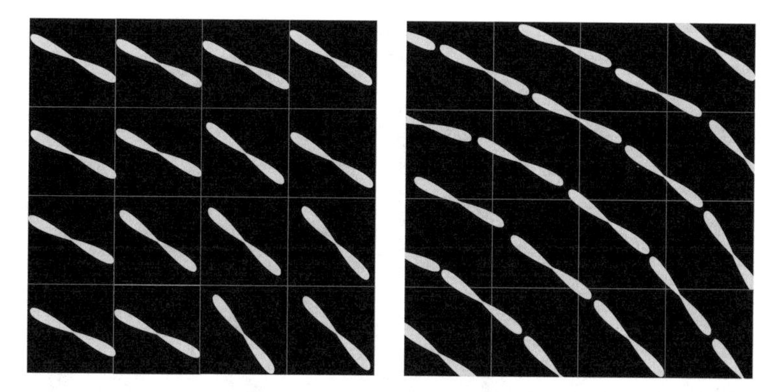

Fig. 5 Comparison of different glyph placement approaches: placement on regular grid (*left*) and continuous placement (*right*) with increased fiber continuity

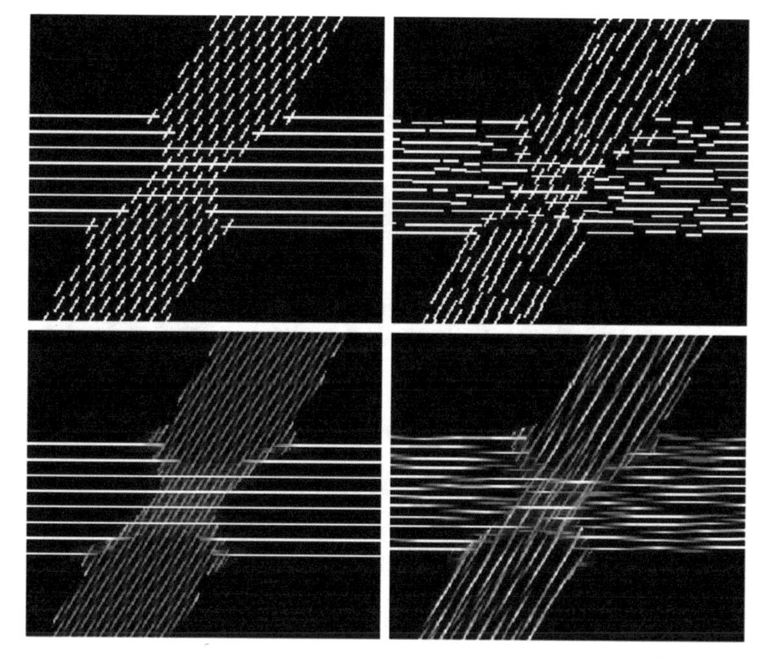

Fig. 6 Comparison of input structures (*top*) and their LIC results (*bottom*) of different cylinder glyph placement approaches: on regular grid (*left*) and continuous placement (*right*)

further stopping criterion is an FA threshold of 0.05 to avoid tracking out of the brain region. To avoid overlapping of glyph samples, samples are rejected, if they intersect with any previously accepted sample. Figures 5 and 6 compare the sample placement strategies. In Fig. 5, FOD glyphs were placed within the voxel raster of the original dataset (left) and alternatively, within the super-resolution grid along tracked integration lines (right). Figure 6 compares the two placement strategies

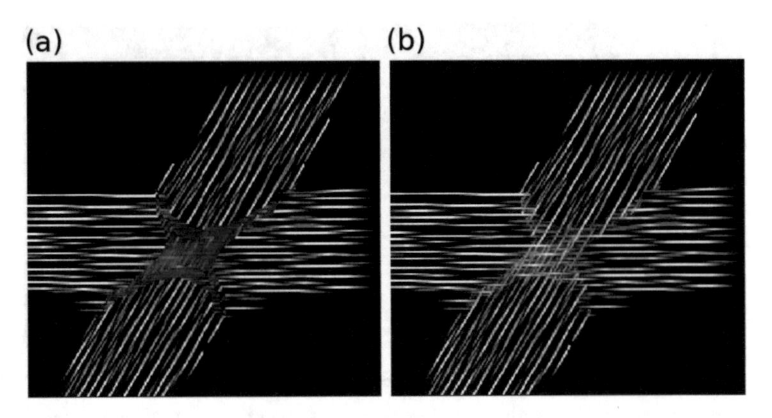

Fig. 7 LIC results from (**a**) HyperLIC and (**b**) multiple-kernel LIC

with cylindrical glyphs with an example of a fiber crossing situation, generated by our diffusion simulation approach. The upper row shows the LIC input patterns, whereas the corresponding LIC results are depicted in the lower row. The parameters used are $F_E = 20$, a FOD length of 12, FOD width of 2 and $n = 20$ integration steps.

For Figs. 6 and 7 we generated synthetic diffusion datasets using a partial volume model similar to the one described in [3]. We chose a gradient b-value of 2000 $\mathrm{s/mm^2}$, a diffusivity of 0.0015 $\mathrm{mm^2/s}$ and a baseline signal S_0 of 100. Furthermore, we used a free volume fraction of 0.35 in voxels completely occupied by fiber segments. We generated datasets with two fibers of 5 mm thickness crossing at an angle of 60°.

4 Multiple-Kernel Line Integral Convolution

The LIC algorithm proposed by Cabral applies a one-dimensional filter kernel with a single flow direction per pixel. In order to take into account more than a single direction the standard LIC method must be adapted. Hotz et. al. [15] suggested generating one LIC result for every eigenvector field, integrating over the field's eigenvalues. All resulting images are overlaid to get a fabric-like texture. With HyperLIC Zeng and Pang proposed a multipass approach, where the LIC algorithm is applied to the principal eigenvector. In a first pass a noise texture is used as the input image to create an intermediate LIC image [43]. In the second pass, the intermediate LIC image is used as the input for LIC of the second eigenvector. Our experiments with an adapted HyperLIC algorithm yielded poor results, particularly in regions of crossing or branching fibers, which could not be clearly depicted (Fig. 7a). In our method, a multi-directional LIC kernel depending on the global and local maxima of the FOD is used. A first integration is carried out along global FOD maxima over the input pattern as described above. This yields the first smoothing result. In a second step, integration is performed along the second local maximum of the candidate voxel, if there is a valid second FOD maximum (see Sect. 4.1).

The resulting value is then combined with that from the first step. Combination of the two values can be performed by averaging. However, it was found that better contrast is achieved by setting the maximum of the two as the voxel's final LIC result. Our multi-kernel approach can be implemented by generating a primary LIC image by application of a deterministic streamline tracking along the start voxel's global FOD maximum, and generation of a second image with the second FOD maximum. The two images are subsequently combined by selecting the maximum of each pixel's two values. In our experiments we used a maximum of two kernels per voxel. The algorithm may be generalized to more than two kernels, if the local FOD maxima can reliably be detected. For tracking a streamline length L of 12 voxels for each direction and a FA threshold of 0.1 was used.

Our method was implemented with the FastLIC algorithm [33], making the process up to 5 times faster. However, since FastLIC exploits redundancies from integration over longer field lines, cumulative integration errors may be introduced and pathways used may be inexact or non-existent.

4.1 FOD Maxima Identification

The global and local maxima of the FOD are used in different steps of the algorithm. To determine the maxima of the FOD we reconstructed the FOD using 606 reconstruction directions, distributed over a hemisphere. A local maximum of the FOD is valid if (1) it reaches at least 50 % of the global maximum, (2) its direction does not deviate less than an angle of 30° from all previously found peaks and (3) all neighboring values on the FOD surface are smaller. This method is used for the definition of the glyphs and the initial tracking directions. During the streamline tracking for every integration step the maximum of the FOD, that fits best with the direction from the previous step, has to be computed. First, we interpolate the 45 spherical harmonics coefficients. Then, we use a Newton-Raphson gradient ascent algorithm. This is the most time-consuming part of the algorithm.

5 Visualization

When applied to a diffusion dataset, the LIC-based approach presented above, results in two highly-resolved volume datasets: the gray-scale LIC volume and the direction volume of averaged anisotropy directions. For visualization of volume data, several approaches could be implemented. In the first method, data is projected into an image plane, for example by volume rendering. In order to generate visualizations with sufficient contrast, the voxel's color and opacity have to be determined by suitable data characteristics. Examples include barycentric opacity mapping [21], threads and halos representations [40] and fiber tract coherence [13]. The second approach involves the generation of slice images, where slices orthogonal to the volume's main axes, oblique slices or even curved slices are defined. Curved slices may be freely defined by the user or by anatomic structures, such as fiber pathways [32].

5.1 Volume Rendering

Since the LIC volume produced by our method depicts fiber structures with good contrast, direct volume rendering by texture mapping is possible. For this purpose the LIC volume, which can be resampled to a lower resolution or clipped to a volume of interest, is loaded to the texture memory of the computer's GPU and rendered by texture mapping. The user can then interactively change the voxel transparency and color in order to highlight structures of interest.

5.2 Directional Color Coding

The easiest way to visualize LIC results is to present them as gray scale slice images depicting the structure of the underlying diffusion field (Fig. 9 left). However, these gray scale texture images do not exactly convey the direction of anisotropic diffusion in a pixel. In DTI, directional encoding is usually provided by color-coded FA maps, which combine the voxel's fractional anisotropy with the principal eigenvector direction of the diffusion tensor. Pajevic and Pierpaoli [28] proposed to assign the components of the eigenvector (x, y, z) to color channels red, green and blue. Using this color model in LIC images results in blocked fiber continuity perception due to abrupt color changes (Fig. 9 middle). Since the in-plane fiber direction is visualized by the structure of the LIC texture, no encoding of left-to-right or up-to-down directions is necessary. As a more appropriate alternative we use a color-coding scheme based on the hue-saturation-brightness (HSB) color model to encode different diffusion properties. HSB is a color representation with a cone-shaped color space. The hue channel is represented by a rotation angle with the main axis of the cone as the rotation axis, spanning from red (0/360) through green (120) to blue (240). The saturation channel is encoded as the fractional distance from the cone's center to its surface and defines the intensity of a color in a value range from 0 to 100. Brightness is measured from the cone's base (0 = black) to its tip (100 = white).

In our color coding scheme the LIC value is scaled from 0 to 100 and it defines the brightness channel. The direction is encoded as follows: when tracking along a streamline we compute the resulting direction vector $\mathbf{v}_{RN}$ as the sum of all the vectors $\mathbf{v}$ used for LIC tracking. The angle γ between the vector $\mathbf{v}_{RN}$ and the normal

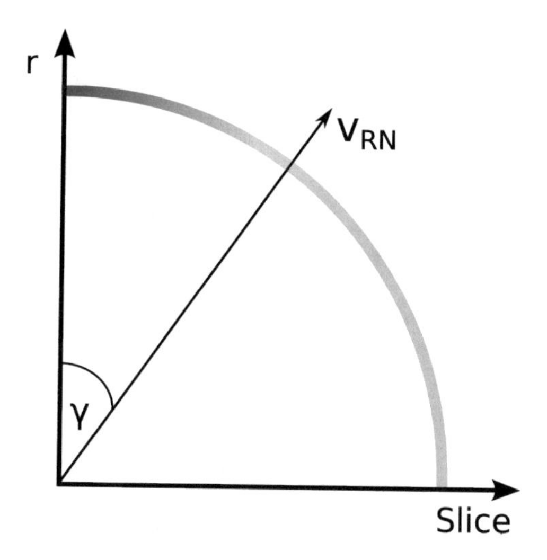

Fig. 8 Determination of hue value by the angle γ between the slice normal $\mathbf{r}$ and the resulting vector $\mathbf{v}_{RN}$

vector $\mathbf{r}$, orthogonal to the image slice, determines the hue value. $|\gamma|$ is mapped from the $[90, 0]$ interval to the $[120, 240]$ color range (Fig. 8). As a result, all streamlines within the slice are green in color and all streamlines orthogonal to the slice are colored in blue, thus avoiding continuity breaks (Fig. 9 right). We only use these two parameters and always set the saturation to 100.

5.3 Slicing

Since the LIC result is a three-dimensional texture, slicing is a major application of the visualization. Orthogonal slices can be visualized separately. But when using a slice thicknesses in the order of the resolution of the LIC volume, e.g. < 0.2 mm within one slice, the amount of information is low (Fig. 10 left). Therefore it is useful to combine adjacent slices (Fig. 10 middle and right). This is carried out by computing the average of the LIC values of adjacent pixels or, with even better contrast, by selecting the maximum value. The resulting slice thickness should be in the order of the resolution of the original diffusion dataset (2 mm). The choice of lower slice thickness leads to visualization results with fiber discontinuity, because only few fibers touch the slice over longer distances.

6 Results

For the in vivo studies two different diffusion data sets were used. The diffusion data set from the brain of a healthy 24 year old human male which was acquired on a 3-T Trio MR scanner (Siemens Healthcare, Erlangen) using a spin-echo echo-planar

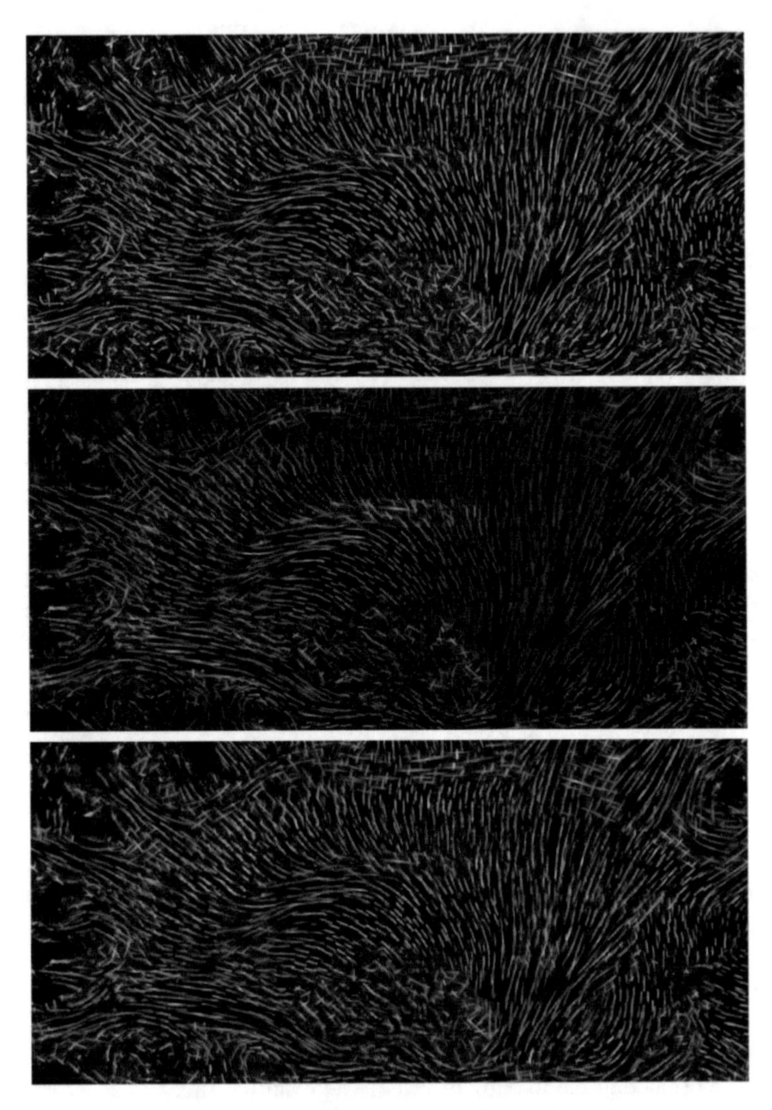

Fig. 9 From *top to bottom*: section of a sagittal LIC slice image with gray scaling (*top*), and directional encoding with RGB (*middle*) and with our HSB color model (*bottom*)

diffusion-weighted sequence (TR = 8000 ms, TE = 105 ms) with 64 diffusion-encoding gradients, a b-value of 2000 s/mm^2 and an isotropic spatial resolution of 2 mm. A data matrix of 108 × 108 was obtained measuring 56 slices. A second dataset from a clinical study of a 6-year old child with a brain tumor in the left central region was acquired on a 1.5 T Vision Scanner using a spin-echo echoplanar diffusion-weighted sequence (TR = 11500 ms, TE = 122 ms) with 60 diffusion-encoding gradients, a b-value of 3000 s/mm^2 and an isotropic spatial resolution of

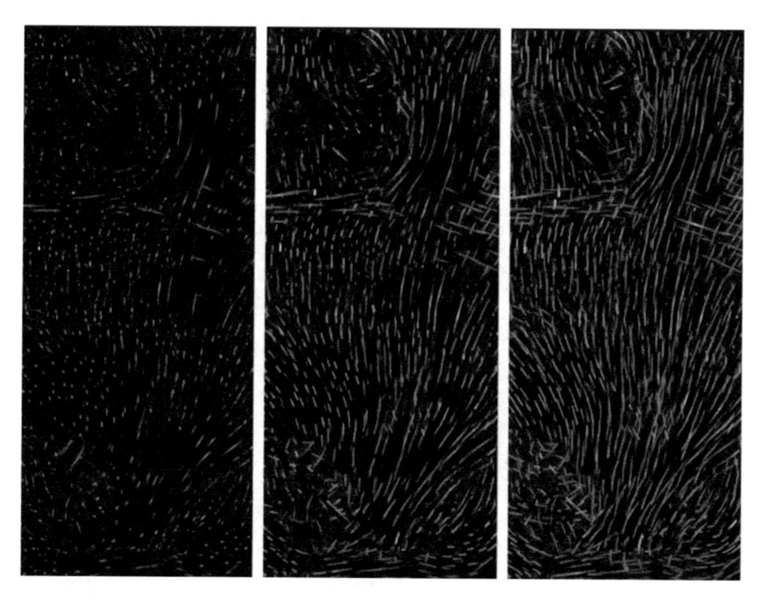

Fig. 10 Combination of adjacent slices (zoomed detail on a sagittal slice of a human brain): single slice (*left*), 5 slices (*middle*) and 10 slices (*right*) of the super-resolution LIC dataset

2.5 mm. A data matrix of 96 × 96 was obtained measuring 60 slices. The child suffered from focal seizures due to this tumor and was scanned as part of a pre-surgical assessment.

We applied a data pre-processing chain on both datasets. First we used a mutual information-based retrospective motion correction scheme in order to remove motion that occurred during the scan. We then computed the diffusion tensor for each voxel and additionally reconstructed the FOD with the constrained spherical deconvolution algorithm using a maximum spherical harmonic order $l_{max} = 8$. We determined one or two main directions for each voxel by detecting the FOD's local maxima. The proposed approaches of input pattern generation, multi-kernel LIC and color-coding were applied to the datasets. For the input pattern generation a resolution factor of F_E of 1/20 and a cylinder length of 12 subvoxels was used. Multi-kernel LIC was performed with a box filter and a maximum length L of 12 integration steps for each direction.

Figure 11 shows the results of all three steps of a coronal slice from the healthy volunteer dataset. The slice thickness used was 1.0 mm. The input structure in Fig. 11 top shows the overall distribution of the glyphs. The placement result already depicts crossing regions. Major fiber pathways, such as pyramidal tract and corpus callosum were depicted with good contrast, including the crossing of callosal projections with pyramidal fibers (Fig. 11 middle). The blue structures in the color coded LIC result represent fibers running orthogonal to the slice plane, such as the cingulum (Fig. 11 bottom).

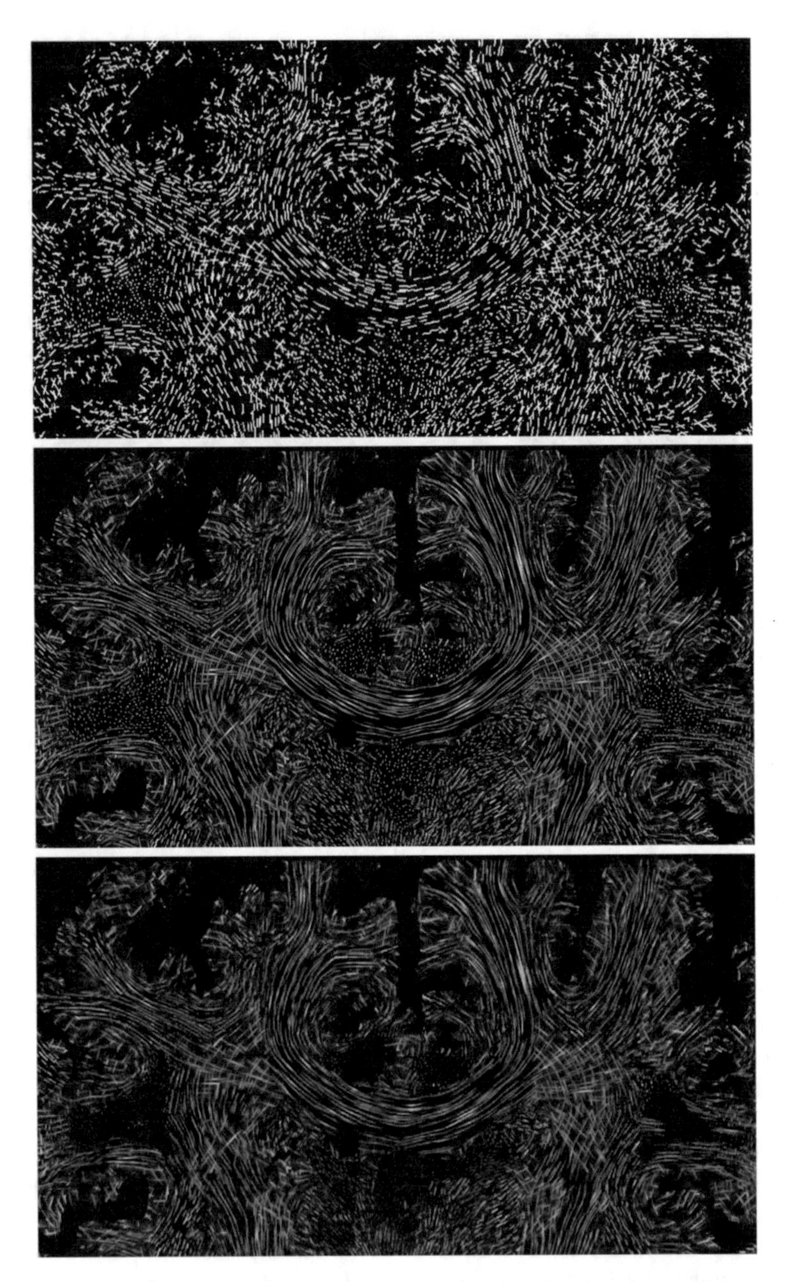

Fig. 11 Coronal slice from healthy volunteer dataset: input pattern (*top*), LIC result (*middle*) and color coded LIC map (*bottom*)

Fig. 12 Combined volume rendering of the LIC result from patient dataset with surface rendering of tumor (*blue*) clearly depicts infiltration of left branch of pyramidal tract (*arrow*)

The LIC result from processing the tumor patient dataset was volume rendered. Additionally the tumor was segmented. LIC and segmented tumor were visualized using the AMIRA software (Fig. 12). The figure indicates a region in which tumorous tissue evidently infiltrated the left branch of the pyramidal tract, without having destroyed anisotropic behavior. Volume rendering of the LIC result was restricted to a subvolume slice of approx. 100 mm thickness with its center in the center of the tumor. By interactive rotation of the 3D visualization the assessment of the relationship between the tumor and the neighboring fiber pathways can be enhanced. This information was of great value for surgeons during operation planning. In addition, the intraoperative electric stimulation confirmed that the fibers within the solid part of this dysembryoblastic neuroepithelial tumor (WHO grade 1) were in fact pyramidal tract fibers. Consequently, a subtotal resection was performed, with only a small part of the tumor remaining. The patient had no postoperative sensorimotor deficit.

The processing and visualization was performed on a Linux (Ubuntu 12.04) workstation with the following specifications: Intel Core i7-3630QM (8 MB cache, 2 GHz $\times$ 8 cores with hyperthreading), 8 GB RAM(DDR3, 1333 MHz), NVIDIA GeForce GT425 (with 1 GByte memory) graphics card. All processing steps were performed using the modular software platform OpenPDT developed by our group.

Since we apply LIC to a highly resolved 3D sample pattern with a grid resolution that surpasses the spatial resolution of the original diffusion dataset by an order of magnitude, the computational workload of the LIC step of our method is quite high. With a typical whole brain diffusion dataset consisting of more than 50 slices and a spatial resolution of approximately 2.0 mm, resulting processing times are of the magnitude of several hours on standard desktop computers.

7 Discussion and Conclusions

We have presented an approach for fiber visualization by application of the LIC algorithm. To generate meaningful LIC maps which reliably represent fiber structures, novel pre- and post-processing methods were proposed. Firstly, the tensor model was dismissed and instead an FOD representation of the local diffusion profile was used, due to its capability of describing multiple anisotropy directions. We used FOD or alternatively multi-cylindrical glyph samples, rasterized by a super-resolution grid, and thus were able to impress anisotropy characteristics into the LIC input pattern and substantially improve the contrast of LIC maps. Secondly, we proposed a sample placement strategy, which uses a uniform random technique to distribute seeds over the data volume and sets samples along integration lines that were produced by deterministic tracking over very short distances. This allowed the continuity of fiber lines in LIC maps to be enhanced. Thirdly, a color-encoding scheme for the directional encoding of LIC maps was presented. The method was applied to synthetic datsets as well as in-vivo datasets of a healthy volunteer and a tumor patient. Results were presented to neuroradiologist and neurosurgeons and were in concordance with their findings. However, our preliminary evaluation results need critical verification by further clinical studies.

7.1 Future Work

As explained above, the method's computational workload might be an obstacle to its integration with clinical protocols. Therefore, algorithmical optimizations, e.g. by exploitation of spatial coherence, as well as usage of computation parallelization have to be implemented. We are already working on the parallelization strategy by transferring the LIC processing to graphic processing units (GPUs). The usage of their high memory bandwidth and high degree of parallelism can substantially reduce the processing time. Since for each voxel of the super-resolution grid the same processing steps have to be performed it is possible to parallelize the algorithm and exploit a GPU's parallel processing architecture. Our preliminary results show that an increase of the processing speed by factors of 100–150 becomes possible, depending on the graphics hardware used.

The visualization strategy of presenting LIC maps has not been widely used in DW-MRI. Therefore, we cannot rely on previous practical experience with this kind of methodology and we are aware of the fact that the method may generate visual stimuli that lead to erroneous interpretations of the diffusion data. With respect to a thorough clinical evaluation we currently focus on diseases which are associated with white matter demyelinization, e.g. multiple sclerosis. Another focus is on the pre-operative assessment of tumor datasets. Application of the method to other 2D and 3D flow fields, e.g. from air or fluid simulations, is another topic of further research.

References

1. Aganj, I., Lenglet, C., Sapiro, G., Yacoub, E., Ugurbil, K., Harel, N.: Reconstruction of the orientation distribution function in single- and multiple-shell q-ball imaging within constant solid angle. Magn. Reson. Med. **64**(2), 554–566 (2010). doi:10.1002/mrm.22365
2. Banks, D.C., Kiu, M.H.: Multi-frequency noise for LIC. In: Yagel, R., Nielson, G.M. (eds.) Proceedings of the 7th Conference on Visualization '96, Lic, pp. 121–126. IEEE CS Press, Los Alamitos (1996)
3. Behrens, T.E.J., Berg, H.J., Jbabdi, S., Rushworth, M.F.S., Woolrich, M.W.: Probabilistic diffusion tractography with multiple fibre orientations: what can we gain? Neuroimage **34**(1), 144–155 (2007). doi:10.1016/j.neuroimage.2006.09.018
4. Cabral, B.: Imaging vector fields using line integral convolution. In: Cunningham, S. (ed.) Proceedings of SIGGRAPH 93, Computer Graphics Proceedings, Annual Conference Series Proceedings of the 20th Annual Conference on Computer Graphics and Interactive Techniques, pp. 263–270. ACM SIGGIGRAPH (1993)
5. Chen, W., Zhang, S., Correia, S., Tate, D.F.: Visualizing diffusion tensor imaging data with merging ellipsoids. In: Eades, P., Ertl, T., Shen, H.W. (eds.) Proceedings on IEEE Pacific Visualization Symposium 2009, pp. 145–151. IEEE CS Press, Los Alamitos (2009). doi:10.1109/PACIFICVIS.2009.4906849
6. Delmarcelle, T., Hesselink, L.: Visualizing second-order tensor fields with hyperstreamlines. IEEE Comput. Graph. Appl. **13**(4), 25–33 (1993)
7. Ehricke, H.H., Otto, K.M., Klose, U.: Regularization of bending and crossing white matter fibers in MRI Q-ball fields. Magn. Reson. Imag. **29**(7), 916–926 (2011)
8. Enders, F., Nimsky, C.: Visualization of white matter tracts with wrapped streamlines. In: Silva, C., Gröller, E., Rushmeier, H. (eds.) Proceedings of IEEE Visualization 2005, pp. 51–58. IEEE CS Press, Los Alamitos (2005). doi:10.1109/VISUAL.2005.1532777
9. Farquharson, S., Tournier, J.D., Calamante, F., Fabinyi, G., Schneider-Kolsky, M., Jackson, G.D., Connelly, A.: White matter fiber tractography: why we need to move beyond DTI. J. Neurosurg. **118**(6), 1–11 (2013). doi:10.3171/2013.2.JNS121294
10. Feng, L., Hotz, I., Hamann, B., Joy, K.: Anisotropic noise samples. IEEE Trans. Vis. Comput. Graph. **14**(2), 342–54 (2008). doi:10.1109/TVCG.2007.70434
11. Goldau, M., Wiebel, A., Gorbach, N.S., Melzer, C., Hlawitschka, M., Scheuermann, G., Tittgemeyer, M.: Fiber stippling: an illustrative rendering for probabilistic diffusion tractography. In: 2011 IEEE Symposium on Biological Data Visualization (BioVis), pp. 23–30 (2011). doi:10.1109/BioVis.2011.6094044
12. Hagmann, P., Jonasson, L., Maeder, P., Thiran, J.P., Wedeen, V.J., Meuli, R.: Understanding diffusion MR imaging techniques: from scalar diffusion-weighted imaging to diffusion tensor imaging and beyond. Radiographics **26**(Suppl. 1), 205–223 (2006). doi:10.1148/rg.26si065510
13. Hlawitschka, M., Garth, C., Tricoche, X., Kindlmann, G., Scheuermann, G., Joy, K.I., Hamann, B.: Direct visualization of fiber information by coherence. Int. J. Comput. Assist. Radiol. Surg. **5**(2), 125–31 (2010). doi:10.1007/s11548-009-0302-5
14. Hoeller, M., Thiel, F., Otto, K., Klose, U., Ehricke, H.: Visualization of high angular resolution diffusion MRI data with color-coded LIC-maps. In: Goltz, U., Magnor, M., Appelrath, H.J., Matthies, H.K., Balke, W.T., Wolf, L. (eds.) Proceedings Informatik 2012, pp. 1112–1124. Gesellschaft für Informik e.V., Braunschweig (2012)

15. Hotz, I., Feng, L., Hagen, H., Hamann, B.: Physically based methods for tensor field visualization. In: Proceedings of the Conference on Visualization '04, pp. 123–130. IEEE CS Press, Los Alamitos (2004)
16. Hsu, E.: Generalized line integral convolution rendering of diffusion tensor fields. In: Proceedings of the International Society for Magnetic Resonance in Medicine (ISMRM), vol. 9, p. 790 (2001)
17. Interrante, V.: Visualizing 3D flow. IEEE Comput. Graph. Appl. **18**(4), 151–53 (1998). doi:10.1109/38.689664
18. Jones, D.K., Knösche, T.R., Turner, R.: White matter integrity, fiber count, and other fallacies: the do's and don'ts of diffusion MRI. NeuroImage **73**, 239–54 (2013). doi:10.1016/j.neuroimage.2012.06.081
19. Kindlmann, G.: Superquadric tensor glyphs. In: Deussen, O., Hansen, C., Keim, D., Saupe, D., Deussen, O., Hansen, C., Keim, D.A., Saupe, D. (eds.) Proceedings of the Sixth Joint Eurographics-IEEE TCVG Symposium on Visualization (2004)
20. Kindlmann, G., Westin, C.F.: Diffusion tensor visualization with glyph packing. IEEE Trans. Vis. Comput. Graph. **12**(5), 1329–35 (2006)
21. Kindlmann, G., Weinstein, D., Hart, D.: Strategies for direct volume rendering of diffusion tensor fields. IEEE Trans. Vis. Comput. Graph. **6**(2), 124–138 (2000)
22. Kratz, A., Kettlitz, N., Hotz, I.: Particle-based anisotropic sampling for two-dimensional tensor field visualization. In: Eisert, P., Hornegger, J., Polthier, K. (eds.) Proceedings of the Vision, Modeling, and Visualization, pp. 145–152. Eurographics Association, Berlin (2011)
23. Mcgraw, T., Vemuri, B.C., Wang, Z., Chen, Y., Rao, M., Mareci, T.: Line integral convolution for visualization of fiber tract maps from DTI. In: Dohi, T., Kikinis, R. (eds.) Proceedings on Medical Image Computing and Computer-Assisted Intervention—MICCAI 2002, pp. 615–622. Springer, Berlin (2002)
24. Merhof, D., Meister, M., Bingol, E., Nimsky, C., Greiner, G.: Isosurface-based generation of hulls encompassing neuronal pathways. Stereotact. Funct. Neurosurg. **87**(1), 50–60 (2009). doi:10.1159/000195720
25. Moberts, B., Vilanova, A., van Wijk, J.: Evaluation of fiber clustering methods for diffusion tensor imaging. In: Silva, C., Gröller, E., Rushmeier, H. (eds.) Proceedings of IEEE Visualization 2005, pp. 65–72. IEEE CS Press, Los Alamitos (2005). doi:10.1109/VIS.2005.29
26. Mori, S., Crain, B.J., Chacko, V.P., van Zijl, P.C.: Three-dimensional tracking of axonal projections in the brain by magnetic resonance imaging. Ann. Neurol. **45**(2), 265–269 (1999)
27. Otto, K.M., Ehricke, H.H., Kumar, V., Klose, U.: Angular smoothing and radial regularization of ODF fields: application on deterministic crossing fiber tractography. Physica Medica : International Journal devoted to the Applications of Physics to Medicine and Biology : Official Journal of the Italian Association of Biomedical Physics (AIFB) **29**(1), 17–32 (2013). doi:10.1016/j.ejmp.2011.10.002
28. Pajevic, S., Pierpaoli, C.: Color schemes to represent the orientation of anisotropic tissues from diffusion tensor data: application to white matter fiber tract mapping in the human brain. Magn. Reson. Med. **43**(6), 921 (2000)
29. Pierpaoli, C., Basser, P.: Toward a quantitative assessment of diffusion anisotropy. Magn. Reson. Med. **36**(6), 893–906 (1996)
30. Schultz, T.: Feature extraction for DW-MRI visualization: the state of the art and beyond. In: Hagen, H. (ed.) Proceedings on Dagstuhl Scientific Visualization: Interactions, Features, Metaphors, vol. 2, pp. 322–345. Schloss Dagstuhl–Leibniz-Zentrum fuer Informatik (2010)
31. Schultz, T., Seidel, H.P.: Estimating crossing fibers: a tensor decomposition approach. IEEE Trans. Vis. Comput. Graph. **14**(6), 1635–1642 (2008). doi:10.1109/TVCG.2008.128
32. Schurade, R., Hlawitschka, M., Scheuermann, B.H.G., Knösche, T.R., Anwander, A.: Visualizing white matter fiber tracts with optimally fitted curved dissection surfaces. In: Bartz, D., Botha, C., Hornegger, J., Machiraju, R. (eds.) Proceedings of Eurographics Workshop on Visual Computing for Biology and Medicine. Eurographics Association, Berlin (2010)
33. Stalling, D., Hege, H.: Fast and resolution independent line integral convolution. In: Mair, S.G., Cook, R. (eds.) Proceedings of the 22nd Annual Conference on Computer Graphics and

Interactive Techniques, SIGGRAPH '95, pp. 249–256. ACM, New York, New York (1995). doi:10.1145/218380.218448
34. Tournier, J.D., Calamante, F., Gadian, D.G., Connelly, A.: Direct estimation of the fiber orientation density function from diffusion-weighted MRI data using spherical deconvolution. Neuroimage **23**(3), 1176–1185 (2004). doi:10.1016/j.neuroimage.2004.07.037
35. Tournier, J.D., Calamante, F., Connelly, A.: Robust determination of the fibre orientation distribution in diffusion MRI non-negativity constrained super-resolved spherical deconvolution. Neuroimage **35**(4), 1459–1472 (2007). doi:10.1016/j.neuroimage.2007.02.016
36. Tuch, D.S., Reese, T.G., Wiegell, M.R., Makris, N., Belliveau, J.W., Wedeen, V.J.: High angular resolution diffusion imaging reveals intravoxel white matter fiber heterogeneity. Magn. Reson. Med. **48**(4), 577–582 (2002). doi:10.1002/mrm.10268
37. Tuch, D.S., Reese, T.G., Wiegell, M.R., Wedeen, V.J.: Diffusion MRI of complex neural architecture. Neuron **40**(5), 885–895 (2003)
38. Wegenkittl, R.: Animating flow fields: rendering of oriented line integral convolution. In: Proceedings of Computer Animation'97, pp. 1–10. IEEE CS Press, Los Alamitos (1997)
39. Weinstein, D., Kindlmann, G., Lundberg, E.: Tensorlines: advection-diffusion based propagation through diffusion tensor fields. In: VIS '99: Proceedings of the conference on Visualization '99, pp. 249–253. IEEE Computer Society Press, Los Alamitos, CA (1999)
40. Wenger, A., Keefe, D.F., Zhang, S., Laidlaw, D.H.: Interactive volume rendering of thin thread structures within multivalued scientific data sets. IEEE Trans. Vis. Comput. Graph. **10**(6), 664–72 (2003). doi:10.1109/TVCG.2004.46
41. Wünsche, B., Linden, J.V.D.: DTI volume rendering techniques for visualising the brain anatomy. In: International Congress Series, Proceedings of the 19th International Computer Assisted Radiology and Surgery Congress and Exhibition, vol. 0, pp. 80–85. Elsevier Science, Berlin (2005)
42. Zhang, S., Demiralp, C., Laidlaw, D.H.: Visualizing diffusion tensor MR images using streamtubes and streamsurfaces. Proc. IEEE Trans. Vis. Comput. Graph. **9**(4), 454–462 (2003)
43. Zheng, X., Pang, A.: HyperLIC. In: Turk, G., van Wijk, J.J., Moorhead II, R.J. (eds.) Proceedings of 14th IEEE Visualization, pp. 249–256. IEEE CS Press, Los Alamitos (2003)

Exploring Crossing Fibers of the Brain's White Matter Using Directional Regions of Interest

Andreas Graumann, Mirco Richter, Christopher Nimsky, and Dorit Merhof

Abstract Diffusion magnetic resonance imaging (dMRI) is a medical imaging method that can be used to acquire local information about the structure of white matter pathways within the human brain. By applying computational methods termed fiber tractography on dMRI data, it is possible to estimate the location and extent of respective nerve bundles (white matter pathways). Visualizing these complex white matter pathways for neuro applications is still an open issue. Hence, interactive visualization techniques to explore and better understand tractography data are required. In this paper, we propose a new interaction technique to support exploration and interpretation of white matter pathways. Our application empowers the user to interactively manipulate manually segmented, box- or ellipsoid-shaped regions of interest (ROIs) to selectively display pathways that pass through specific anatomical areas. To further support flexible ROI design, each ROI can be assigned a Boolean logic operator and a fiber direction. The latter is particularly relevant for kissing, crossing or fanning regions, as it allows the neuroscientists to filter fibers according to their direction within the ROI. By precomputing all white matter pathways in the whole brain, interactive ROI placement and adjustment are possible. The proposed fiber selection tool provides ultimate flexibility and is an excellent approach for fiber tract selection, as shown for some real-world examples.

1 Introduction

The human brain is a highly interconnected organ comprising about 100 billion neurons and 150 trillion synapses. In order to gain insight into the complex architecture, diffusion magnetic resonance imaging (dMRI) and suitable reconstruction

A. Graumann (✉) • M. Richter
AG Visual Computing, University of Konstanz, Konstanz, Germany
e-mail: andreas.graumann@uni-konstanz.de; mirco.richter@uni-konstanz.de

C. Nimsky
Department of Neurosurgery, University Hospital Marburg, Marburg, Germany
e-mail: nimsky@med.uni-marburg.de

D. Merhof
Institute of Imaging & Computer Vision, RWTH Aachen University, Aachen, Germany
e-mail: dorit.merhof@lfb.rwth-aachen.de

L. Linsen et al. (eds.), *Visualization in Medicine and Life Sciences III*, Mathematics and Visualization, DOI 10.1007/978-3-319-24523-2_8

techniques have emerged (see [1] for an overview). dMRI measures the diffusion of water, which originates from the random motion of molecules due to thermal energy. In fibrous tissue such as the white matter of the nervous system, diffusion is restricted to a preferred direction. Therefore, conclusions about the underlying tissue structure can be drawn from the diffusion measurements. This makes it possible to analyze the structure of the human brain in vivo, which is of interest for different areas in medical research and applications.

The acquired diffusion characteristics support the diagnosis of certain pathological disorders, e.g. acute ischemic stroke, degenerative diseases such as Alzheimer's disease or psychiatric disorders such as schizophrenia. In neuroanatomy and neurosurgery, the diffusion measurements serve as a basis for the reconstruction of white matter structures.

However, the processing and visualization of dMRI data is a non-trivial task due to the complexity of the acquired data, which comprehensively describes the local diffusion properties. The inherent complexity of the diffusion imaging data has triggered research efforts towards the reconstruction and visualization of white matter tracts, which comprise the design of meaningful scalar metrics, adequate reconstruction algorithms, as well as comprehensive visualization techniques.

In recent years, there has been a steady progress, both on the imaging side as well as on the fiber reconstruction side. The advent of more sophisticated diffusion models based on high angular resolution diffusion imaging (HARDI) has led to more detailed information regarding the underlying diffusion profiles. This is of particular interest in case of regions with a complex white matter architecture, such as fanning, kissing or crossing fibers. Based on this imaging data, fiber tractography algorithms have emerged that are able to address such complex fiber configurations. The pathways produced by tractography are abstract representations of possible routes through the white matter of the brain.

Recent tractography approaches comprise global and probabilistic methods, as well as approaches that are designed to resolve fiber crossings by regularized HARDI tractography. In Reisert et al. [2], a global approach to generate optimally distributed fibers along the white matter structures is proposed. The probabilistic approach proposed in Jeurissen et al. [3] uses a residual bootstrap tractography to account for orientations of multiple intravoxel fiber populations. An approach to simultaneously estimate the local fiber orientations and perform multi-fiber tractography is introduced in Malcolm et al. [4], which proved to significantly improve the angular resolution at crossings and branchings. Jeong et al. [5] combine independent component analysis tractography with a ball-stick model to isolate intravoxel crossing fibers even for small numbers of fibers with the same orientation. In order to better resolve complex fiber configurations, Rowe et al. [6] propose a tractography approach that takes into account fiber dispersion which aids their approach in finding connectivity commonly missed by other methods.

However, visualizing these complex white matter pathways for neuro applications is still a non-trivial task, especially in the light of recent developments in tractography for complex fiber configurations. Hence, interactive visualization techniques to explore and better understand tractography data are required.

First attempts for selecting fiber tracts comprise filtering [7] and regions of interest (ROIs) [8]. However, since these approaches do not provide any direct user control, interactive fiber selection techniques have been proposed [9–11] which make it possible to interactively create and modify ROIs and associated properties. However, these interactive approaches do not specifically address complex fiber configurations which are of increasing interest in advanced fiber tracking approaches.

In this paper, we propose a new interaction technique to support exploration and interpretation of white matter pathways, particularly in regions with complex fiber configurations. Our application makes it possible to interactively manipulate manually segmented, box- or ellipsoid-shaped ROIs to selectively display pathways that pass through specific anatomical areas. To further support flexible ROI design, each ROI can be assigned a Boolean logic operator. Our key contribution is a new interaction technique which allows specifying a fiber direction to assist in the exploration and identification of fiber tracts from dMRI tractography. This technique is particularly relevant for kissing, crossing or fanning regions, as it allows the neuroscientist to filter fibers according to their direction within the ROI in order to resolve complex fiber configurations.

2 Related Work

Tractography algorithms are commonly used to reconstruct fibers from dMRI data for further visualization. However, due to the vast amount of generated fibers, selection techniques are required in order to make this information accessible.

To further process dMRI tractography results, Zhang et al. [7] proposed to filter reconstructed fibers based on length, average linear anisotropy, and distance separating neighboring fibers. In Conturo et al. [8], ROIs are introduced to select fiber bundles that connect anatomically or functionally predefined regions. A combination of ROIs with Boolean AND, OR, and NOT operations to isolate particular fiber tracts is proposed in Wakana et al. [12]. Although these groups anticipated the potential value of selection techniques for dMRI tractography, the aforementioned approaches do not comprise any interactive filtering technique.

Interactive fiber selection by placing and interactively manipulating box- or ellipsoid-shaped regions to selectively display pathways was initially proposed in [9, 10]. Apart from specifying ROIs, their user interface makes it possible to assign a Boolean operator to a ROI as well as to further specify pathway properties such as length or average fractional anisotropy. In Blaas et al. [11], a similar technique is presented which is computationally more efficient due to dedicated data structures for speed-up. The box- or ellipsoid-shaped ROIs are overcome in Akers et al. [13], where arbitrarily shaped ROIs are generated using a special bimanual interface for 3D pathway selection which employs a pen and a trackball. In merhof10, a segmentation user interface is provided in order to generate user-defined ROIs, and the ROI behaviour can be further specified using

Boolean operators. The approach proposed in Cai et al. [14] extends the previously presented box- and sphere-shaped ROI selectors by a multi-view interface to enable comparative visualizations. In [15], a 2D embedding of the fiber tracts is displayed along with the 3D view, with the aim of removing visual clutter which can be helpful when selecting fiber tracts.

New interaction mechanisms were explored in [16], where a framework for real-time selection of neuronal fiber bundles using a Wii remote control (a wireless controller for Nintendo's gaming console) is employed. In order to achieve a smooth interaction, a novel space partitioning data structure is proposed, which allows for queries that are much faster than previous state-of-the-art approaches.

An automated approach for fiber tract selection is presented in [17]. Given a ROI, the approach locates the point on the fiber that is closest to the centroid of the ROI and the fiber direction at this point and performs a k-means fiber clustering based on these parameters. In this way, a delineation of major tract systems can be achieved without user interaction (e.g. by using ROIs from an atlas).

A broad overview addressing some of these approaches but also covering related fields such as streamline clustering, edge detection and segmentation, topological methods, and extraction of anisotropy creases is provided in [18].

However, none of these approaches is specifically dedicated towards complex fiber configurations such as kissing, crossing or fanning fibers, which may emerge when using state-of-the-art fiber tracking algorithms (Figs. 1 and 2).

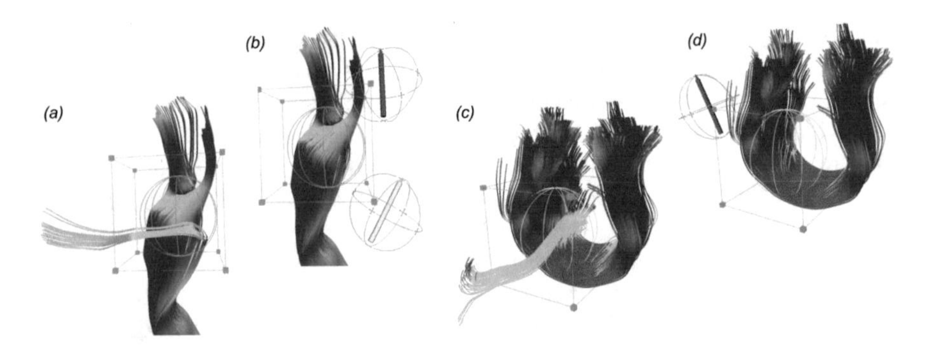

Fig. 1 Illustration of filtering method based on directional regions of interest (ROIs). *Left:* Deactivated directional filtering yields an imperfect result for motor pathway which also comprises lateral fibers in green (**a**). Optimal result after directional filtering, providing the motor pathway without any confounding fibers (**b**). *Right:* Fiber selection based on ROI (directional filtering deactivated) yields imperfect result containing fibers of both the cingulum and the corpus callosum (**c**). Activation of directional filtering allows to either extract the corpus callosum (**d**) or the cingulum (not shown here)

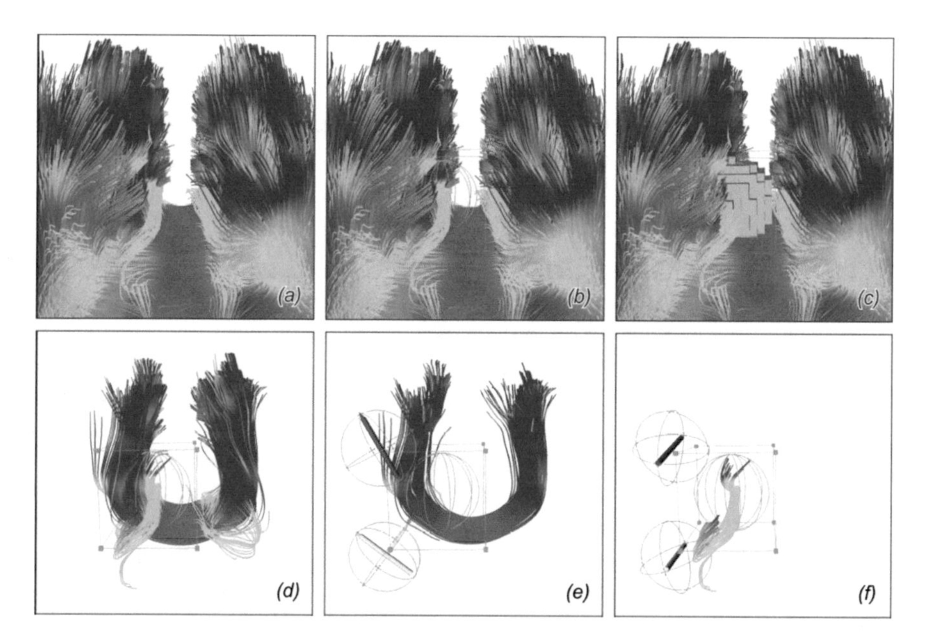

Fig. 2 (**a**) Fiber tractography result of the whole brain showing a view across the corpus callosum. (*b*) Positioning of ROI (wireframe representation) with the aim to segment the cingulum (*green tract*). (**c**) Voxel representation of ROI to clearly see where ROI intersects the fibers. (**d**) Fiber selection based on ROI (directional filtering deactivated) yields imperfect result containing fibers of both the cingulum and the corpus callosum. Activation of directional filtering allows to either extract the corpus callosum (**e**) or the cingulum (**f**)

3 Data Acquisition and Preprocessing

3.1 Data Acquisition

dMRI and anatomical MRI data was acquired on a 3T MRI scanner Tim Trio (Siemens, Erlangen, Germany). For anatomical reference, T1-weighted 3D images were acquired (3D magnetization-prepared rapid gradient echo; repetition time, 1900 ms; echo time, 2.26 ms; field of view, 256 mm; matrix, 256×256; slice thickness, 1 mm; 176 slices, sagittal).

dMRI data was acquired using a single shot echo-planar imaging sequence with 30 noncollinear diffusion-encoding gradients (repetition time, 7800 ms; echo time, 90 ms; field of view, 256 mm; matrix, 128×128; slice thickness, 2 mm; numbers of excitations, 1; $b = 1000\ \text{s/mm}^2$; voxel size $2 \times 2 \times 2\ \text{mm}^3$).

Acquisition of all data sets took about 15 minutes per subject.

3.2 Precomputing Pathways

For fiber tractography, a selection of different methods is available in our toolbox, which comprise both standard tractography algorithms for diffusion tensor imaging (DTI) data as well as more advanced techniques for HARDI data. The following methods can be selected:

STT: Streamline tracking (STT) [8, 19, 20] is a common method for fiber tracking based on DTI data. This method follows the principal diffusion direction throughout the volume, where the integration is typically performed numerically with Runge-Kutta schemes of order 1 to 4. A user-defined threshold based on fractional anisotropy (FA) [21] is applied to stop fiber propagation if the direction of dominant diffusion is not well defined.

TEND: Tensor deflection (TEND) [22–24] takes advantage of the whole diffusion tensor and computes the new fiber direction as a product of the local diffusion tensor and the incoming fiber direction. Similarly to STT, a termination threshold based on FA is used in order to stop propagation if the fiber leaves regions with anisotropic diffusion.

Maximum tracking: For fiber reconstruction based on HARDI data, a deterministic multidirectional tracking proposed by Descoteaux et al. [25] is used, with the generalized fractional anisotropy (GFA) index [26] as the anisotropy measure to terminate fiber propagation. This tractography approach is able to describe crossing and splitting fiber bundles based on the fiber orientation distribution function (ODF). It extends the classical STT approach by taking into account multiple ODF maxima in every integration step.

GFT: A global fiber tracking (GFT) approach with acceptable computing times for a broad class of practical applications was proposed [2]. The approach generates optimally distributed fibers along the white matter structures and is able to model noise in diffusion data through global optimization with particle simulation.

The first three methods are local approaches which have several steps such as seed point selection, fiber propagation and fiber termination in common. Starting from predefined seed points, fibers are propagated until a termination criterion (FA or GFA) is reached. Other commonly applied criteria for streamline termination are a predefined maximum length in order to assure termination of the algorithm after a finite number of steps, or a maximum bending angle [20] which assumes that anatomical fibers do not take sharp turns.

In order to interactively select fiber bundles for exploring fiber tracts, a tracking of the entire white matter region of the brain needs to be generated in the first place. For this purpose, the first three tractography approaches (STT, TEND and maximum tracking) require seed points to initialize fiber tracking. Accordingly, seed points are generated for all voxels above a user-defined FA or GFA threshold similarly to Conturo et al. [8].

To summarize, in our tool the neuroscientist can make different selections to control the outcome of the initial tractography result: choice of different tracking algorithms, FA/GFA-based threshold for seed point generation, FA/GFA-based threshold for fiber termination, maximum bending angle, threshold for minimum and maximum fiber length.

4 Method

In order to extract individual fiber tracts, the neuroscientist needs to define ROIs that are employed in the fiber selection process. In the following sections, all features of the proposed ROI tool are outlined in detail, which comprise functionality to define the ROI (Sect. 4.1), to assign a Boolean operator to the ROI (Sect. 4.2) and functionality to control and filter the direction of fibers within a ROI (Sect. 4.3). Implementation details are provided in Sect. 4.4.

4.1 Definition of ROI

User-defined ROIs based on manual segmentation—For manually segmented ROIs, a voxel-based segmentation tool is used which makes it possible to create ROIs of arbitrary shape and extent. For this purpose, the dMRI dataset that is measured without diffusion gradient is loaded into the segmentation ROI tool and used as anatomical reference. The user-defined ROI based on manual segmentation is drawn manually on axial, sagittal or coronal slices based on anatomical knowledge and is finally imported into the fiber tracking tool for further processing. In this way, it is possible to define non-trivial ROI shapes, e.g. in the vicinity of a tumor.

Box-shaped ROIs—Box-shaped ROIs can be either created by using a control widget where the user can manually specify the position according to the axial, sagittal and coronal slice number and the desired size of the box, or by dragging the cursor directly within the dataset displayed in the 3D viewer to the desired position. After creating such a ROI, it is possible to adjust the size of the ROI interactively by dragging the edges of the ROI (see Sect. 4.4 for details). This provides the possibility to edit the ROI directly and interactively within the 3D viewer according to the user's needs.

Sphere-shaped ROIs—Analogously to box-shaped ROIs, sphere-shaped ROIs can be created in the two same ways, i.e. by defining the ROI position and size within a control widget or by dragging the cursor to the desired ROI position within the 3D viewer. The size of a sphere-shaped ROI is defined by its radius, which can also be adjusted interactively by dragging the displayed transformer nodes besides the sphere (see Sect. 4.4 for details).

Furthermore, each ROI can be visualized in two different ways: *wire frame view* and *voxel view*, as illustrated in Fig. 3. The wire frame view allows to better position

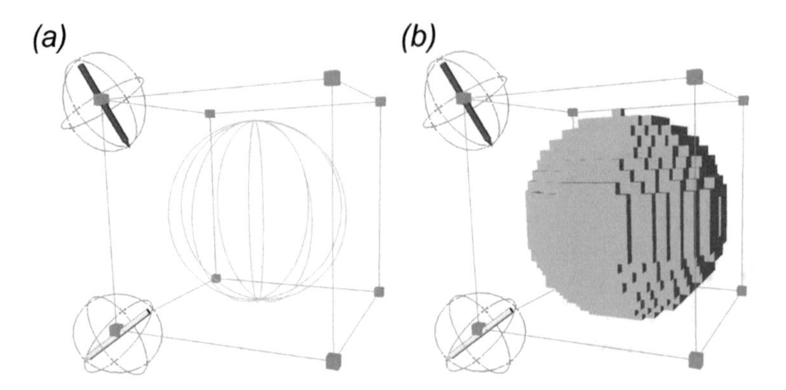

Fig. 3 Sphere-shaped ROI and Open Inventor manipulators. The ROI can either be displayed as wireframe *(left)* or as voxels *(right)*

the ROI in space but is less suitable to determine whether individual fibers are crossing the ROI or not, whereas the voxel view allows to exactly estimate all fibers which are crossing the ROI.

4.2 Assigning Boolean Operators to ROIs

Boolean logic has many applications and is generally used in order to define conditions on sets to retrieve specific elements. Combining ROIs with Boolean operations to isolate particular fiber tracts was first proposed by Wakana et al. [12] and proved to be of great value in several interactive ROI selection tools [9, 10, 14]. In our tool, the three basic Boolean operators are implemented, which comprise logical OR and logical AND to perform the union and intersection of sets of fibers, and logical NOT to exclude fibers.

More specifically, a ROI with associated Boolean operator AND requires any fibers to cross this ROI, i.e. multiple AND ROIs result in a fiber bundle that traverses all AND ROIs. A ROI with assigned Boolean operator OR requires all fibers to traverse either this ROI or any other OR ROI. In order to exclude fibers from the extracted fiber tract, a NOT ROI may be defined which deletes any fibers that cross the ROI. Effectively, the functionality of the NOT ROI is identical to the sculpture-based removal proposed in Cai et al. [14], where a box or sphere-shaped sculpture widget is used to sculpt away undesirable fibers.

As already demonstrated [9, 10, 12, 14], more flexibility for tract selection can be achieved by combining ROIs with different associated Boolean operators. For this reason, this approach has also been integrated into our tool for ROI based selection of fiber tracts.

4.3 Assigning Directions to ROIs

In addition to the assignment of Boolean operators to ROIs, we further propose to assign directions to ROIs. This allows to only extract fibers with a given direction within the ROI. Especially when exploring regions with kissing, crossing or fanning fibers, this gives the user the possibility to extract only those fibers which are relevant and important for a given purpose.

For this purpose, each ROI has two direction manipulators that can be separately activated and deactivated, respectively. These manipulators allow to visually adjust two different user-defined directions. For each fiber, the average direction of all fiber segments within the ROI is used to compare it to the desired user-defined direction given by the directional manipulators. Furthermore, the user can define a parameter to control the allowed deviation between the user-defined and actual average direction of the fiber.

The implementation of the box- and sphere-shaped ROIs with direction selection is performed in Open Inventor Version 2.1.5-10-16 and is described in detail in Sect. 4.4.

4.4 Implementation

In this section, the implementation of the ROIs and the directional filtering of the fibers is described. The box- and sphere-shaped ROIs are constructed with an Open Inventor scene graph which is outlined in detail in Sect. 4.4.1. The implementation of the filtering of fibers according to a user-defined direction is explained in Sect. 4.4.2. Details about an efficient implementation of real-time filtering are provided in Sect. 4.4.3.

4.4.1 Open Inventor Scene Graph

Open Inventor provides a large library of objects, including geometric primitives and interactive manipulators, which can be modified and extended according to own requirements [27]. All objects are represented as nodes and can be connected with each other. The resulting graph is called Open Inventor scene graph.

In Fig. 4 (left), the basic scene graph for a sphere-shaped ROI is displayed. The node *SoSphere* represents the sphere. To adjust the size of the sphere interactively, the manipulator *SoTransformBoxDragger* is added to the scene graph. The transformation node *SoTransform* applies the scaling caused by the manipulator to the sphere.

In addition to the manipulator for adjusting the size, each ROI provides two arrows to assign a direction to the ROI, which can be activated and deactivated via the user interface and appear or disappear in the 3D view accordingly. For this

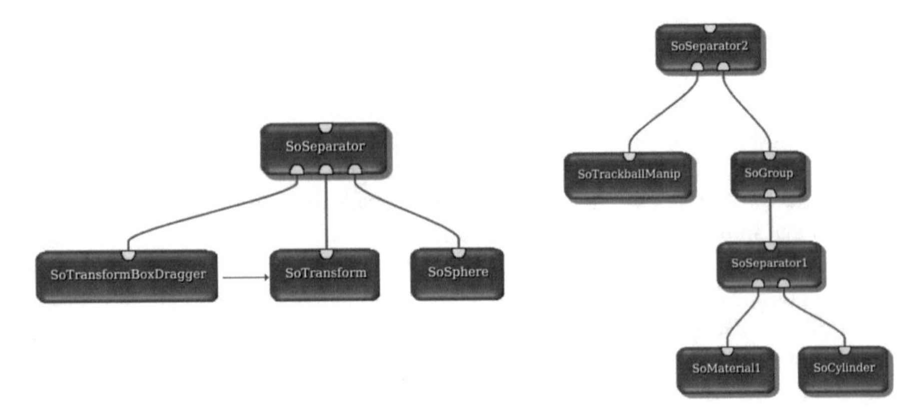

Fig. 4 *Left*: Open Inventor scene graph for sphere-shaped ROI. *Right*: Open Inventor scene graph to define a ROI direction

purpose, the basic scene graph is extended with an arrow represented as a long cylinder. The manipulation node *SoTrackballManip* enables to rotate the cylinder in each direction to define a direction associated to the ROI. Figure 4 (right) shows this extension of the basic scene graph.

The resulting sphere-shaped ROI is shown in Fig. 3. The radius of the ROI can be adjusted by dragging one of the manipulator nodes besides the sphere. Furthermore, the user can also activate and adjust up to two directional manipulators (yellow and red arrow), which have their own trackball for manipulation. A ROI can be displayed in two different ways: Fig. 3a shows the wireframe representation of a sphere-shaped ROI, whereas in Fig. 3b the ROI voxels are displayed.

Box-shaped ROIs are implemented analogously to sphere-shaped ROIs. In case of box-shaped ROIs, width and length can be adjusted with the Open Inventor manipulator (rather than the radius as in sphere-shaped ROIs).

4.4.2 Implementation of Directional Filtering

After creating or adjusting a ROI, all fibers are filtered automatically according to the position and size of the ROI and its assigned Boolean operator. If the directional filtering option is also activated, the given directions are taken into account as well to extract fibers with a desired direction.

For this purpose, a callback method is invoked directly after adjusting the ROI. This callback method updates the ROI voxels automatically and calls a function to filter the fibers with the given ROIs and fiber directions.

When filtering the fibers with a direction, this callback method calculates for each fiber the average direction of all fiber segments within the ROI, and compares this average direction with the user-defined direction given by the manipulator. All fibers with a direction close to the desired direction (the user can define a maximum deviation from the initially defined direction) are extracted and visualized.

4.4.3 Real-Time Fiber Selection

For efficiency reasons, the process of filtering fibers is accelerated by an octree implementation, similarly to Blaas et al. [11]. For this purpose, each ROI is encompassed by a cube which is subdivided into eight child cubes of equal size. This subdivision is recursively applied to each child cube until the smallest cube comprises a pre-defined amount of voxels (256 voxels in our case). Each leaf cube contains a list of those voxels within the boundaries of the leaf cube that are marked by the ROI. With this method, each point of a fiber can be efficiently checked for containment in the ROI by recursively searching through the octree based on the coordinate of the fiber point.

5 Results and Discussion

For benchmarking, a PC equipped with an Intel Core i7 8×3.4 GHz processor, 32 GB RAM and an NVidia Geforce GTX 660 Ti graphics card with 2 GB graphics memory was used. The fiber tracking results within the whole brain consisted of approximately 10,000 fibers comprising about 2,000,000 fiber segments in total.

Table 1 gives a detailed overview of the performance results of the presented fiber selection tool. For each of the use-case scenarios presented in Figs. 2, 5 and 6, the number of fibers of the whole-brain tracking is provided as well as the timings for extracting individual white matter tracts based on the ROIs displayed in the respective subfigures.

In order to illustrate the functionality of our tool, different use-case scenarios are shown in Figs. 2, 5 and 6, which are discussed in the following:

Scenario 1: The cingulum is a thin, arching fiber bundle that forms the white matter core of the cingulate gyrus, following it from the subcallosal gyrus of the frontal lobe beneath the rostrum of corpus callosum, to the parahippocampal gyrus and uncus of the temporal lobe. As one of the connecting parts of the limbic

Table 1 Timings in seconds for filtering fibers of the whole brain in the use-case scenarios presented in Figs. 2, 5 and 6

Figure	Number of fibers to filter	Time
2 d	4050	0.07 s
2 e	4050	0.22 s
2 f	4050	0.20 s
3 a	10,687	0.05 s
3 b	10,687	0.05 s
3 c	10,687	0.05 s
3 d	10,687	0.70 s
4 b	10,017	0.04 s
4 c	10,017	0.05 s
4 d	10,017	0.80 s

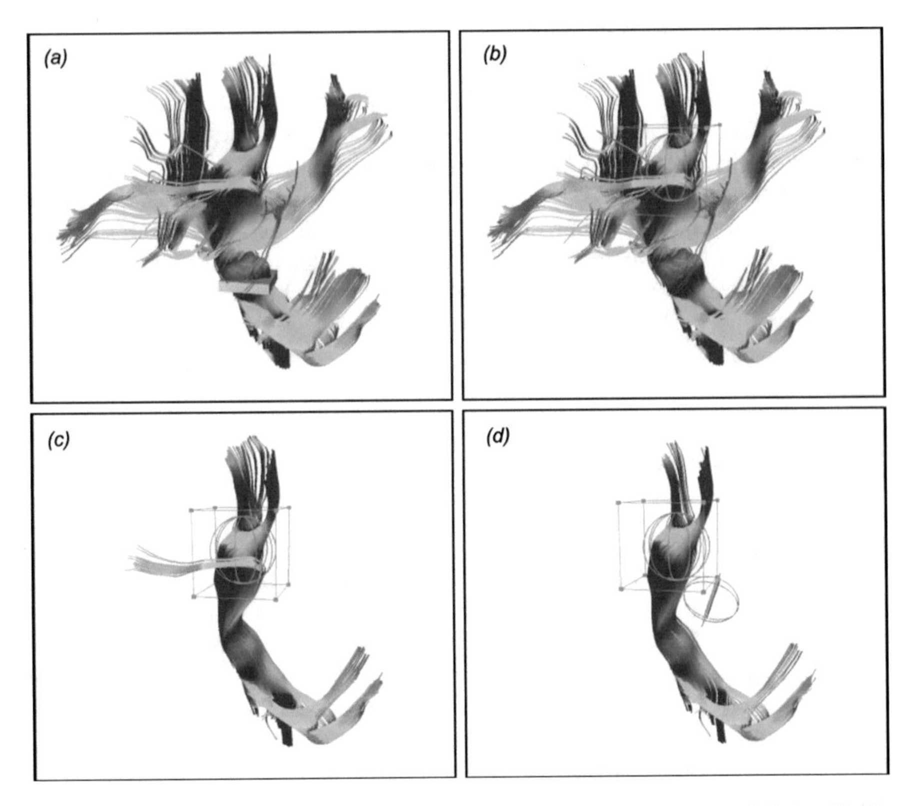

Fig. 5 (**a**) Corona radiata of the left hemisphere after tractography with a box ROI located in the brainstem. (**b**) Positioning of ROI in region that refers to the motor pathway. (**c**) Without directional filtering, the motor pathway is displayed with some remaining lateral fibers (*green*). (**d**) Directional filtering provides optimal result for the motor pathway, without any confounding fibers as in (**c**)

system, its anterior part is linked to emotion, whereas the posterior section is involved in cognitive functions. In Fig. 2, a fiber tractography result of the whole brain is provided that shows a view across the corpus callosum (red) with the cingulum in green *(a)*. In *(b,c)*, a ROI is positioned that fully encompasses the cingulum with the aim to segment this fiber tract. In this scenario, fiber selection of the cingulum based on standard ROI tools would be challenging and would require multiple ROIs in order to reliably extract this structure. This can be seen in *(d)*, where fiber selection based on a standard ROI without any directional filtering yields an imperfect result containing fibers of both the cingulum and the corpus callosum. However, after activating the directional filtering, either the corpus callosum *(e)* or the cingulum *(f)* can be easily extracted.

Scenario 2: The corona radiata comprises both descending and ascending neuronal fibers that carry almost all of the neural traffic from and to the cerebral cortex. The corona radiata can be depicted by positioning a ROI in the brainstem as shown in Fig. 5a. In order to select the fraction of fibers of the corona radiata

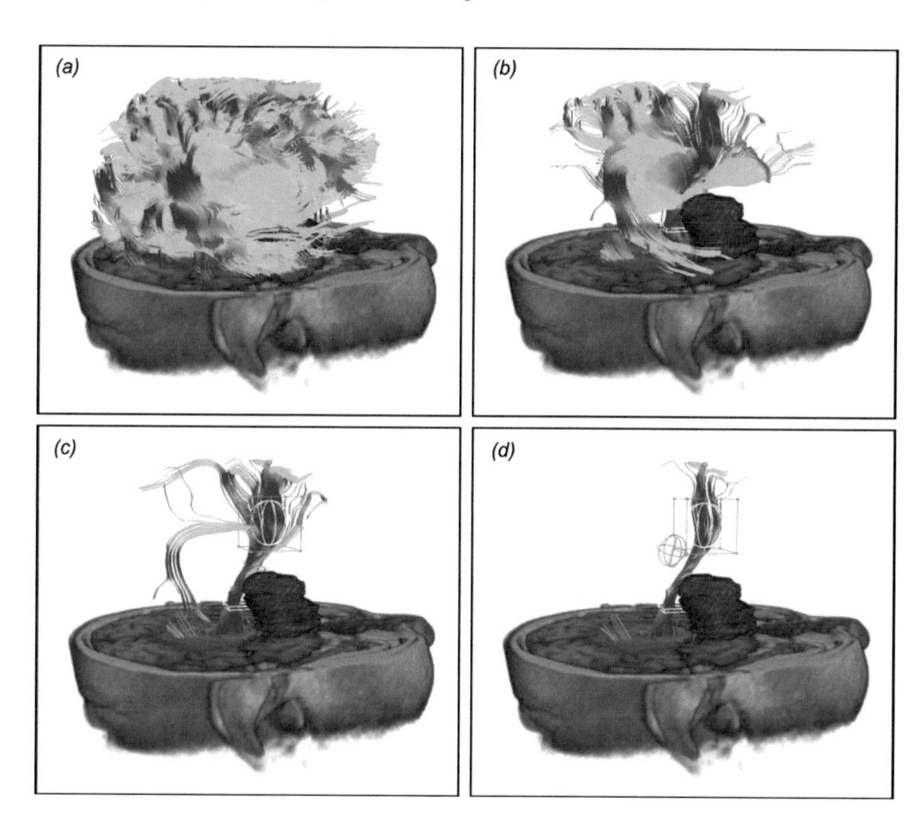

Fig. 6 Brain tumor patient (tumor delineated in red). (**a**) Tractography result of the whole brain. (**b**) Corona radiata of the right hemisphere after tractography with a box ROI located in the brainstem, with projection fibers passing nearby the tumor. (**c**) Filtering with ROI in region that refers to the motor pathway. Without directional filtering, the motor pathway is displayed with some remaining lateral fibers (*green*). (**d**) Directional filtering provides optimal result for the motor pathway, without any confounding fibers as in (**c**)

which refers to the motor tract, another ROI is positioned in that respective region *(b)*. In this scenario, fiber selection based on standard ROI tools would provide an imperfect filtering for the motor pathway with remaining lateral fibers (green), as illustrated in *(c)*. However, with the directional filtering tool proposed in this work, an optimal result can be achieved as shown in *(d)* without any confounding fibers.

Scenario 3: Fiber tractography results are of interest for different neuro applications, which also comprises planning for brain tumor surgery as illustrated in Fig. 6. In order to provide an anatomical reference, the anatomical MRI dataset is rendered together with the fibers. The tumor which has been segmented by a medical expert is delineated in red. In order to interactively explore the fibers in the vicinity of the tumor, ROI-based tools are of great interest. Starting with a tractography result of the whole brain *(a)*, the corona radiata of the right hemisphere can be displayed by filtering with a box ROI located in the brainstem *(b)*. Note the projection fibers passing nearby the tumor. Filtering with a spherical ROI in the region that refers to the motor pathway provides an imperfect result with remaining lateral fibers (green) if directional filtering is deactivated *(c)*, and an optimal result if directional filtering is used *(d)*.

According to the experience of the neuroscientists who have explored our tool, the wireframe representation is preferable in order to position a ROI, as this rather transparent representation of the ROI provides better spatial orientation and more accurate positioning with respect to a target white matter structure. The voxel view on the other hand was preferred in order to exactly see and cross-check which fibers are actually intersected by the ROI.

As regards the directional manipulators, either one manipulator (possibly with a greater directional tolerance) or two manipulators were used. In general, two directional manipulators provide more flexibility, e.g. for crossing fibers or in case of fiber branchings, and were preferably used in regions with more diverging fibers such as in Fig. 1 *(left)*.

Considering practical application in neurosciences, a flexible and interactive tool for white matter fiber bundle selection is an important prerequisite to explore and interact with the tractography data. However, in order to explore and view fibers within their anatomical context, a multimodal visualization combining function and surrounding anatomy is required as shown in Fig. 6. By investigating such a multimodal representation, the neuroscientist is able to e.g. verify whether important tract systems are located in the vicinity of the tumor. Our software application including the newly developed tool for directional ROIs provides interaction and visualizations of high quality and at interactive frame rates.

In comparison to previously presented techniques that do not aim at interactive manipulation [7, 8, 12] or that are intended to operate fully automatically [17], interactive adjustment of ROIs offers high flexibility and makes it possible to extract fiber bundles with a complex architecture. This can be essential in many medical applications related to neurosciences or neurosurgery, e.g. for selection of displaced fiber bundles in the presence of a brain tumor, or in case of crossing fiber bundles in the area of the centrum semiovale. For this reason, previous research focused on interactive ROI-based techniques [9–11, 13, 14, 28]. In addition to dedicated data structures that enable interactive processing times, these approaches share the idea of supporting user control by means of automated ROI placement, flexible ROI design and Boolean operators assigned to ROIs. The directional ROIs proposed in this paper add an additional degree of freedom by providing a means of filtering fibers within the ROI according to a user-defined fiber direction. The presented approach could potentially also combined with advanced user interfaces, such as the framework presented in [16].

6 Conclusion

In light of recent advances in diffusion imaging and tractography to better capture the complex architecture of white matter fiber tracts, appropriate tools for selecting fiber structures are required likewise, as proposed in this work. For this purpose, we have presented an interface to interactively explore dMRI fiber tractography results. As opposed to previous applications, our interface allows to interactively manipulate ROIs to selectively display pathways with dragging operations directly in the 3D viewer. Our key contribution, a new interaction technique which allows to specify a fiber direction to assist the exploration and identification of fiber tracts, proved to be of great value in case of complex fiber configuration. By assigning a fiber direction to a ROI, it becomes possible to resolve kissing, crossing or fanning regions as illustrated by several real-world examples. Overall, the presented tool overcomes the limitations of current tract selection tools and offers ultimate flexibility in tract selection.

References

1. Jones, D.K. (ed.): Diffusion MRI: Theory, Method, and Applications. Oxford University Press, Oxford (2011)
2. Reisert, M., Mader, I., Anastasopoulos, C., Weigel, M., Schnell, S., Kiselev, V.: Global fiber reconstruction becomes practical. Neuroimage **54**(2), 955–962 (2011)
3. Jeurissen, B., Leemans, A., Jones, D.K., Tournier, J.D., Sijbers, J.: Probabilistic fiber tracking using the residual bootstrap with constrained spherical deconvolution. Hum. Brain Mapping **32**(3), 461–479 (2011)
4. Malcolm, J.G., Michailovich, O., Bouix, S., Westin, C.F., Shenton, M.E., Rathi, Y.: A filtered approach to neural tractography using the Watson directional function. Med. Image Anal. **14**, 58–69 (2010)
5. Jeong, J.W., Asano, E., Yeh, F.C., Chugani, D.C., Chugani, H.T.: Independent component analysis tractography combined with a ball-stick model to isolate intravoxel crossing fibers of the corticospinal tracts in clinical diffusion MRI. Magn. Reson. Med. (2012). Epub ahead of print. doi: 10.1002/mrm.24487
6. Rowe, M., Zhang, H., Alexander, D.: Utilising measures of fiber dispersion in white matter tractography. In: Proceedings of the MICCAI Workshop on Computational Diffusion MRI, pp. 2–12 (2012)
7. Zhang, S., Demiralp, C., Laidlaw, D.: Visualizing diffusion tensor MR images using streamtubes and streamsurfaces. IEEE Trans. Vis. Comput. Graph. **9**(4), 454–462 (2003)
8. Conturo, T.E., Lor, N.F., Cull, T.S., Akbudak, E., Snyder, A.Z., Shimony, J.S., et al. Tracking neuronal fiber pathways in the living human brain. Proc. Natl. Acad of Sciences of the United States of America (PNAS). 1999;96(18):10422–10427.
9. Akers, D., Sherbondy, A., Mackenzie, R., Dougherty, R., Wandell, B.: Exploration of the brain's white matter pathways with dynamic queries. In: IEEE Visualization, Proceedings of the Conference on Visualization 2004, pp. 337–384 (2004)
10. Sherbondy, A., Akers, D., Mackenzie, R., Dougherty, R., Wandell, B.: Exploring connectivity of the brains white matter with dynamic queries. IEEE Trans. Vis. Comput. Graph. **11**(4), 419–430 (2005)

11. Blaas, J., Botha, C.P., Peters, B., Vos, F.M., Post, F.H.: Fast and reproducible fiber bundle selection in DTI visualization. IEEE Visualization, Proceedings of the Conference on Visualization, pp. 59–64 (2005)
12. Wakana, S., Jiang, H., Nagae-Poetscher, L.M., van Zijl, P.C., Mori, S.: Fiber tract-based atlas of human white matter anatomy. Radiology **230**(1), 77–78 (2004)
13. Akers, D.: CINCH: a cooperatively designed marking interface for 3D pathway selection. In: 19th Annual ACM Symposium on User Interface Software and Technology (UIST), pp. 33–42 (2006)
14. Cai, H., Chen, J., Auchus, A.P., Correia, S., Laidlaw, D.H.: InShape: in-situ shape-based interactive multiple-view exploration of diffusion MRI visualizations. In: International Symposium on Visual Computing (ISVC), pp. 706–715 (2012)
15. Chen, W., Ding, Z., Zhang, S., MacKay-Brandt, A., Correia, S., Qu, H., et al.: A novel interface for interactive exploration of DTI fibers. IEEE Trans. Vis. Comput. Graph. **15**(6), 1433–1440 (2009)
16. Klein, J., Scholl, M., Köhn, A., Hahn, H.K.: Real-time fiber selection using the Wii remote. In: Proceedings of the SPIE Medical Imaging, pp. 76250N 1–8 (2010)
17. de Luis-Garcia, R., Alberola-Lopez, C.: Tractography clustering for fiber selection in ROI-based diffusion tensor studies. In: Proceedings of the IEEE Engineering in Medicine and Biology Society (EMBC), pp. 5665–5668
18. Schultz, T.: Feature extraction for DW-MRI visualization: the state of the art and beyond. In: Hagen, H. (ed.) Scientific Visualization: Interactions, Features, Metaphors. Vol. 2 of Dagstuhl Follow-Ups. Schloss Dagstuhl – Leibniz-Zentrum für Informatik, pp. 322–345 (2011)
19. Mori, S., van Zijl, P.C.M.: Fiber tracking: principles and strategies - a technical review. NMR Biomed. **15**(7–8), 468–480 (2002)
20. Basser, P.J., Pajevic, S., Pierpaoli, C., Duda, J., Aldroubi, A.: In vivo fiber tractography using DT-MRI data. Magn. Reson. Imag. **44**(4), 625–632 (2000)
21. Basser, P.J., Pierpaoli, C.: Microstructural and physiological features of tissues elucidated by quantitative diffusion tensor MRI. J. Magn. Reson. Ser. B **111**(3), 209–219 (1996)
22. Westin, C.F., Maier, S., Khidhir, B., Everett, P., Jolesz, F.A., Kikinis, R.: Image processing for diffusion tensor magnetic resonance imaging. In: Proceedings of the Medical Image Computing and Computer Assisted Intervention (MICCAI), pp. 441–452 (1999)
23. Weinstein, D., Kindlmann, G., Lundberg, E.: Tensorlines: advection-diffusion based propagation through diffusion tensor fields. In: Proceedings of the IEEE Visualization (VIS), pp. 249–253 (1999)
24. Lazar, M., Weinstein, D., Tsuruda, J., Hasan, K., Arfanakis, K., Meyer, E., et al.: White matter tractography using diffusion tensor deflection. Hum. Brain Mapping **18**(4), 306–321 (2003)
25. Descoteaux, M., Deriche, R., Knosche, T.R., Anwander, A.: Deterministic and probabilistic tractography based on complex fibre orientation distributions. IEEE Trans. Med. Imag. **28**(2), 269–286 (2009)
26. Tuch, D.S.: Q-ball imaging. Magn. Reson. Med. **52**(6), 1358–1372 (2004)
27. Wernecke, J.: The Inventor Mentor: Programming Object-Oriented 3D Graphics with Open Inventor, Release, vol. 2, 1st ed. Addison-Wesley Longman Publishing, Boston, MA (1993)
28. Merhof, D., Greiner, G., Buchfelder, M., Nimsky, C.: Fiber selection from diffusion tensor data based on Boolean operators. In: Proc. Bildverarbeitung für die Medizin (BVM), pp. 147–151 (2010)

Multi-Modal Visualization of Probabilistic Tractography

Mathias Goldau and Mario Hlawitschka

Abstract Neuroscientists use visualizations of diffusion data to analyze neural tracts of the brain. More specifically, probabilistic tractography algorithms are a group of methods that reconstruct tract information in diffusion data and need proper visualization. One problem neuroscientists are facing with probabilistic data is putting this information into context. Neuroscience experts already successfully utilized several techniques together with structural MRI to detect neural tracts in the living human brain which were previously only known from tracer studies in macaque monkeys. Whereas the combination with structural MRI, i.e., T1 and T2 images, has been important for these studies, new challenges ask for an integration of other imaging modalities. First, we provide an overview of the currently used visualization techniques. Then, we show how probabilistic tractography can be combined with other techniques, trying to find new and useful visualizations for multi-modal data.

1 Introduction

The human connectome and its visualization are one of the hottest topics in the field of neuroscience and medical scientific visualization [35, 51, 58]. Hereby, the term *connectome* describes the entirety of all neural connections of the brain, addressed at different types [57] and scales. We focus on anatomical connectivity analysis at the scale of millimeters within the *living* brain, for which magnetic resonance imaging (MRI) and diffusion-weighted MRI (dMRI) are excellent imaging techniques. dMRI measures the strength of the diffusion of water molecules along multiple orientations and allows to estimate major white matter tracts, as diffusion is directed along cell structures defining those tracts. The process of reconstructing such tracts is

M. Goldau (✉)
Leipzig University, Leipzig, Germany
Max-Planck-Institute for Neurological Research, Cologne, Germany
e-mail: math@informatik.uni-leipzig.de

M. Hlawitschka (✉)
Leipzig University, Leipzig, Germany
e-mail: hlawit@informatik.uni-leipzig.de

L. Linsen et al. (eds.), *Visualization in Medicine and Life Sciences III*, Mathematics and Visualization, DOI 10.1007/978-3-319-24523-2_9

called tractography and became an important tool in neuroscience over the past years. Many approaches have been proposed to extract neural tracts at high accuracy from dMRI data [32]. In addition, neuroscience experts need anatomical context to evaluate tractography data.

Among tractography algorithms, probabilistic diffusion tractography is a widely used and promising tool to probe for white matter tracts in diffusion data [32, 50]. Probabilistic tracts are volumetric scalar maps of *connectivity scores* ranging from zero to one, which relate a seed region or seed point to other areas in the brain. If the score is close to one there is good evidence that there is a physical connection of this point to the seed region. Otherwise, if the score is close to zero, there is high evidence that there is no physical connection between this point and the seed region. Therefore, the existence of a physical connection between every point of the brain to this seed region is expressed by this connectivity score. For a review on tractography and probabilistic tractography, including its limitations, see [6, 32].

Margulies et al. recently presented a very valuable overview of visualization techniques currently used in the neuroscience community for connectome visualization [39]. He state that there is a serious lack of visualizations combining image modalities. To fill this gap, we analyze the combination of various imaging modalities and probabilistic tractography data. Therefore, we survey typical datasets neuroscientists are working with and typical visualization techniques used in anatomical space. We conclude by giving recommendations on how to combine techniques to better analyze structural connectivity data.

2 Background and Related Work

We first want to provide an overview of visualization techniques for probabilistic tracts. Then, we relate our aim to recent survey articles and discuss techniques in terms of perception useful for our analysis. Finally, we discuss recent work based on the different data types, which is the basis for our classification and provides the structure of this paper.

2.1 Visualizing Probabilistic Tracts

The most complete and up to date overview of visualization techniques for probabilistic tractography data is given by Margulies [39]. The techniques working in anatomical space can be classified into two groups: three-dimensional or slice-based. Three-dimensional techniques typically employ three-dimensional structures, such as surface estimations, and slice-based techniques, such as axis planar cutting planes, using only two dimensions for representation.

Slice based techniques typically superimpose some texture or colormap of the tract over anatomical context. A common technique is the heat-map (also known

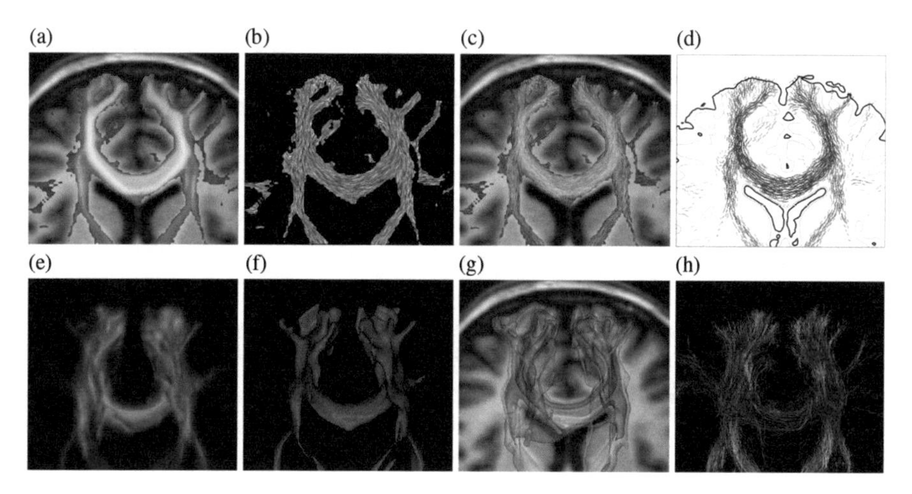

Fig. 1 Overview of visualization techniques used for visualizing probabilistic diffusion tractography data. All figures render the same tract using the same camera position. (**a**) Heat-map of the probabilistic tract over T1 image. (**b**) LIC of major diffusion orientation filtered by the probabilistic tract. (**c**) Same as (**b**) but with T1 background and colored LIC with heat-map. (**d**) Red Fiber-Stipples with boundary curves as context. (**e**) DVR of the tract, with transfer function mapping low scores to white shadow and the core of the tract to opaque red. (**f**) Isosurface, with isovalue set to 50 %. (**g**) Three nested isosurfaces. (**h**) Probabilistic fibers defining the connectivity scores

as hot-iron colormap), which can be used as overlay over structural MRI data [49], see Fig. 1a. There, the color is made transparent in case of no connectivity, to dark red for low connectivity and then gradually to bright yellow for highest connectivity.

Line integral convolution (LIC) [11], generates a texture from a vector field by smearing a noise texture along integral lines. This can be used to depict major diffusion directions in regions of the tract (Fig. 1b and c). Tract density imaging (TDI) [12] produces similar results by partially projecting fibers onto a slice.

Fiber-Stippling [26, 29], generates small line stipples to depict diffusion direction. The connectivity score is encoded in stipple density and opacity. Anatomical context is given as contours as well as usual structural MRI underlay, see Fig. 1d. We consider Fiber-Stipples an important tool because it was used in a Technical Spotlight in the European Journal of Neuroscience to reveal the structural bases of cerebellar networks within the basal ganglia which were previously only known from studies using transneuronal virus tracers in macaque monkeys [50].

In the category of three-dimensional techniques, direct volume rendering (DVR) assigns a certain color and opacity to each volume sample which are then compiled into one image. Hereby, the color mapping is modeled by a transfer function [60], see Fig. 1e for an example. Another three-dimensional technique uses isosurfaces to show outlines of the volumetric connectivity score. The isosurfaces are typically represented as triangle meshes and depict only a particular connectivity score (isovalue) of the tract, see Fig. 1f. Berres et al. [7] proposed nested isosurfaces to display multiple values, see Fig. 1g. Brecheisen et al. [10] propose similar results, but with an illustrative rendering. Some probabilistic tracking algorithms repeatedly start a randomized deterministic tracking from the seed region. The number of tracts hitting a voxel is similar to the connectivity score. Polylines from such trackings are called probabilistic fibers [16], see Fig. 1h.

2.2 *Surveys*

For identifying important image modalities and most commonly used visualization techniques we consulted recent surveys [1, 38, 39, 55]. As probabilistic diffusion tractography originates from dMRI and structural MRI, especially other MRI modalities, such as functional MRI (fMRI) or magnetic resonance angiography (MRA) [59], might be of interest. While fMRI gives information about functional activity and connectivity, MRA data provides structural information of blood vessels. Each image acquisition technique generates different raw data from which additional data might be derived. The surveys state that most common visualization techniques in anatomical space use textured (axis planar) slices, glyphs, polylines, or surfaces.

2.3 *Color and Perception*

When combining different techniques, the perception of shape and color plays an important role. Ware [61] provides a good overview on the use and perception of colors. In addition, he also provides useful guidelines on choosing colors in various settings. For example, Ware suggests to use Bauer's method [5] to compute the most discriminative color to a given set of fixed colors, cf. Paragraph "Fixed Colors". Additionally Ware propose to enhance luminescence when chromatic contrast is exhaused, cf. Sect. 4.1.1, or outlining glyphs, cf. Fig. 6, to further increase visual contrast. Additionally, Kapri et al. [60] evaluates optimal coloring for probabilistic tract in virtual environment regarding uncertainty.

2.4 *Data*

For a systematic investigation, we first need an overview of what data is available and of which visualization techniques are typically applied. In the following paragraphs we briefly introduce various imaging modalities and types of data neuroscientists are working with. We then collect information on typical visualization techniques and summarize the information in Table 1 which serves as basis for our study. Even though they have proven to be very helpful to neuroscientists, to focus this paper, we exclude visualization techniques outside of anatomical

Table 1 Each row corresponds to a specific data type, named in the first column. The remaining columns then identify common visualization techniques used for this data type

	Common visualization techniques					
	Textured slices					
Data	Monochrome	Color	Glyph	Contours	Lines	Surface
CT	x			x		
Structural MRI: T1,T2,	x			x		
Diffusion images	x					
DTI			x			
Complex diff. models			x			
Major diff. orientation		x			x	
Diffusion indices	x					
Probabilistic tracts		x				x
Deterministic tracts					x	
Functional activity		x				
Functional connectivity			x		x	
Segmentation		x		x		x
PET/SPECT		x				
Surface data						x

space such as dendrograms for hierarchical agglomerative clusterings, colored matrices depicting functional correlation, and radial graphs with force-directed edge bundling, cf. [9, 34, 39].

2.4.1 CT and Structural MRI

Usually computed tomography (CT) [36] and structural MRI data (T1,T2) [14] are displayed on axis-aligned slices (coronal, sagittal, or axial) using a monochrome texture representing the measured data. Since small nuances of gray are important (see Fig. 2), contrast and brightness manipulators are often used for contrast enhancement. Besides monochrome textures, different tissues (e.g. gray matter) or structures (e.g. lesions, tumors, blood vessels, ventricular system) need to be highlighted. Contrast improvement can be done using contrast agents during the measurement, but some structures that are visible in the raw data can be extracted in a postprocessing step. T1 data, for example, is also rendered on inflated cortical surface maps or used to segment brain structures such as the ventricular system. Furthermore, MRA data is used to reconstruct three-dimensional surfaces as well. As segmentations and surfaces are derived data, we will discuss them separately in Sects. 2.4.6 and 2.4.8.

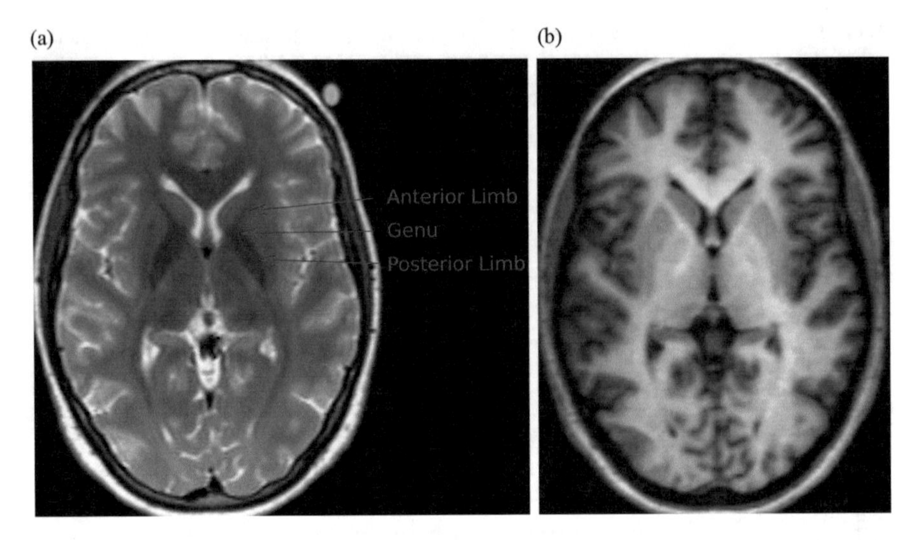

Fig. 2 These figures show the difficulty of reading structural MRI data at the example of the *internal capsule* (IC) which consists of mainly three parts: Its *anterior limb*, *posterior limb*, and the *genu*. In (**a**) *red* labels indicate their positions on an axial slice of a T2 image. (**b**) Same subject, same axial slice position but with T1 data textured. With prior knowledge from (**a**) you hopefully grasp the IC in (**b**) as well

2.4.2 Diffusion Images

Diffusion imaging measures diffusion strength along multiple orientations, also called gradients or angles. For a typical high angular diffusion image (HARDI), there are more than 30 different gradients, each resulting in a three-dimensional scalar image representing the diffusion strength along its orientation [67]. Such data is rarely visualized directly (if so, then on slices) and is typically used for image registration processes, artifact reduction schemes, and, most importantly, to construct data representing some model of the diffusion process.

The diffusion tensor is one of the oldest and most popular diffusion models [4, 32] and founded the diffusion tensor imaging (DTI). Although there are limitations to this model, for example not being able to resolve crossings of even major white matter bundles, it still has a very big (clinical) importance [1]. Compared to other diffusion models, this model is a rather simplistic approach, which also might be a reason for its ongoing success. The most common visualization of all diffusion models is using glyphs [3, 30], often placed on axis-aligned slices. One of the biggest limitations of DTI is that it is not able to resolve fiber crossings locally. Even worse, Wedeen found white matter organization to be highly affected by fiber crossings [63]. This triggered a new demand of more complex diffusion models such as diffusion spectrum imaging (DSI) or q-ball imaging (QBI), cf. [32]. Nonetheless their *direct* visualizations remain glyph centric.

One of the most important parts of DTI is the major diffusion orientation computed from the eigensystem of the second order tensor. A straight-forward visualization technique is to map orientations into a color space. The most popular coloring technique is the RGB-coloring scheme [48]. It maps a three-dimensional orientation vector to RGB color space, so superior–inferior orientations mapped to blue, left–right to red and finally anterior–posterior orientations are mapped to green. Besides this, other coloring schemes have been developed [15]. Other approaches depict the major diffusion orientation using gray-scale textures generated from LIC or by a small line per voxel, which is called *vector-plot* [24].

Diffusion indices or biomarkers are important scalar values generally derived from diffusion models. They have been successfully used for certain diagnoses and brain development studies and hence are important for brain analysis. The most popular diffusion indices are derived from the rather simplistic diffusion tensor and have been used in various clinical studies to study specific diseases, aging, and developmental processes, e.g. [17, 53]. For example the fractional anisotropy (FA), apparent diffusion coefficient (ADC), mean diffusivity (MD) or radial diffusivity (RD). Beside DTI based diffusion indices, new generalized diffusion indices, such as peak fractional anisotropy (PFA), generalized anisotropy (GA), and generalized fractional anisotropy (GFA), have been conceived [23].

2.4.3 Tractography

As described earlier, tractography is the process of reconstructing neural tracts out of diffusion data. The algorithms can be classified as follows: Deterministic tractography methods generate polylines [21] whereas probabilistic tractography methods often generate volumetric data. For an overview on probabilistic tractography, we refer to Sects. 1 and 2.1. In the following we focus on deterministic tractography.

Deterministic tracts are often visualized as pure lines or tubes [39] and might be colored globally or locally. Global coloring assigns the color of its *main* orientation to a polyline, while local coloring schemes assign individual colors to each line segment. Typical coloring schemes are the same as for directional color coding mentioned earlier in Sect. 2.4.2. To increase depth-perception, polylines may be enhanced with illumination and shadows [20]. Besides rendering the lines directly, there are also techniques super-sampling the density of deterministic tracts producing LIC-like textures such as TDI.

2.4.4 Functional Activity

Functional MRI (fMRI) is used to locate neural activity indirectly through metabolism activity measured by the blood oxygen level dependent (BOLD) factor [52]. The resulting data are scalar maps indicating regions of brain activity. This activity is often rendered in the same way as probabilistic tracts by using heat-map or rainbow colormaps on slices with structural MRI background.

Electroencephalography (EEG) [47] and Magnetoencephalography (MEG) [2] are other very popular techniques for measuring brain activity. These techniques capture time-dependent electric or magnetic field strengths, respectively, with sensors placed along the scalp. The measurement delivers plots of electric or magnetic activity over time, induced by neural activity inside of the brain. The spatial location of the activation areas at a time can be computed by so-called source reconstruction. This reconstruction leads to scalar activation maps that, from a visualization point of view, can be treated similarly to fMRI activation maps and are therefore not listed separately in Table 1.

2.4.5 Functional Connectivity

Functional connectivity typically refers to a statistical concept of the correlation between spatially distinct units e.g. between cortex areas. The most common visualization within anatomical space is to connect cortex regions with a strong correlation by a straight line or splines. Other approaches try to sample glyphs on the gray matter–white matter interface denoting the anatomical direction of their functional correlations [9].

2.4.6 Segmentation

One of the big goals of neuroscience is to classify brain regions based on functional units, which are important on the cortex as well as in white matter tissues. For example finding a cortex parcellation that decomposes the cortex into functional units or finding white matter clustering schemes that decompose white matter tracts into major white matter fiber bundles are both hot topics in the community. The visualization techniques labeling different brain regions are usually limited to employ most discriminative colors to each single unit [27, 66]. Besides coloring, the visualization of segmentation data uses three-dimensional surfaces [25, 31] as well as slice-based techniques [62]. For highlighting segmentation results sometimes also contours are used to outline cortex areas or the gray matter–white matter interface.

2.4.7 PET and SPECT

Positron emission tomography (PET) or single-photon emission computed tomography (SPECT) measures photons from the radioactive decay of an injected radionuclide, which has been aggregated by different metabolic systems. Other applications are the analysis of brain hemodynamics by tracking chemicals in blood flow [13] or so called amyloid imaging, which visualizes the presence of senile plaques linked to Alzheimer's disease (AD) [45]. The so gained three-dimensional images localize areas of high tracer presence and are typically visualized in anatomical space on slices using rainbow colormaps [56].

2.4.8 Surface Data

Neuroscientists frequently use surface data, ranging from cortex estimations [22, 65], over boundaries of lesions or tumors, the ventricular system, or the blood system [46] to boundary surfaces on brain segmentations. Most visualization systems represent these surfaces using triangular mesh structures or solid surface renderings using colored and illuminated facets.

3 Methods

The data and visualization techniques we are working with in this paper are summarized in Table 1. Even though the list of data types might be incomplete and might lack data used to answer specific questions, to the best of our knowledge, it lists the *most popular* data types. Other imaging techniques may fall into a similar category as one of the listed modalities. Also please note that this table only lists *common* visualization techniques, not all possible techniques. This is also why there is no column for direct volume rendering, as it is rarely used in the neuroscience community. We compiled the table based on recent publications, which have been already discussed in the related work.

We implemented methods for the combined visualization of probabilistic tractography (as seen in Fig. 1) with state-of-the-art visualization techniques for other modalities as shown in Table 1 and compared the results.

4 Results and Discussion

In this section, we discuss possible combinations of probabilistic tractography with data from Table 1. For each column in this table we will discuss the scenarios in separate subsections. This groups similar data and simplify the overview on how can it combined with probabilistic tractography data. Please note, other data not listed in Table 1 may also be assigned to one of the columns and thus may be threatened similarly.

4.1 Textured Slices

Textures are very often used with slice based visualization techniques. Although other geometry may be used for texturing, for example inflated cortical surface maps textured with T1 data, slices are still very prominent. One reason for that is, that slices deliver detailed context. This does not imply that three-dimensional visualizations are useless, other visualizations may still profit from their strength in communicating the overall structure. However, most multi-modal visualizations using textures on slices, destroy this valuable context by using opaque overlays. Also semi-transparent renderings introduce the problem of color mixture, which may lead to wrong interpretation of both foreground and background color. All techniques for probabilistic tract visualization share the same problem: Rendering the tracts along with other data will occlude it more or less. Fiber-Stipples is a (slice-based) technique for rendering probabilistic tracts, which tries to minimize exactly this occlusion and hence is in that point superior to all other techniques. This is the reason why we suggest Fiber-Stipples for visualizing probabilistic tracts in combination with other textures on slices, except for directional textures as discussed in Sect. 4.1.3.

4.1.1 Monochrome Backgrounds

There are many monochrome images providing important information for neuroscientists. CT, structural MRI, angiography data, diffusion MRI, and diffusion indices are data sources for such images. As mentioned earlier (cf. Sect. 2.4.2), direct visualization of the diffusion gradients is rare and neuroscientists so far focus on providing gray-scale images as anatomical context to their technique, either taking measured data (such as T1 or T2 images) or by projecting their data into a reference frame and using reference data as context [60]. Axis aligned slices of T1 and T2 images are the most commonly used reference data for all techniques displaying probabilistic tracts. This is mainly because probabilistic tractography and T1 and T2 images are MRI based techniques.

A problem when using such underlays, and especially with monochrome underlays, is, that they typically have bright and dark regions. Hence the luminescence of the background conflicts with the luminescence of the foreground color. Ware [61] (pp. 111–112) states that fine details may then be hard to perceive and proposes to increase luminance contrast. This is of special importance to Fiber-Stipples, which encode most information in such fine details of the texture. This can be solved by controlling (i.e., reducing) brightness and contrast of the background image. Of course this will have the drawback of loosing detailed features in the background. Another possibility is to adapt the luminescence of the stipples depending on the local background. This removes the possibly of using quantitative color-coding. Fortunately, Fiber-Stipples provides two channels to communicate the connectivity score: the stipple coloring, typically by controlling opacity, and secondly, the stipple density, which remains unchanged. An example of such a contrast enhanced stipple coloring is given in Fig. 3. However, such an adaptive color coding may lead to misinterpretation of the data and needs further investigation. In addition, fully opaque stipples might distract the user in regions of low connectivity scores. The problem is less severe when the background data provides sufficient contrast along the probabilistic tract. As an example, fractional anisotropy (FA) which is a prominent scalar value derived from diffusion information (diffusion index), is also used as termination criteria for many tractography algorithms or as a indicator on bundle integrity [41]. While major white matter bundles will often follow regions

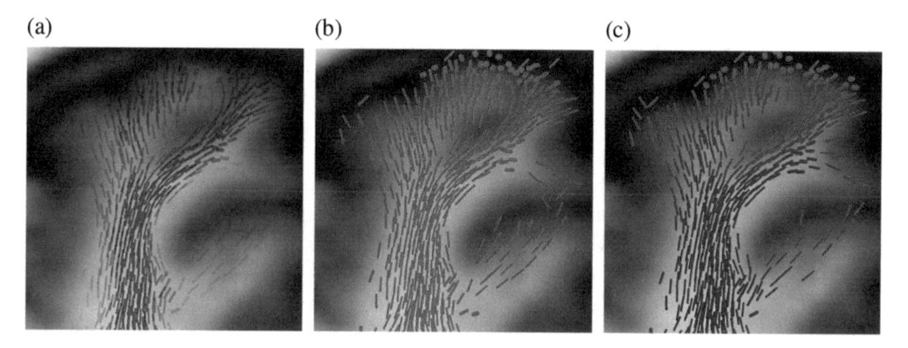

Fig. 3 Fiber-Stipples on T1 images. (**a**) Close up view of original Fiber-Stipples. (**b**) Full opaque stipples with fixed maximal luminescence. (**c**) Adaptive luminescence with full opacity. This way, the stipples have higher contrast

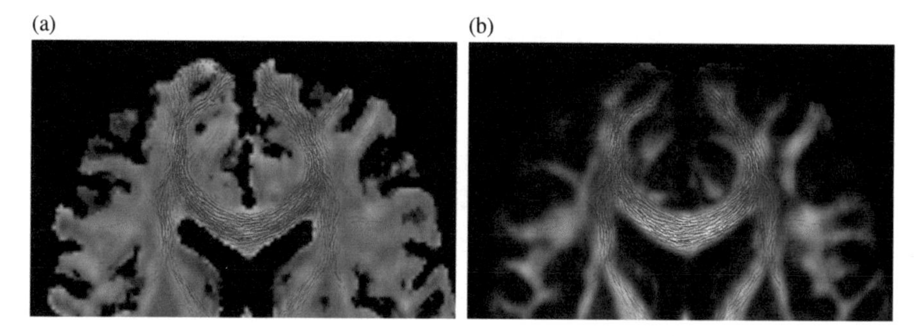

Fig. 4 Fiber-Stipples over diffusion indices: axial diffusivity (**a**) and fractional anisotropy (**b**). Other prominent indices such as radial diffusivity or mean diffusivity look very similar and are not depicted here

of high FA, *most* of the stipples will pass bright regions and the color can be chosen accordingly. When white matter tracts also project into cortex, which is a region of low FA and thus rendered in dark colors, a uniform coloring might be suboptimal then, see Fig. 4 for images dealing with diffusion indices.

Beside, the combination of structural MRI or diffusion indices with probabilistic tracts, we think angiography data (from CT or MR) might be worth to combine with probabilistic tracts. An application might be surgical planning, where the location of important blood vessels and white matter tracts may become critical. On the other side, pure CT data might not provide enough soft tissue contrast for probabilistic tracts. The optimal visualization of probabilistic tracts in such a scenario is not known and needs further investigation. Contextual information from angiography data can be represented in slices as contours or colormaps, as well as three dimensional surfaces or volume renderings.

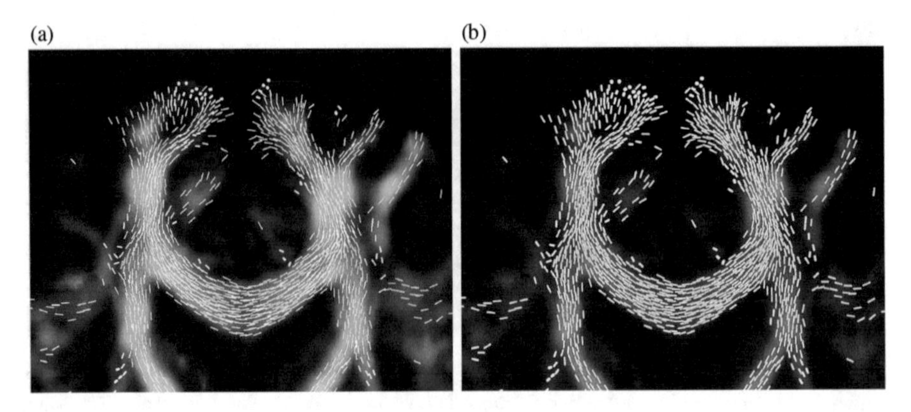

Fig. 5 Fiber-Stipples on colored background representing RGB-coloring scheme of the major diffusion direction filtered by FA. In all images the stipples are fully opaque. (**a**) White stipples on unmodified RGB-colored FA background. (**b**) Outlined white stipples with decreased background luminance from 100 % to 80 %

4.1.2 Colored Background

Beside monochrome textures, neuroscientists are also working with colored textures for superimposed data. As an example, fMRI data is colored with the heat-map as overlay for T1 data as context. Another example is the RGB-coloring of diffusion orientations (cf. Sect. 2.4.2) which may be used directly as texture, but also with additional filters such as the fractional anisotropy (FA), see Fig. 5.

Some colored textures may use the whole color space (e.g. rainbow colormap) and some are limited to a fixed number of colors or transitions between only two or three colors (e.g. heat-map). As visualization techniques for probabilistic tracts usually use some color, a combination with colored background generally depends on whether there are unused colors left which also provide discernible contrast. Techniques not using color but texture are LIC or TDI, which are discussed separately in Sect. 4.1.3. Due to the same motivation as given in Sect. 4.1 regarding the occlusion, Fiber-Stipples are suggested for combinations of probabilistic tracts with colored backgrounds as well. However, the encoding of connectivity score in stipple opacity may be of no use, as the color blending might change colors and hence lead to misinterpretation.

Arbitrary Colored Backgrounds

When a texture is arbitrarily colored it may use almost the whole colorspace and any color used for rendering the probabilistic tract might fail. Prominent examples are the rainbow colormap [8] or directional color codings [15, 48] (filtered or unfiltered). With disabling the opacity, black or white may sound as a good choice then for coloring the stipples, see Fig. 5. While the luminescence of the background may

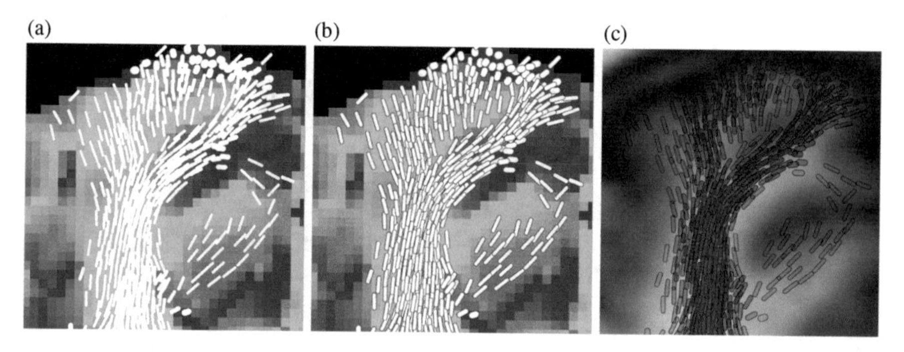

Fig. 6 White Fiber Stipples over arbitrary colored background. To ease the perception of directionality of Fiber-Stipples, a small outline around each stipple is used. In (**a**) no such outline is present on fully opaque white Fiber-Stipples, while in (**b**) the outline substantially improves the shape. This outlining technique can also be applied to Fiber-Stipples with variable opacity encoding the connectivity score (**c**)

still influence the visualization, we further suggest to limit or increase (depending on data) the background luminescence as well, see Fig. 5b. To further increase contrast we propose to outline (cf. also Ware [61] page 123) the stipples as depicted in Figs. 5b and 6. This way the combination of probabilistic tract data with underlying colored texture may still be perceived well.

Colormaps Defined by Few Colors

When working with colormaps defined by only two or three colors and their continuous transitions, a most complementary color may be practical for coloring the probabilistic tract. For example we believe, when working with the heat-map, used e.g. for indicating BOLD activity in fMRI, blue might be a good choice as it provides good chromatic contrast, see Fig. 7. The combination of fMRI data with probabilistic tracts is a very interesting one, as fMRI data communicates functional connectivity and the probabilistic tract communicates structural connectivity. This contributes to the visualization of effective connectivity [19, 57]. Other combinations with PET data are possible as well [44], but, from a visualization point of view, the combinations are then very much the same as for fMRI data, due to the same type of data.

Another possibility to combine visualizations of probabilistic tracts with colormap overlays using only few colors, is to apply the colormap directly on Fiber-Stipples, see Fig. 8 for an example. Although it seems a natural solution for combining Fiber-Stipples with such colormaps, important topological information may not be rendered. A tract not crossing a center of neural activation may then need additional topological information for better interpretation. Another solution would be to combine two three-dimensional techniques, e.g. use nested isosurfaces [7] to combine the modalities. One surface highlights the probabilistic tract while another

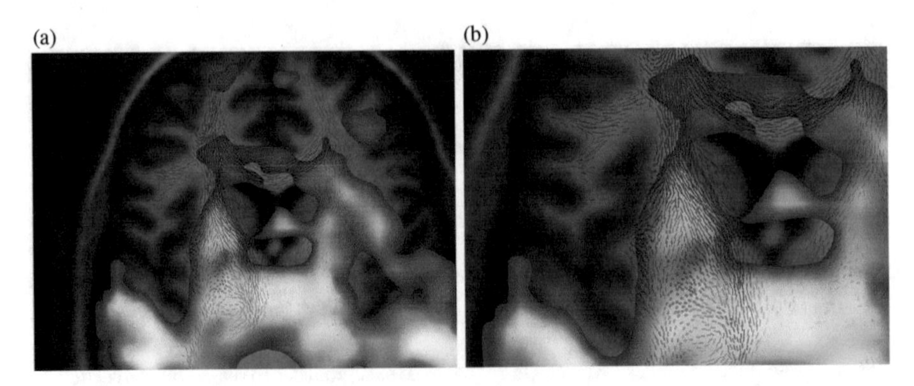

Fig. 7 FMRI Activation Maps are often visualized as colormap overlay onto T1 images as in (**a**), and (**b**) shows a close up view. For such activation maps the so called heat-map is very common

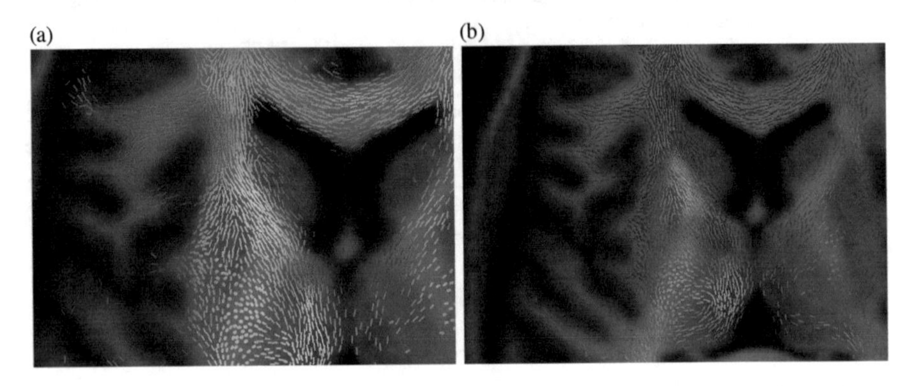

Fig. 8 Fiber-Stipples color coded with the fMRI BOLD activity signal. (**a**)—*bright yellow* stipples indicate not high connectivity score, as this is encoded in stipple density, but a high BOLD signal activity. In (**b**), the BOLD signal is thresholded, so *blue* colored stipples represent stipples below this BOLD threshold

may highlight the fMRI activity region. However, when anatomical context is very important we suggest Fiber-Stipples again for visualization.

Fixed Colors

Finally, textures with few fixed colors are very common in neuroscience to label segmented structures. For example several white matter atlases label segmented structures with different colors, see Fig. 9. As same as in the previous section, it would be practical to have a most discriminative color, used for Fiber-Stipples representing the probabilistic tract. Bauer et al. [5] revealed, that the most discriminative colors should lie outside of the convex hull defined by the given colors in CIE color space [64]. Unfortunately, the more colors are used, the more difficult it is, to find such a color. Furthermore, it is also important to adapt luminescence contrast as

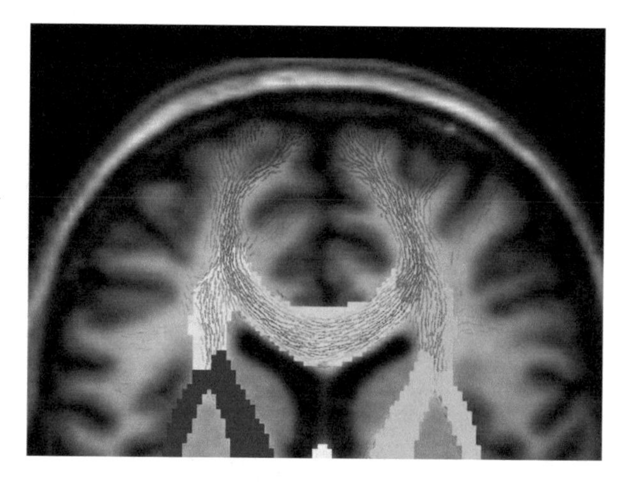

Fig. 9 Segmentation from the JHU white matter tractography atlas [42] labeling certain white matter regions. Here parts of the *forceps minor*, *internal capsule*, *external capsule* and parts of the *corpus callosum* in the frontal lobe are labeled with different colors

well for better color discrimination, as the colors used for labeling structures might have very high or low luminance as well. Although it sound possible to change color labels of white matter structures, so that the set of fixed colors allows a better choice for the stipple color, we do not encourage such a change, as scientists may be used to the given colors for specific atlases. Last but not least, the visualization technique depicting the probabilistic tract may also employ the given set of colors as labels for the structures (the very same way as outlined in Paragraph "Colormaps Defined by Few Colors") which imposes the same problem of missing topological information. This would be important to answer specific questions such as, to determine if a given tract just barley hits the structure or if it is projecting right into its center.

4.1.3 Directional Textures

A special case of textures are textures indicating directional information using fine patterns, with line integral convolution (LIC) being the best known representative of this group, cf. Fig. 1b. Beside LIC, tract density imaging (TDI) [12] and other fabric-like visualizations have been proposed to display tensor information.

In order to combine such textures with probabilistic tracts we think semi-transparent colormaps are most useful here, see Fig. 1c. The directional textures are rather versatile when it comes to applying the pattern to different planes or surfaces. Their main disadvantage is the restless pattern that makes it harder to see small structures in front of it (cf. [61]). These disadvantages are depicted in Fig. 10c.

However, these textures transport some directionality (from vector field) and eventually also some density (cf. TDI). Both properties may be depicted with Fiber-Stipples again, leaving room for anatomical context which would be occluded by such textures otherwise, see Fig. 10d. There, two crossing tracts are depicted by the major diffusion orientation. The FSL software package offers a technique to estimate a second diffusion direction out of DTI data by using Bayesian estimators

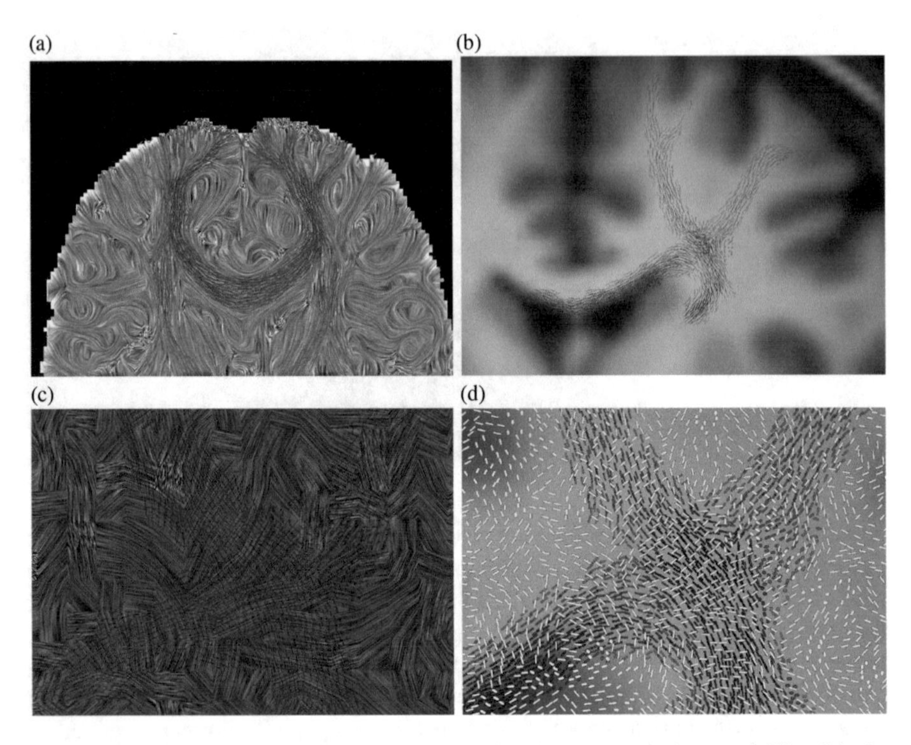

Fig. 10 Combinations of Fiber-Stipples with directional textures. (**a**) shows red Fiber-Stipples over LIC background computed from DTI's major diffusion orientation. (**b**) shows two different tracts which cross. The *red* tract is part of the *corpus callosum* while the *blue* is part of the *cortico spinal tract*. T1 is used as background. (**c**) Close up view of (**b**) but with LIC of the second diffusion orientation (computed by FSL BedpostX) as background instead of T1. (**d**) Same as (**c**) but LIC replaced with (*yellow*) Fiber-Stipples

implemented in BedpostX [33], which helps to better understand crossings. Such a second direction is then depicted as LIC in Fig. 10c and with yellow Fiber-Stipples instead in Fig. 10d. Please note that, other visualization techniques, such as the vector-plot, cannot communicate density.

4.2 Glyphs

Glyphs have been used in many settings before, and can be combined with surface and volume representations [40]. Special care has to be taken when glyphs are not sampled along a plane or surface, but express volumetric information. Then, occlusion leading to a perceived change in glyph density may distract the viewer.

So, combining glyphs with probabilistic diffusion tractography would be best on a plane or surface. When placing glyphs on an arbitrary surface, e.g. on cortex [9] or on an isosurface of a probabilistic tract, glyph placement should be performed as proposed by Kratz et al. [37] to get optimal results. In short, they employ an anisotropic Voronoi cell rendering to obtain best packaging on two-manifold domains.

When combining glyphs with slice based techniques, they will typically occlude some visualization parts. As there are also slice based techniques using glyphs for the visualization of the probabilistic tract (Fiber-Stipples), mixing different glyph types might be very confusing, especially when they need to maintain a density property. On the other hand, glyphs of the same type *might* be combined. For example, Fiber-Stipples can also be regarded as a specific glyph-based technique to render probabilistic tract information. As the technique resembles manual drawings of stipples [54], they may be safety combined (up to a certain extend) with other Fiber-Stipples, as long as they have discernible colors.

Last but not least, integrating the tract into the given glyph might also be an option. To summarize this, it is hard to tell from which combination such a visualization would benefit and we vote for keeping it simple and use either Fiber-Stipples or a plain colormap to communicate the tract as an underlay for the glyphs.

4.3 Line Data

Neuroscientists use various line data in their visualizations. Examples are, polylines or tubes representing white matter fiber tracts, outlines of structures, splines as connections between regions and so on.

4.3.1 Contours

Contours outline specific parts of T1 data, such as cortex parcellations, white matter segmentations, or lesions. Often only one or two contour lines are used, e.g. Fiber-Stipples uses one contour line for outlining the gray matter and another for outlining the gray matter–white matter interface, see Fig. 11a.

As contours are often used on surfaces we focus on combinations of slice based techniques for probabilistic tracts with contours. Such combinations might be useful to see if a tract is affected by a lesion, or if it projects into outlined regions. The combination of colormaps as overlays and contours would clearly lead to desired results, although the overlays occlude anatomical context. On the other hand, contours might be hard to perceive along with directional patterns such as LIC or TDI. Likewise, Fiber-Stipples with boundary curves have problems (to many contours with different meaning), see Fig. 11b. If we could find a way to depict contours, probabilistic tract and context at once, this would then be the optimal combination for such data. As an example we look back at the combination of Fiber-Stipples with fMRI data, where stipples are colored with the heat-map of BOLD activity data (cf. Paragraph "Colormaps Defined by Few Colors" and Fig. 8). There was a lack of topological information, which could be closed by using additional

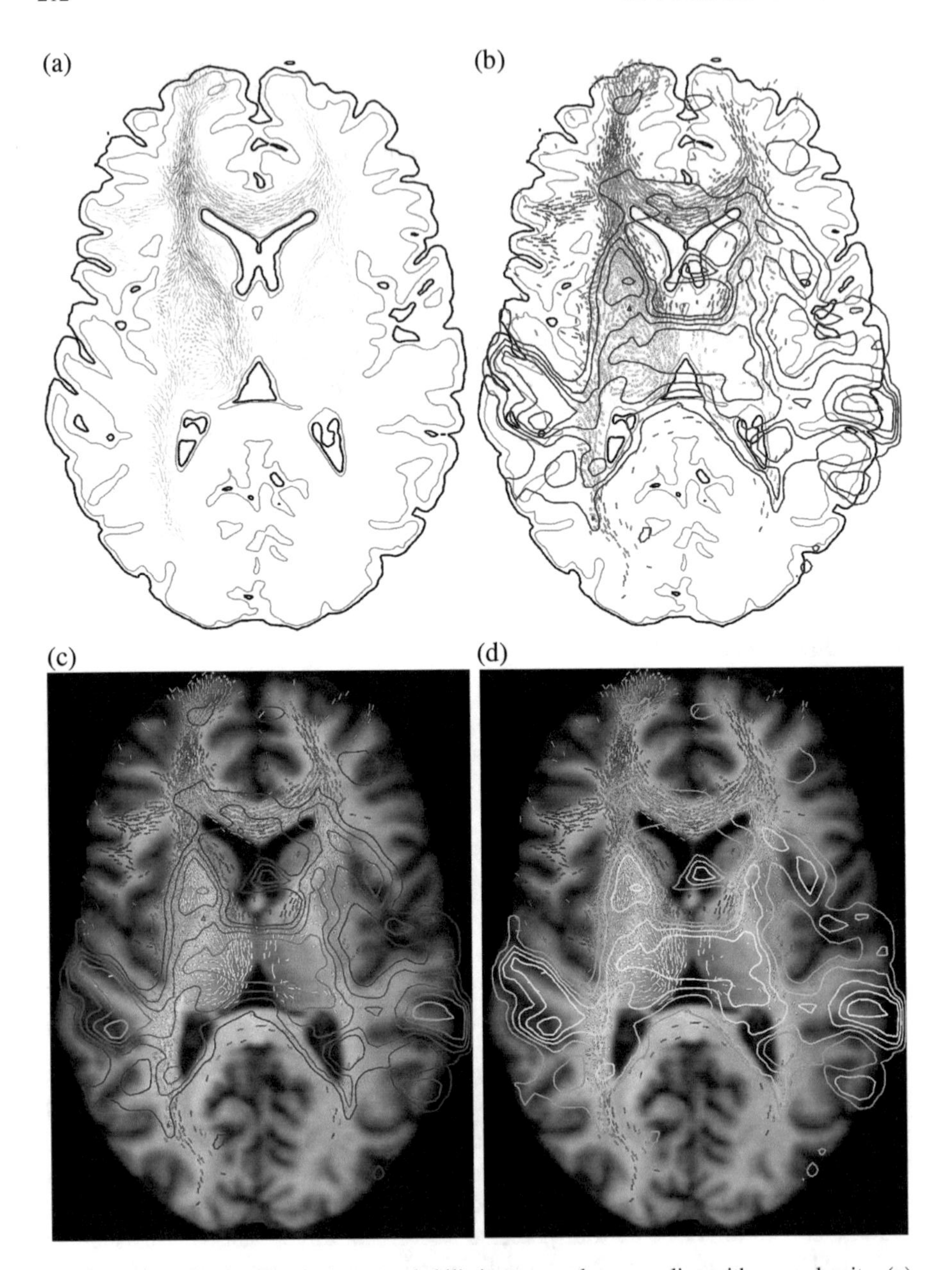

Fig. 11 All images render the same probabilistic tract at the same slice with same density. (**a**) shows the original Fiber-Stipples technique with enabled opacity encoding which is disabled in (**b**)–(**d**) (**b**) combines the fMRI activation from Fig. 7 with Fiber-Stipples and renders contour lines (90 %, 80 %, 70 %, 60 %, and 50 %) of the activity field in *blue*. (**c**) same as (**b**) but with T1 background and contours colormapped from light blue (90 %) to *dark blue* (50 %). (**d**) same as (**c**) but using the heat-map for contours as well

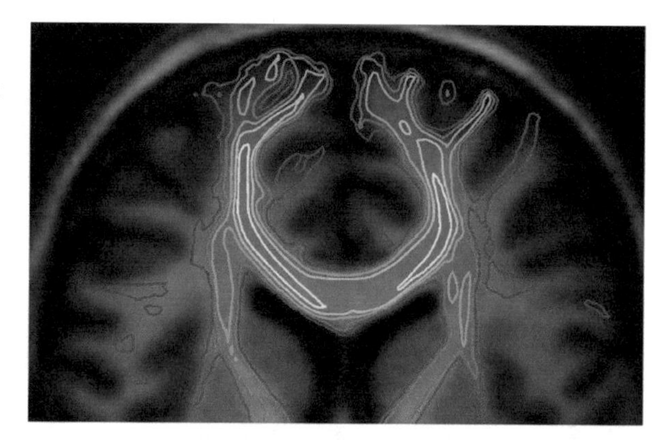

Fig. 12 Probabilistic tract rendered with contour lines, 100 %, 85 %, 70 %, etc. using the heat-map

information, such as contours. This leads to a visualization of Fiber-Stipples without boundary curves over T1 background, where the Fiber-Stipples are colored with fMRI texture (heat-map) and additional contours, for the fMRI data, provide the missing topological information. We must admit, it is not easy to select the right colors to avoid the whole scene to become visually cluttered. Examples of such visualizations are given in Fig. 11. Please note that this introduces a new and alternative visualization for probabilistic tracts as well, see Fig. 12. In summary, we would here for Fiber-Stipples again, as it preserves valuable context on slices while still being able to work with contour data arising from other modalities.

4.3.2 3D Polylines

In neuroscience, three-dimensional polylines are often used for representing neural tract data, cf. deterministic tractography [32, 43]. As such polylines visualize a three-dimensional course through the brain, we believe it best to use three-dimensional visualizations of probabilistic tracts to combine them. Solutions are semi-transparent isosurfaces Fig. 13a, or illustrative techniques for confidence intervals [10].

Other applications of such polylines or splines are representations of functional connectivity [9]. However, such a combination should be considered carefully as though the line's endpoints reside in anatomical space, their course does not have an anatomical foundation.

4.4 Surfaces

Neuroscientists work a lot with surfaces from brain structures, blood vessels but most importantly, cortex estimations. Due to the same fact, that those surfaces typically represent three-dimensional objects we recommend to use three-dimensional tract representations as well. As an example see Fig. 13b with a semi-transparent iso-

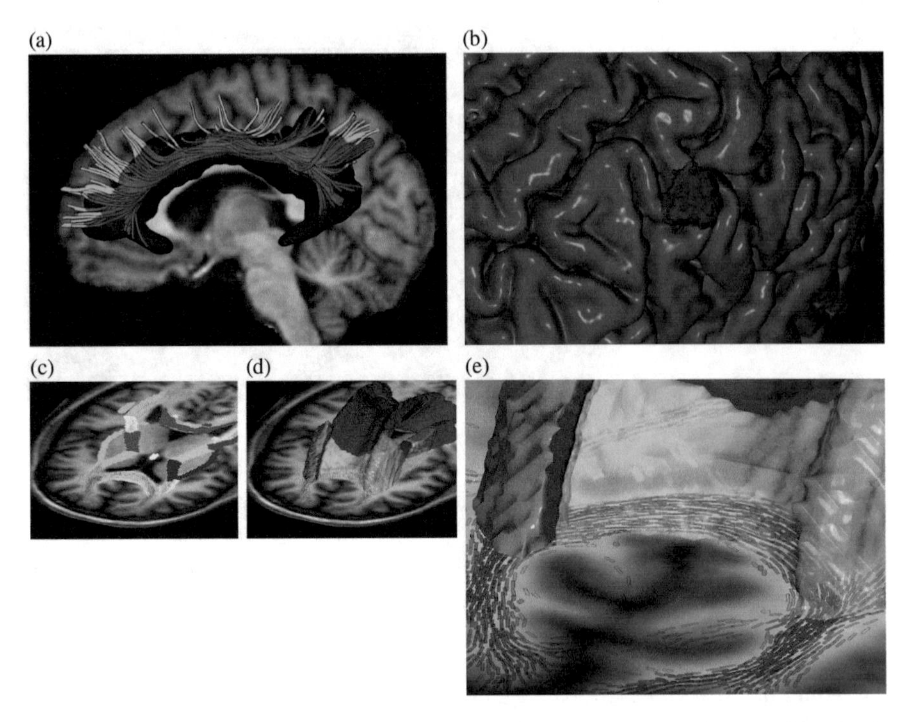

Fig. 13 Combinations of probabilistic tractography with three-dimensional visualizations. (**a**) 3D polylines (rendered as tubes) from deterministic tractography representing *left cingulum* (CNG) bundle, partially enclosed by an isosurface. (**b**) Cortex approximation surface (*semi-transparent gray*) and probabilistic tract (*red*) with seed region in gray matter–white matter interface of the precentral gyrus. (**c**) The same white matter atlas [42] as used for segmentation as in Fig. 9 but using a different viewing perspective, (**d**) with the segmentation represented by surfaces, some of them are opaque and some are semi-transparent, and (**e**) a close up view of (**d**)

surface for the cortex and an opaque isosurface for the tract. Although, it is possible to combine slice based visualizations, such as Fiber-Stipples (Fig. 13c–e), the interpretation of the data will remain very difficult. For example the stipple orientation and density might not be perceived correctly from different viewing perspectives. A better slice based communication of the tract with the surface is to use cutting planes with colormaps, see Gorbach [28] (Fig. 1). Please note, in case of using those cutting planes the three-dimensional course might be lost. As an example, the *fornix* tract is mostly diagonal to all three (sagittal, axial and coronal) cutting planes.

5 Conclusion

In this work we gave an overview of current visualization techniques for probabilistic diffusion tractography data and its combination with other image modalities from the neuroscience domain. We summarized common data types and visualization techniques from the neuroscience community to systematically investigate multi-modal visualization with probabilistic tractography data.

In Fig. 2, we pointed out why anatomical context is of utmost importance when analysing white matter tracts. It turned out, that the Fiber-Stipples technique is a very flexible visualization technique for probabilistic tracts as it supports various superimposed data while still maintaining precise and detailed context. The resulting images typically incorporate a lot of information, thus, often the simple sparse and effective visualizations will stand out from their competitors. We showed that a combination of Fiber-Stipples with contour lines may communicate complex multi-modal visualizations. As an example we show the combination with fMRI activation regions. Additionally, we showed, that contour lines serve as simplistic visualizations of probabilistic tracts as well. Furthermore, we observed that color plays an important role when combining modalities and gave certain guidelines regarding contrast and color as well as improvements such as outlining to enhance the resulting visualization.

However, slice-based visualization techniques have drawbacks when three-dimensional objects, such as cortex surface approximations, demand for a combined visualization. In such cases we believe that three-dimensional techniques, e.g. isosurface renderings, may be superior to slice-based visualizations.

6 Future Work

The presented techniques should be considered a guideline for the creation of meaningful and useful visualizations focusing on probabilistic tractography data. With improved or new acquisition techniques arising daily, this list should not be considered complete but should be enhanced in the future. Whereas we focus on probabilistic data, an overview including other specific tasks would be beneficial. Although we reasoned design decisions wherever possible, we will evaluate them in a user study.

Acknowledgements The authors thank the OpenWalnut project [18] for providing the software framework supporting neuroscience visualization and for distributing the Fiber Stipple source code. The software platform was used for conducting this study. We thank C. Heine and A. Wiebel for the fruitful discussions and for commenting on drafts of this article. Last but not least we want to thank the reviewers which gave us very helpful comments and advises.

References

1. Assaf, Y., et al.: Diffusion tensor imaging (DTI)-based white matter mapping in brain research: a review. J. Mol. Neurosci. 34(1), 51–61 (2008)
2. Barnes, G., et al.: Magnetoencephalogram. Scholarpedia **5**(7), 3172 (2010)
3. Barr, A.H.: Superquadrics and angle-preserving transformations. IEEE CGA **1**(1), 11–23 (1981)

4. Basser, P.J., et al.: MR diffusion tensor spectroscopy and imaging. Biophys. J. **66**, 259–267 (1994)
5. Bauer, B., et al. Distractor heterogeneity versus linear separability in colour visual search. Perception (1996)
6. Behrens, T., et al.: Probabilistic diffusion tractography with multiple fibre orientations: what can we gain? Neuroimage **34**(1), 144–155 (2007)
7. Berres, A., et al.: Tractography in context: multimodal visualization of probabilistic tractograms in anatomical context. In: VCBM, pp. 9–16 (2012)
8. Borland, D., et al.: Rainbow color map (still) considered harmful. IEEE CGA **27**(2), 14–17 (2007)
9. Bottger, J., et al.: Three-dimensional mean-shift edge bundling for the visualization of functional connectivity in the brain. IEEE TVCG **20**(2), 471–480 (2013)
10. Brecheisen, R., et al.: Illustrative uncertainty visualization of DTI fiber pathways. Vis. Comput. **29**(4), 297–309 (2013)
11. Cabral, B., et al.: Imaging vector fields using line integral convolution. In: Proceedings of the 20th Annual Conference on Computer Graphics and Interactive Techniques. SIGGRAPH '93, pp. 263–270 (1993)
12. Calamante, F., et al.: Track-density imaging (TDI): super-resolution white matter imaging using whole-brain track-density mapping. NeuroImage **53**(4), 1233–1243 (2010)
13. Chen, J.J., et al.: Cerebral blood flow measurement using fMRI and PET: a cross-validation study. Int. J. Biomed. Imag. (2008)
14. Dawson, J., et al.: Magnetic resonance imaging. Scholarpedia **3**(7), 3381 (2008)
15. Demiralp, C., et al.: Coloring 3D line fields using Boy's real projective plane immersion. IEEE Transactions on Visualization and Computer Graphics **15**(6), 1457–1464 (2009)
16. Descoteaux, M., et al.: Deterministic and probabilistic tractography based on complex fibre orientation distributions. IEEE Trans. Med. Imag. **28**(2), 269–286 (2009)
17. Douaud, G., et al.: DTI measures in crossing-fibre areas: increased diffusion anisotropy reveals early white matter alteration in MCI and mild Alzheimer's disease. Neuroimage **55**(3), 880–890 (2011)
18. Eichelbaum, S., et al.: OpenWalnut. http://www.openwalnut.org
19. Eichelbaum, S., et al.: Visualization of effective connectivity of the brain. In: VMV, pp. 155–162 (2010)
20. Eichelbaum, S., et al.: LineAO – improved three-dimensional line rendering. IEEE TVCG **19**(3), 433–445 (2013)
21. Fillard, P., et al.: Quantitative evaluation of 10 tractography algorithms on a realistic diffusion MR phantom. NeuroImage **56**(1), 220–234 (2011)
22. Fischl, B.: FreeSurfer. NeuroImage **62**(2), 774–781 (2012)
23. Ghosh, A.: High Order Models in Diffusion MRI and Applications. Ph.D. thesis. INRIA Sophia-Antipolis Méditerranée (2011)
24. Golay, X., et al.: High-resolution isotropic 3D diffusion tensor imaging of the human brain. MRM **47**(5), 837–843 (2002)
25. Goldau, M., et al.: Visualizing DTI parameters on boundary surfaces of white matter fiber bundles. In: IASTED, pp. 53–61 (2011)
26. Goldau, M., et al.: Fiber stippling: an illustrative rendering for probabilistic diffusion tractography. In: IEEE BioVis, pp. 23–30 (2011)
27. Gorbach, N.S., et al.: Hierarchical information-based clustering for connectivity-based cortex parcellation. FNINF **5**(18) (2011)
28. Gorbach, N.S., et al.: Information-theoretic connectivity-based cortex parcellation. In: MLINI, pp. 186–193 (2012)
29. Hlawitschka, M., et al.: In: Linsen, L., et al. (eds.) Hierarchical Poisson-Disk Sampling for Fiber Stipples. Visualization in Medicine and Life Sciences, vol. II. Springer, pp. 19–23 (2013)
30. Hlawitschka, M., et al.: Interactive glyph placement for tensor fields. In: Advances in Visual Computing: Third International Symposium, ISVC, vol. LNCS 4841 and LNCS 4842, pp. 331–340 (2007)

31. Jainek, W.M., et al.: Illustrative hybrid visualization and exploration of anatomical and functional brain data. CGF **27**(3), 855–862 (2008)
32. Jbabdi, S., et al.: Tractography: where do we go from here? Brain Connect. **1**(3), 169–183 (2011)
33. Jenkinson, M., et al.: FSL. NeuroImage **62**(2), 782–790 (2012)
34. Jianu, R., et al.: Exploring 3D DTI fiber tracts with linked 2D representations. IEEE TVCG **15**(6), 1449–1456 (2009)
35. Johansen-Berg, H.: Human connectomics—what will the future demand? NeuroImage **80**, 541–544 (2013)
36. Kalender, W.A.: X-ray computed tomography. Phys. Med. Biol. **51**(13), R29 (2006)
37. Kratz, A., et al.: Anisotropic sampling of planar and two-manifold domains for texture generation and glyph distribution. TVCG **18**, 1563 (2013)
38. Le Bihan, D., et al.: Diffusion MRI at 25: exploring brain tissue structure and function. NeuroImage **61**(2) , 324–341 (2012)
39. Margulies, D.S., et al.: Visualizing the human connectome. NeuroImage **80**, 445–461 (2013)
40. Meyer-Spradow, J., et al.: Voreen: a rapid-prototyping environment for ray-casting-based volume visualizations.IEEE CGA **29**(6), 6 (2009)
41. Mielke, M.M., et al.: Fornix integrity and hippocampal volume predict memory decline and progression to Alzheimer's disease. Alzheimer's Dementia **8**(2), 105–113 (2012)
42. Mori, S., et al.: MRI atlas of human white matter, Elsevier Science (2005)
43. Mori, S., et al.: Three-dimensional tracking of axonal projections in the brain by magnetic resonance imaging. Ann. Neurol. **45**(2), 265–269 (1999)
44. Neuner, I., et al.: Multimodal imaging utilising integrated MR-PET for human brain tumour assessment. Eur. Radiol. **22**(12), 2568–2580 (2012)
45. Nordberg, A.: PET imaging of amyloid in Alzheimer's disease. Lancet Neurol. **3**(9), 519–527 (2004)
46. Nowinski, W., et al.: Three-dimensional interactive and stereotactic human brain atlas of white matter tracts. Neuroinformatics **10**(1), 33–55 (2011)
47. Nunez, P., et al.: Electroencephalogram.Scholarpedia **2**(2), 1348 (2007)
48. Pajevic, S., et al.: Color schemes to represent the orientation of anisotropic tissues from diffusion tensor data: application to white matter fiber tract mapping in the human brain. MRM **42**, 526–540 (1999)
49. Parker, G.J., et al.: A framework for a streamline-based probabilistic index of connectivity (PICo) using a structural interpretation of MRI diffusion measurements. JMRI **18**(2), 242–254 (2003)
50. Pelzer, E.A., et al.: Cerebellar networks with basal ganglia: feasibility for tracking cerebello-pallidal and subthalamo-cerebellar projections in the human brain. Eur. J. Neurosci. **38**(8), 3106–3114 (2013)
51. Pfister, H., et al.: Visualization in connectomics. In: CoRR abs/1206.1428 (2012)
52. Rogers, B.P., et al.: Assessing functional connectivity in the human brain by fMRI. In: Magn. Reson. Imag. **25**(10), 1347–1357 (2007)
53. Rose, S.E., et al.: Loss of connectivity in Alzheimer's disease: an evaluation of white matter tract integrity with colour coded MR diffusion tensor imaging. JNNP **69**(4), 528–530 (2000)
54. Schmahmann, J.D., et al.: Fiber Pathways of the Brain. Oxford University Press (2006)
55. Shenton, M., et al.: A review of magnetic resonance imaging and diffusion tensor imaging findings in mild traumatic brain injury. Brain Imag. Behav. **6**(2), 137–192 (2012)
56. Small, G.W., et al.: PET scanning of brain tau in retired national football league players: preliminary findings. Am. J. Geriatr. Psychiatry **21**(2), 138–144 (2013)
57. Sporns, O.: Brain connectivity. Scholarpedia **2**(10), 4695 (2007)
58. Sporns, O., et al.: The human connectome: a structural description of the human brain. PLoS Comput. Biol. **1**(4), e42 (2005)
59. Vaphiades, M.S., et al.: Management of intracranial aneurysm causing a third cranial nerve palsy: MRA, CTA or DSA? Semin. Ophthalmol. **23**(3), 143–150 (2008)

60. von Kapri, A., et al.: Evaluating a visualization of uncertainty in probabilistic tractography. Proc. SPIE **7625**(1), 762534 (2010)
61. Ware, C.: Information visualization: perception for design. Morgan Kaufman (2012)
62. Wassermann, D., et al.: Unsupervised white matter fiber clustering and tract probability map generation: applications of a Gaussian process framework for white matter fibers. NeuroImage **51**(1), 228–241 (2010)
63. Wedeen, V.J., et al.: Diffusion spectrum magnetic resonance imaging (DSI) tractography of crossing fibers. Neuroimage **41**(4), 1267–1277 (2008)
64. Wyszecki, G., et al.: Color science. John Wiley and Sons (1982)
65. Yotter, R.A., et al.: Local cortical surface complexity maps from spherical harmonic reconstructions. NeuroImage **56**(3), 961–973 (2011)
66. Zhang, S., et al.: Identifying white-matter fiber bundles in DTI data using an automated proximity-based fiber-clustering method. IEEE TVCG **14**, 1044–1053 (2008)
67. Zhan, L., et al.: How many gradients are sufficient in high-angular resolution diffusion imaging (HARDI). In: 13th OHBM (2008)

Part IV
Cohort Studies and Time-Varying Phenomena

Visual Analytics of Image-Centric Cohort Studies in Epidemiology

Bernhard Preim, Paul Klemm, Helwig Hauser, Katrin Hegenscheid, Steffen Oeltze, Klaus Toennies, and Henry Völzke

Abstract Epidemiology characterizes the influence of causes to disease and health conditions of defined populations. Cohort studies are population-based studies involving usually large numbers of randomly selected individuals and comprising numerous attributes, ranging from self-reported interview data to results from various medical examinations, e.g., blood and urine samples. Since recently, medical imaging has been used as an additional instrument to assess risk factors and potential prognostic information. In this chapter, we discuss such studies and how the evaluation may benefit from visual analytics. Cluster analysis to define groups, reliable image analysis of organs in medical imaging data and shape space exploration to characterize anatomical shapes are among the visual analytics tools that may enable epidemiologists to fully exploit the potential of their huge and complex data. To gain acceptance, visual analytics tools need to complement more classical epidemiologic tools, primarily hypothesis-driven statistical analysis.

1 Introduction

Epidemiology is a scientific discipline that provides reliable knowledge for clinical medicine focusing on prevention, diagnosis and treatment of diseases [14]. Research in epidemiology aims at characterizing *risk factors* for the outbreak of diseases and at evaluating the efficiency of certain treatment strategies, e.g., to compare a new treatment with an established gold standard. This research is strongly hypothesis-driven and statistical analysis is the major tool for epidemiologists so

B. Preim • P. Klemm (✉) • S. Oeltze • K. Toennies
Otto-von-Guericke University Magdeburg, Universitätsplatz 2, 39106 Magdeburg, Germany
e-mail: bernhard.preim@ovgu.de; paul.klemm@ovgu.de; steffen.oeltze@ovgu.de; klaus@isg.cs.uni-magdeburg.de

H. Hauser
University of Bergen, Høyteknologisenteret (HiB), Thormøhlensgate 55, 5008 Bergen, Norway
e-mail: Helwig.Hauser@UiB.no

K. Hegenscheid • H. Völzke
University of Greifswald, Institut für Community Medicine, Walter Rathenau Str. 48, 17475 Greifswald
e-mail: hegenscheid@uni-greifswald.de; voelzke@uni-greifswald.de

L. Linsen et al. (eds.), *Visualization in Medicine and Life Sciences III*, Mathematics and Visualization, DOI 10.1007/978-3-319-24523-2_10

far. Correlations between genetic factors, environmental factors, life style-related parameters, age and diseases are analyzed. The data are acquired by a mixture of interviews (self-reported data, e.g., about nutrition and previous infections) and clinical examinations, such as measurement of blood pressure. Statistical correlations, even if they are strong, may be misleading because they do not represent *causal relations*. As an example, the slightly reduced risk of heart infarct and cardiac mortality for elderly people reporting to drink one glass of wine every evening (compared to people drinking no alcohol at all) may be due to the involved low level of alcohol but may also be a consequence of a very regular and stress-free lifestyle [14]. When something happened *before* an event, it is an indicator for a causal relationship. However, care is necessary, since many things happen in the life of an individual before, e.g., a heart attack, but do *not* cause it.

Thus, statistical correlations are the starting point for investigating *why* certain factors increase the risk of getting diseases. Epidemiology is not a purely academic endeavor but has huge consequences for establishing and evaluating preventive measures even outside of medicine. The protection of people from passive smoking, recommendations for various vaccinations and the introduction of early cancer detection strategies, e.g., mammography screening, are all based on large-scale epidemiological studies. Also the official guidelines for the treatment of widespread diseases, such as diabetes, are based on *evidence* from epidemiological studies [14]. While this all may sound obvious, it is a rather recent development. *Evidence-based* medicine often still has to "fight" against recommendations of a few opinion leaders arguing based on their personal experience only.

The analysis techniques used so far are limited to investigating hypotheses based on known or suspected relations, e.g. hypotheses related to observations or previous publications. The available tools support the analysis of a few dimensions, but not of the hundreds of attributes acquired per individual in a cohort study. Both typical visualization techniques as well as analysis techniques, e.g., support vector machines, do not scale well for hundreds of attributes [41]. While we are not able to describe *solutions* for these challenging problems, we give a survey on recent approaches aiming also at *hypothesis generation*.

Organization This chapter is organized as follows. In Sect. 2 we describe important concepts and terms of epidemiology including observations from epidemiologic workflows. This discussion is restricted to those terms that are crucial for communicating with epidemiologists, understanding requirements and for designing solutions that fit in their process. In Sect. 3, we discuss how (general) information visualization and data analysis techniques may be used for epidemiologic data. Section 4 describes the analysis of image data from cohort studies and how this analysis is combined with the exploration of non-image attribute data. This section represents the core of the chapter and employs a case study where MRI data of the lumbar spine are analyzed along with attributes characterizing life-style, working habits, and back pain history.

2 Background in Epidemiology

Population-Based Studies Epidemiological studies are based on a *sample* of the population. The reliability of the results obviously depends on the size of that sample but also strongly on the selection criteria. Often, data from patients treated in one hospital are analyzed. While this may be a large number of patients, the selection may be heavily biased, e.g., since the hospital is highly specialized and diseases are often more severe or in a later stage compared to the general population.

Population-based studies, where representative portions of a population (without known diseases) are examined, have the potential to yield highly reliable results. The source population may be from a city, a region or a country. Individuals are randomly selected, e.g., approaching data bases of population registries. The higher the percentage of people who accept the invitation and actually take part in the study, the more reliable the results are.

In this chapter, we focus on longitudinal population-based studies. The sheer amount and diversity in terms of type of data makes it difficult to fully identify and analyze interesting relations. We will show that information visualization and visual analytics techniques may provide substantial support that complements the statistical tools with their rather simple statistical graphics. Most epidemiological studies were restricted to nominal (often called categorical) and scalar data, e.g., related to alcohol consumption, and body mass index as one measure of obesity.

Image-Centric Epidemiological Studies More recently, for example, in the Rotterdam study [22], also non-invasive imaging data, primarily ultrasound and MRI data, are employed. Petersen and colleagues [32] report on six studies involving cardiac MRI from at least 1000 individuals in population-based studies. These high-dimensional data enable to answer analysis questions, e.g., how does the shape of the spine changes as a consequence of age, life style and diseases? We focus on such *image-centric* epidemiological studies.

Epidemiology and Public Health There are different branches of epidemiology. One branch deals with predictions to inform public health activities. These include measures in case of an epidemic—an acute *public health* problem, mostly related to infectious diseases. The recent article "computational epidemiology" [29] was focussed on this branch of epidemiology. Another branch of epidemiology aims at long-term studies and at findings primarily essential for prevention. Image-centric cohort studies, the focus of this article, belong to this second branch. The target user group consists of epidemiologists who can be expected to have a high level of expertise in statistics. Thus, their findings involve statistical significance, confidence intervals and other measures of statistical power.

Healthy Aging and Pathologic Changes An essential problem in the daily clinical routine is the discrimination between healthy age-related modifications (that may not be reversed by treatment) and early stage diseases (that may benefit from immediate treatment). As a consequence, elderly people are often not adequately treated. As a general goal for epidemiological studies, better and more reliable markers for early stage diseases are searched for. The cardiovascular branch of the Rotterdam study, for example, aims at an understanding of atherosclerosis, coronary heart disease and "cardiovascular conditions at older age" [22].

Modern Epidemiology Epidemiology faces new challenges due to the rapid progress, e.g., in genetics and sequencing technology as well as medical imaging. Acquisition of health data thus becomes cheaper and more precise. In cohort studies, as much potentially relevant data as possible are acquired as a basis for an as broad as possible spectrum of analysis questions. This includes blood, urine and tissue samples, information about environmental conditions and the social milieu.

Visual Analytics for Modern Epidemiology In the past, epidemiology primarily dealt with hypotheses aiming to prove them, e.g., the efficiency of early cancer detection programs in terms of mortality and long term survival [14]. Since recently, more and more data mining is performed to identify correlations. Results of such analyses, however, need to be very carefully interpreted. If thousands of potential correlations are analyzed automatically, just by chance some of them will reach a high level of statistical significance.

An essential support for epidemiology research is to define relevant subgroups. To perform separate analyses for women and men as well as for different age groups is a common practice in epidemiology. However, relevant subgroups may be defined by a non-obvious combination of several attributes that may be detected by a combination of cluster analysis and appropriate visualization.

Since the information space is growing with each examination cycle, Pearce and Merletti [31] pointed out in 2006 that methods are needed which can cope with this complexity and enable the analysis of underlying causes of a certain disease. Visual analytics (VA) methods can support epidemiological data assessment in different ways, e.g. by defining subgroups based on a multitude of attributes that exhibit a certain characteristic. For the analysis of scalar and categorical data, established information visualization techniques combined with clustering and dimension reduction are a good starting point, but need to be tightly integrated with statistic tools epidemiologists that are more familiar with. For image-centric studies, however, new visualization, (image) analysis and interaction techniques are needed.

In the following, we define essential terms in epidemiology and give an overview on cohort studies that employ medical image data as an essential element. Finally, we describe how image data, derived information and other data complement each other to identify and characterize risks.

2.1 Important Terms

Prevalence and Incidence Epidemiology investigates how often certain diseases or *clinical events*, such as a cerebral stroke or sudden heart death, occur in the population. Two terms are important to characterize this frequency. The *prevalence* indicates the portion of people suffering from a disease at a given point in time. The *incidence* represents how many people suffer from a disease or event in a certain interval, usually one year. High prevalence is usually associated with high economic costs. Population-based studies focus on diseases with a high prevalence, such as diabetes, coronary heart disease or neurodegenerative diseases. Even these diseases do not occur frequently in a random population including many younger people (where the prevalence of these diseases is low). A rare disease, such as amyotrophic lateral sclerosis, may have a prevalence of 5 from 100,000. Thus, even in a large population-based study probably no individual suffers from this disease.

Absolute and Relative Risks Another essential epidemiological term is the *risk* for a clinical event, such as outbreak of a certain disease, severity (stage) or death. As an example, a study related to cardiac risk may investigate angina pectoris, myocardial infarction, atrial fibrillation depending on attributes such as age and sex. The *absolute risk* characterizes the likelihood of getting a disease in life time. The absolute risk for a woman to develop breast cancer in the Western world is particularly high for women aged 50–60 (2.6 %) and 60–70 (3.7 %). Therefore, for these age groups, mammography screening—aiming at early detection and thus optimal treatment—was introduced.

The *relative risk* (RR) characterizes the increased risk if an individual is exposed to a certain risk factor, e.g., smoking, excessive weight, or alcohol abuse. It is based on a comparison with a control group not exposed to that risk factor. A value of $RR < 1$ represents a factor that protects, e.g., moderate physical activity. Exciting observations are often the combined effects of several parameters. A certain factor may be protective for some people (younger, slim women) and is involved with an increased risk for others. The combined risk may be significantly smaller or larger than could be expected from individual factors.

Moreover, relationships are often distinctly non-linear or even non-monotonic. Dose-response relationships are often non-linear. RR increases slowly (almost no effect for a small dose) and increases much faster for higher levels of a dose, e.g., exposure to toxicity. A typical non-monotonic relation is *U-shaped*, that is both very low and very high instances of an attribute involve an increased risk, whereas values in between are associated with a reduced risk. Examples are weight (both very low and very high weight are associated with an increased risk for mortality) and sleeping time (both very short and very long sleepers have an increased risk for developing psychiatric disorders [22]). Such relations cannot be characterized by a global RR value. Instead, tools are necessary that support the hypothesis of a U-shaped relation by estimating their parameters with some kind of best-fit algorithm.

2.2 *Image-Centric Cohort Studies*

Image Data in Epidemiology The acquisition of image data is determined by the available time, by financial resources, by the epidemiological importance and by ethic considerations. Epidemiological studies require approval by a local ethics committee. As a consequence, healthy individuals in a cohort study should not be exhibited to a risk associated to the examinations carried out. Thus, MRI should be preferred over X-ray or CT imaging for its non-radiation nature. Petersen and colleagues [32] explain why cardiac CT is less feasible in a cohort study and even MR is only used without a contrast agent in their study due to ethical reasons. MRI data and ultrasound data are the prevailing modalities in both the SHIP as well as the Rotterdam study. Unfortunately, MRI and ultrasound data do not exhibit standardized intensity values (in contrast to CT data). Moreover, MRI and ultrasound data suffer from inhomogeneities and various artifacts. Thus, they are more difficult to interpret for humans and more difficult to analyze with computational means. These data are used to measure, e.g., the thickness of vessel walls, the abdominal aorta diameter and plaque vulnerability in the coronary vessels [22]. The intensive use of MRI in epidemiological research also explains to some extent which questions are analyzed: MRI is the best modality for the analysis of brain structures and thus serves to explore early signs of Parkinson's, Alzheimer's and other neurodegenerative diseases. Epidemiological research aims at identifying such brain pathologies in a pre-symptomatic stage. Among the sources for such investigations are MR Diffusion Tensor Imaging data that enable an assessment of white matter integrity [22].

The selection of imaging parameters is always a trade-off between conflicting goals related to quality, e.g., image resolution, signal-to-noise ratio, patient comfort, e.g., examination time and associated costs. As a consequence, to shorten overall examination times in cohort study examinations, not the highest possible quality is available, i.e., a slice distance of 4 mm is more typical than 1 mm. A great advantage of MRI is that this method is very flexible and enables to display different structures in different sequences, such as T1-, T2- and proton density-weighted imaging. MRI data in cohort studies often comprise more than ten different sequences.

Standardization in Image Acquisition Due to the rapid progress in medical imaging, sequences, protocols and even (MR) scanners are frequently updated in clinical routine (similar to the update frequency on a computer). These updates would severely hamper the comparison of imaging results and thus the assessment of natural changes and disease outbreak. Thus, differences in acquisition parameters are essential *confounding variables*. Therefore, for one cohort and examination cycle that may last up to several years, no updates are allowed. Moreover, all involved physicians and radiology technicians are carefully instructed to use the same standardized imaging parameters. This point is even more important for longitudinal studies with repeated imaging examinations. Even if MR scanners and protocols are not updated, the life cycle of MR coils leads to changes of image quality that need to be monitored and compensated.

2.3 *Examples for Image-Centric Cohort Study Data*

In the following, we describe selected comprehensive and on-going longitudinal cohort studies. Both use a number of (epidemiologic) *instruments* that are innovative in cohort studies and thus lead already to a large number of insights documented in hundreds of (medical) publications. A considerable portion of these publications employ results from imaging data. However, the full potential of analyzing organ shapes, textures and spatial relations quantitatively is not exploited so far.

The Rotterdam Study A prominent example is the *Rotterdam* Study,[1] initiated in 1990 in the city of Rotterdam, in the Netherlands. Similar to later studies, it was motivated by the demographic change with more and more elderly people suffering from different diseases and their interactions. After the initial study involving almost 8000 men and women, follow-ups at four points in time were performed—the most recent examinations took place in the 2009–2011 period. In the later examination cycles, also new individuals were involved leading to datasets from almost 15,000 patients [22].

The original focus of the Rotterdam Study was on neurological diseases, but meanwhile it has been extended to other common diseases including cardiovascular and metabolic diseases. The study has an enormous impact on epidemiological and related medical research, documented in 797 journal publications registered in the pubmed database (search with keyword "Rotterdam Study", January 30, 2014). Among them are predictions for the future prevalence of heart diseases and many studies on potential risk factors for neurodegenerative diseases. For a comprehensive overview of the findings, see [22] that summarizes the findings of more than 240 papers related to the Rotterdam Study. In a similar way, [23] is a significant update of these findings with more recent data.

Norwegian Aging Study A long-term study in Norway investigates the relations between brain anatomy (as well as brain function), cognitive function, and genetics in normally aging people.[2] In total 170 individuals (120 of them female), aged between 46 and 77 (mean 62), were examined in Bergen and Oslo in by now three waves (first wave in 2004/2005, next in 2008/2009, and most recently in 2011/2012) [47]. While naturally not all of these subjects could be followed through all three waves, still most of them were subjected to an extensive combination of

1. neuropsychological tests, including tests of the intellectual, language (memory), sensory/motor, and attention/executive function,
2. MRI data, including co-registered T1-weighted anatomical imaging, diffusion tensor imaging, and—from the second wave on—also resting-state functional MRI, as well as
3. genotyping (first wave only) [46].

[1] http://www.erasmus-epidemiology.nl/research/ergo.htm, accessed: 1/31/2016.

[2] http://org.UiB.no/aldringsprosjektet/, accessed: 1/31/2016.

The substantially heterogeneous imaging and test data are used to study aging-related questions about the modern Norwegian population, for example, how anatomical and functional changes in the human brain possibly relate to the later development of Alzheimer and dementia. Important findings include the relation between hippocampal volumes and memory function in elderly women [47] and the relation between subcortical functional connectivity and verbal episodic memory function in healthy elderly [48].

SHIP The Study of Health in Pommerania (SHIP) is another cohort study broadly investigating findings and their potential prognostic value for a wide range of diseases. The SHIP tries to explain health-related differences after the German reunion between East and West Germany. It was initiated in the extreme northeast of Germany, a region with high unemployment and a relatively low life expectancy.

In the first examination cycle (1997–2001) 4308 adults of all age groups were examined, followed by a second and a third cycle that was finished at the end of 2012. The instruments used changed over time with some initial image data (liver and gallbladder ultrasound) available already in the first cycle and others, in particular whole body MRI, added later. The use of whole body MRI was unique in 2008 when the third examination cycle started. Breast MRI for women is performed, whereas for men MR angiography data are acquired, since men suffer from cardiovascular diseases significantly earlier than women [43]. In addition, a second cohort (SHIP-Trend) was established comprising 4420 adult participants.

Diagnostic reports are created by two independent radiologists who follow strict guidelines to report their findings in a standardized manner. The pilot study to discuss the viability and potential of such a comprehensive MR exam is described by Hegenscheid and colleagues [20]. The overall time for the investigation is two (complete) days with 90 min for the MR exam. The SHIP helped to reliably determine the prevalence of risk factors, such as obesity, and diseases. Major findings of the SHIP are increased levels of obesity and high blood pressure (compared to the German population) in the cohort. The MR exams alone identified pathological findings in 35 % of the sample population. More than 400 publications in peer-reviewed journals are based on SHIP data (January 2014).

UK Biobank The UK Biobank represents a comprehensive approach to study diseases with a high prevalence in an aging society, such as hearing loss, diabetes and lung diseases. Half a million individuals will be investigated in one examination cycle from which 100,000 receive an MRI from 2014 onwards. The rationale for the number of individuals to be included is explained by Peterson and colleagues [32]: they aim at a reliable identification of even moderate risk factors (RR between 1,3 and 1,5) for diseases with a prevalence of 5 %. The prospective study should have a comprehensive protocol of cardiac MRI, brain MRI and abdominal MRI. This prospective cohort study also involves genetic information.[3]

[3]http://www.ukbiobank.ac.uk, accessed: 1/31/2016.

The German National Cohort The recently started "German National Cohort" in Germany is based on experiences with a number of moderate-size studies, such as SHIP, and examines some 200,000 individuals over a period of 10–20 years. Individuals will be invited in three waves to characterize changes. Due to the large-scale character, imaging is distributed over five cities. Thus, the subtle differences in imaging within different scanners have to be considered.[4] It explicitly aims at improvements in the treatment of chronic diseases and involves a variety of tissue samples, e.g., lymphocytes. Imaging in 30,000 individuals is again performed with MRI, comprising whole body, brain and heart.

2.4 *Epidemiological Data*

Epidemiological data are huge and very heterogeneous. As an example, in the UK biobank 329 attributes relate to physical measures, such as pulse rate, systolic and diastolic blood pressure, and various measures relate to vision or hearing. 471 attributes relate to interviews (socio-demographics, health history, lifestyle, ...).

The data that are stored per individual is standardized but not completely the same, e.g., childbirth status and menstrual period are available for women only. Image data and derived information, e.g., segmentation results, significantly increased both the amount and complexity of data. Longitudinal cohort study data are time-dependent. While some *instruments*, such as blood pressure measurements, are available for all examination cycles, others were added later or removed. Individuals drop out, because they move, die or just do not accept the invitation to a second or third examination cycle. It is important to consider also such incomplete data but to be aware of potentially misleading conclusions.

The great potential of image-centric studies is that image data and associated laboratory data as well as data from interviews are available. An epidemiological study, such as the SHIP, has a large *data dictionary* that precisely defines all attributes and their ranges. While laboratory data are scalar values, most data from interviews are nominal or ordinal values. In particular, data from interviews exhibit an essential amount of uncertainty. Self-reports with respect to alcohol and drug use, cigarette smoking and sexual practices may be biased towards "expected or socially accepted" answers. Epidemiologists are not only aware of these problems but developed strategies to minimize the negative effects, e.g., by asking redundant questions. After data collection, experts spend a lot of effort to improve the quality of the data. Despite these efforts, visual analytics techniques have to consider outliers, missing and erroneous data.

Geographic Data Geographic data play a central role in public health where the dynamics of local infections are visualized and analyzed (*disease mapping*).

[4] http://www.nationale-kohorte.de/, accessed: 1/31/2016.

Chui and colleagues [8] presented a visual analytics solution directly addressing this problem by combining three dedicated views. Also in cohort study data, geographic data are potentially interesting to understand local differences in the frequency and severity of diseases as an interaction between environmental factors and genetic differences. This branch of epidemiology is referred to as *spatial epidemiology*. Beale and colleagues [4] investigated differences between rural and urban populations. In their comprehensive survey, Jerrett and colleagues [24] considered spatial epidemiology as an emerging area. However, we do not focus on spatial epidemiology since cohort study data typically comprise rather narrow regions and thus may not fully support such analysis questions.

2.5 Analysis of Epidemiological Workflow

The following discussion of observations and requirements for computer support is largely based on discussions with epidemiologists as well as the inspiring publication by Thew and colleagues. According to [40]

- epidemiological hypotheses are mostly observations made by physicians in clinical routine,
- corresponding attributes are chosen based on the observations and further experience, and
- regression analysis is frequently used to determine whether the investigated attribute is a risk factor or not.

Major requirements for an epidemiological workflow (again based on [40]) are:

- Results have to be reproducible. Due to the iterative data assessment, methods need to be applied to new data sets as well and the results need to be comparable between different assessment times to characterize the change. User input needs to be monitored all the time to enable reproducible results.
- A major result of an epidemiological analysis is whether certain factors influence a disease significantly. Relative risk (as a measure of effect size) and p-values as statistical significance level are particularly important.

Although these requirements neither consider image data nor visual analytics, they have to be considered also in these more innovative settings. Reproducibility, for example, means that clustering with random initialization is not feasible. Moreover, reports must be generated that clearly reveal all settings, e.g., parameters of clustering algorithms that were used for generating the results.

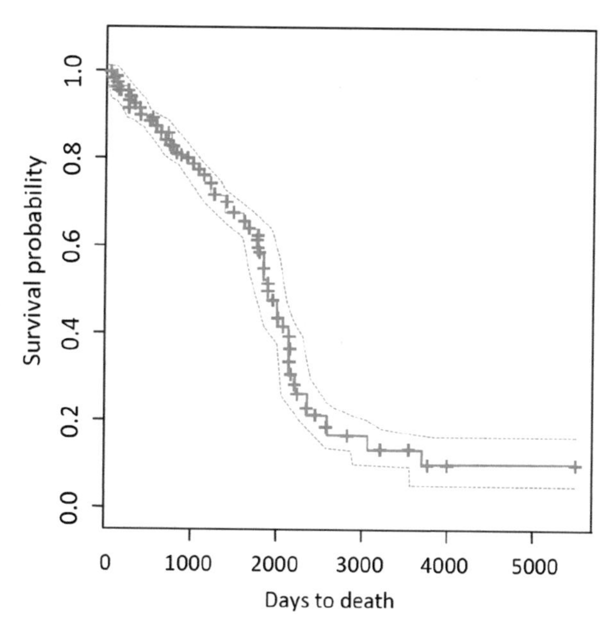

Fig. 1 A Kaplan-Meier curve indicates how many patients survive at least a certain time. The more patients pass away, the larger is the confidence interval indicated by the *dotted lines*. The *crosses* mark each time a patient dies to further provide information on the reliability of the data that decreases over time (courtesy of Petra Specht, University of Magdeburg)

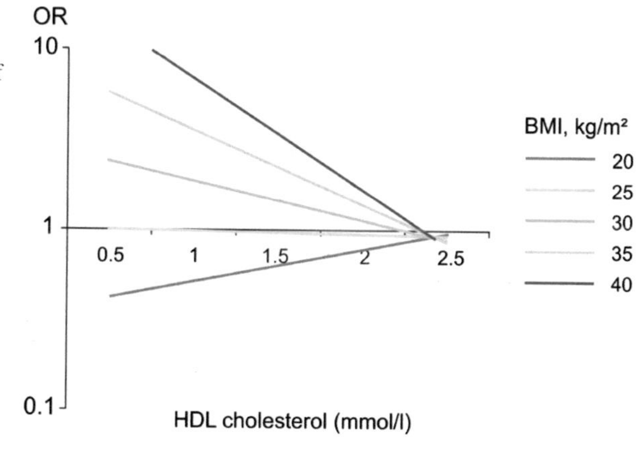

Fig. 2 The relative risk for cholelithiasis in men associated with a high level of a certain type of cholesterol slightly increases with a low BMI, but decreases for individuals with high or very high BMI. This multifactorial situation is depicted in an *interaction term* (inspired by [42])

Since statistical analysis plays such an important role, statistics packages, such as SPSS,[5] R[6] and STATA[7] dominate in epidemiology. They provide various statistical tests also in cases where assumptions, such as a normal distribution, are not valid. Also the peculiarities of categorical data are considered. Visual analysis, so far, plays a minor role. As an example, Figs. 1 and 2 illustrate two graphical representations frequently used in epidemiology: *Kaplan-Meier curves* and *interaction terms.*

[5]http://www-01.ibm.com/software/analytics/spss/products/statistics/, accessed: 1/31/2016.

[6]http://www.r-project.org/ accessed: 1/31/2016.

[7]http://www.stata.com/, accessed: 1/31/2016.

A Kaplan-Meier curve shows the survival of patients, often as a comparison between different treatment options.

3 Visual Analytics in Epidemiology

The visualization of correlations in the epidemiological routine is largely restricted to scatterplots with regression lines and box plots to convey a distribution. Scatterplots may be enhanced, e.g., by coloring items according to certain characteristics, e.g., a diagnosis or by adding results from cluster or Principal Component Analysis (PCA) [38]. The frequency related to a particular combination of values is often encoded by adapting saturation or darkness of colors.

Visual analytics methods can complement the statistical analysis and provide methods to *explore* the data. Efficient methods are essential to cope with the large amount of data and provide rapid feedback that is essential for any exploration process.

One of the first attempts to employ information visualization techniques for medical (image and non-image) data was realized in the WEAVE system [19]. The system incorporated parallel coordinates as well as real time synchronization between different views. Another essential tool, inspired by the WEAVE system, was presented by Blaas and colleagues [6]. They enabled feature derivation techniques and incorporated segmentation techniques providing a powerful framework for heterogeneous medical data. Later work by Steenwijk and colleagues was more focussed on epidemiology. They provided an exploratory approach to analyze heterogeneous epidemiological data sets, including MRI [38]. They consider parameters on normalized and not normalized domains, while only normalized domains are comparable between subjects. Normalization means, for example, to register MRI brain data to an atlas to compare individual differences.

Mappers are used to project data into normalized domains. As an example, in brain analysis, a mapper defines the relation between an individual brain and a brain atlas that contains normalized and averaged information derived from many individual data. Feature extraction pipelines can be build visually by using a pipeline of mappers. The visualization is realized through multiple coordinated views which either represent scalar data or volumetric images. Different techniques to color code data, to align them and add further information are provided to enhance scatterplots. Steenwijk and colleagues evaluated their tools with specific examples from neuroimaging and questions related to a neurological disease where relations between clinical data (anxiety-depression scales, mental state scales) and MR-related data are analyzed (Fig. 3). Normalized data domains are represented using scatterplots and parallel coordinated views. Dynamic changes are visualized using a time plot. The selection is linked between views and allows for multi-parameter comparison of clusters.

Zhang and colleagues [49] build a web-based information visualization framework for epidemiological analysis through different views. They divide the analytics

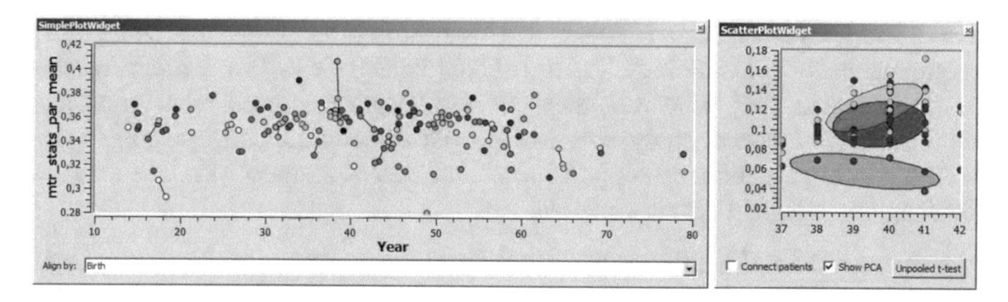

Fig. 3 *Left*: A scatterplot relates magnetic transfer ratios to age. Items relating to the same patient (over time) are connected via a *line*. *Colors* indicate a diagnosis. *Right*: An enhanced scatterplot with PCA results for three subgroups overlaid (courtesy of Martijn Steenwijk, VU University Medical Center Amsterdam)

process in *batch analytics* and *on-demand analytics*. Batch analytics steps are performed automatically as a new subject is added to the data set and aim to create groups by means of a certain condition. On-demand analytics are performed by user requests. Subjects are visualized using treemaps, histograms, radial visualizations and list views. However, neither filtering and grouping nor the interaction between the views are explained.

Recently, Turkay and colleagues [41] described a framework to analyze the data of the Norwegian aging study, aiming particularly at *hypothesis generation*. For this purpose, they give an overview on the dimensions in their dataset that conveys statistical properties, such as mean, standard deviation, skewness and kurtosis. The two latter measures characterize how asymmetric the data distribution is. Scatterplots display pairwise measures related to *all* dimensions. *Deviation plots*, a new technique, enables grouping and supports a comparison of measures for a subgroup to the whole cohort.

The possibilities of VA tools can be summarized as follows:

- Manual/automatic definition (brushing) of interesting parameters and ranges of values in attribute views,
- Linking of attribute views for identifying relations,
- Analysis across aggregation levels, parameters and subjects,
- Definition of groups either interactively by means of (complex) brushes, or semi-automatically by means of clustering, and
- Visual queries and direct feedback enable easy exploration

While having different applications in mind, the Polaris system of Stolte and colleagues also employs multiple coordinated views to validate hypotheses [39]. It uses a variety of different information visualization techniques to map ordinal/nominal or quantitative data of a relational data base. The system itself formulates data base queries and the mapping to create visualizations for the requested attributes. They choose the visualization mapping automatically based on the attribute types that are viewed in context with each other. This allows for a fast visualization of different attribute combinations in order to drill down to the information of interest. General visual analytics tools, such as Polaris and Weaver, in principle support some of the requirements for epidemiology. The use of coordinated views, brushes and switches is advantageous [44]. However, they are not designed to cope with the special requirements of cohort study data and do not directly support epidemiological workflows. In particular, no support for a combined assessment of image data and other epidemiological data is available.

Commercial Visual Analytics Systems There are a number of commercial software tools specialized on data visualization. As an example, Tableau [28] provides an interface for creating different visualizations based on attribute drag-and-drop.[8] These approaches deliver fast results with respect to visualizing different attributes. However, it is not supported to derive new data, such as scores for diseases, body mass index and other data that is relevant in epidemiology. QLikView is a similar tool for visualizing data associations. The user can design a frontend where associations can be assessed using multiple information visualizations. Thus, the user can drill down to the desired information. Statistics features regarding epidemiological key figures are limited.

Spotfire/IVEE [1] is able to handle more complex analysis of data sets and allows for interactive filtering of attributes. It can be linked to the statistical computing programming language `R`, which makes it versatile in comparison with its competitors. However, users need to be familiar with the `R` syntax.

Commercial systems cannot be enhanced or embedded in another system with hassle-free data exchange. The focus of commercial data visualization tools is business intelligence yielding a focus on quantitative data sources. At the same time they excel at incorporating user collaboration by including comment sections and share filters or entire setups of a dashboard.

4 Analysis of Medical Image Data for Epidemiology

Medical images are not by themselves useful for epidemiological analysis, since the semantics of image elements (pixel, voxels) is too low. The resulting extremely high-dimensional feature space would be unsuitable for visual analysis. Hence, image data is sequentially aggregated and reduced. The different steps of this process i.e. image analysis, shape analysis of extracted objects and subsequent clustering are characterized and discussed in this section. Throughout the section, we often refer to one case study, where MRI data from the lumbar spine is analyzed. The representation of assumptions on the lumbar spine shape and location as well as the object detection scheme used are examples of viable and common approaches. We do not claim that these techniques are better than any other approaches.

[8]http://www.tableausoftware.com/, accessed: 1/31/2016.

4.1 Medical Image Analysis

One of the major purposes of image analysis for a cohort study is to quantify anatomy, e.g., by volume, shape or spatial relations between structures. Quantification may be used to establish a range of *normal values* for different age groups and to characterize variations. Such variations may also confirm a disease and thus add to data derived from clinical tests. Thus, the use of image data enables more reliable conclusions w.r.t. the incidence and prevalence of diseases. As discussed in Sect. 2.2, MRI is particularly interesting because of the wide range of different information represented in these data. Thus, the examples discussed in the following relate to MRI data although the principles are more general.

Due to the wide range of image analysis tasks, the techniques should be adaptable to different analysis goals. The parameterization of an image analysis technique should be intuitive and the interaction should be kept to a minimum. The latter aspect is particularly important, since often several thousand datasets need to be analyzed. While largely manual approaches are acceptable in some clinical settings, such as radiation treatment planning, they are not feasible in the evaluation of cohort study data. The reduction of interactivity is not only a matter of effort, but also to meet the essential goal of *reproducible approaches*.

Detection and Segmentation of Anatomical Structures A modular system is a possible means to meet the central requirement of image analysis in cohort study data. An example of a cohort study is the liver segmentation of [17], where concurrent detection and localization processes are combined for initial segmentation that is then fine-tuned in a model-driven segmentation step and finalized by a data-driven correction process. It has been shown that processes can be re-used and re-combined to solve a different segmentation task on similar data (kidney segmentation in MRI [18]). Alternatively, the necessary *domain knowledge*, related to expected size, basic shape, position and grey values, can be separated from the detection and segmentation module. This strategy is attractive, since the user has not to care about the detection process when changing the application. Two problems have to be addressed in this case:

- What is the expectation about the data support integrated to fit a model?
- How is the with-class variation of the object in search separated from the between-class variation?

Point distribution models (PDM) [30] address the second question by training on sample segmentations. Model fitting is realized by a registration step. When training is not feasible, a prototypical model may be used instead. It is associated with restricted input about variation (a few parameters only) and qualitative knowledge about configuration or part-relationship.

In the following, we describe the linear elastic deformation of a finite element model (FEM) as a common method to model shape variation. The user specifies the average shape and two elasticity parameters: Young's modulus defines how much external force is needed for a deformation and Poisson's ratio describes how the deformation is transferred orthogonal to the direction of an incident force [35]. The decomposition of the prototypical shape into finite elements bounded by nodes and specification of the elasticity parameters results in a stiffness matrix K that relates the node displacement u to incident forces f [Eq. (1)]:

$$Ku = f \tag{1}$$

Different kinds of nodes may be specified that are attracted by different kinds of forces. Boundary nodes are attracted by the intensity gradient and inner nodes are attracted from expected intensity or texture. For letting an FEM move and deform into an object in an image, deformation is made dependent on time t. Behavior then also depends on mass M of the FEM and object-specific damping D [Eq. (2)].

$$M\ddot{u}(t) + D\dot{u}(t) + Ku(t) = f(t) \tag{2}$$

M represents the resistance of the moving FEM to external forces and allows the model to move over spurious image detail (e.g., gradients caused by noise). Damping D avoids oscillation of the FEM. The system of differential equations is decoupled by solving the following generalized eigenproblem.

$$KE = ME\Lambda \text{ with } E^T KE = \Lambda \text{ and } E^T ME = I$$

where Λ is the diagonal matrix with real-valued eigenvalues and I is the identity matrix.

After projecting the data on the eigenvector matrix, the differential equations can be solved fast and in a stable manner. Moreover, the projection on the eigenvectors (called *modes of vibration*, see Fig. 4) separates deformation into components representing rigid transformation, major deformation modes and remaining minor deformation modes. The vibration modes can be used similar to the variation modes of an ASM to derive a quality-of-fit formulation for a fitted model instance. Since only a few anatomical objects have such a specific shape that it can be described by a simple deformation model and since training of additional information should be avoided, it is useful to complement simple deformation with pre-specified information on part-relationships. Part-relationships may describe the configuration of the object of interest w.r.t neighboring structures or may represent decomposition into parts (see [33]).

Extending the FEM to a hierarchical model requires the introduction of a second layer FEM. Each sub-shape is represented by an FEM on the first layer and sub-shape FEMs are connected to the second layer. The type of connection regulates dependencies between sub-shapes and may range from distance constraints to co-deformation. FEMs for the first and second layer are created and assembled in the same fashion than elements are assembled for the sub-shape FEM [35].

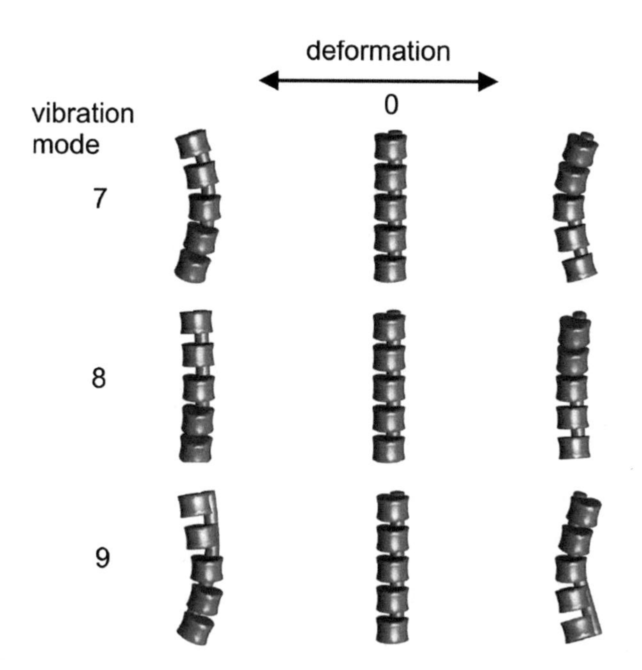

Fig. 4 Vibration modes 7–9 of a lower spine model. Vibration modes 1–6 represent rigid transformations (courtesy of Marko Rak, University of Magdeburg)

Case Study: Analysis of Vertebrae Back pain and related diseases exhibit a high prevalence and are thus a focus of the SHIP (recall [43]). Specific goals are

- to define the prevalence of degenerative changes of the spine,
- to identify risk factors for these changes,
- to correlate degenerative changes with actual symptoms, and
- to better understand the progress from minor disease to a severe problem that requires medical treatment.

Epidemiologists hypothesize that smoking, heavy physical activity and a number of drugs that are frequently used are risk factors for back pain. Based on clinical observations, epidemiologists suggested to focus on the lumbar spine—the lowest part of the spine comprising five vertebrae. As a first step, the spine and lumbar vertebrae should be detected in T1-weighted and T2-weighted MRI data from SHIP.

Although local optimization could be complemented by stochastic global optimization [11], Rak and colleagues used only local optimization, since the initialization is simple for the given data. The user places a model instance in a sagittal view on the middle slice of the image sequence which is then transformed based on local image attributes. The model is constructed according to the appearance of vertebrae and spine in a sample image sequence. Vertebra sub-shapes were connected with a spine sub-shape by a structural model on the second level.

The spine model supported proper localization of the vertebrae. Since its most discriminate aspect was the cylindrical shape, it was represented by a deformable

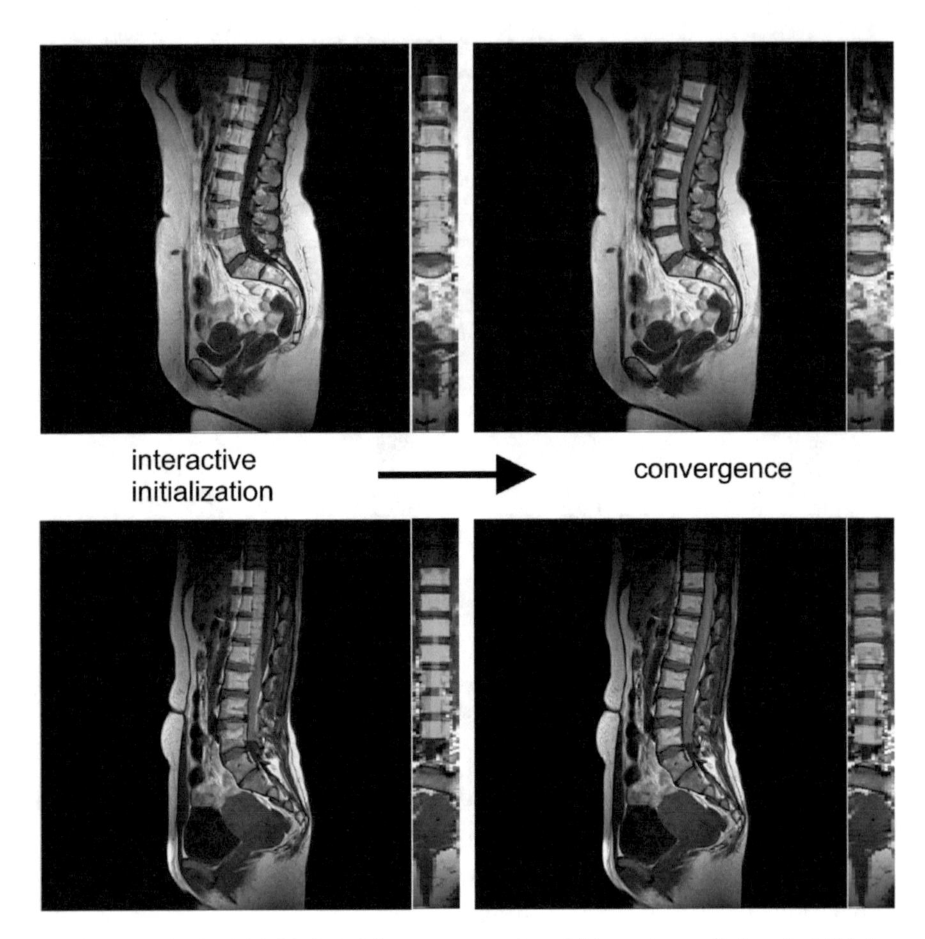

Fig. 5 Examples for initialization and convergence of model instances applied to the MRI data (courtesy of Marko Rak, University of Magdeburg)

cylinder consisting of inner nodes only. The vertebrae shape was represented indirectly by inner nodes as well, since reliability of the intensity gradient was low. For each of the two shape models, the vertebra and the spine, a weighted combination of the T1-weighted and the T2-weighted image was computed as appearance input. Weights for each of the two models were determined a priori and produced a clearly recognizable local minimum for vertebra and spine appearance, respectively. The user placed a model instance in the vicinity of the object on a sagittal slice. Computation time until convergence was between 1.1 and 2.6 s per case. The method was evaluated on 49 data sets from the SHIP. The detection was considered successful if the center of each vertebra sub-shape was in the corresponding vertebra in the image data, which was achieved in 48 of the 49 cases (see Fig. 5 for examples and [35] for further details).

4.2 Shape Analysis

Epidemiologists are used to work with numerical and categorical data which then is tested for statistic validity. Medical image data also allows to consider characteristic object shapes. As an example, the shape of the liver may depend in a characteristic manner on infections (hepatitis), alcohol consumption, or obesity. Eventually, shape characteristics may change even before a disease becomes symptomatic. If this turns out, shape changes may be employed as an early stage indicator.

While the quantitative analysis of shapes or parts thereof (*morphometrics*) is a recent trend in epidemiology due to the availability of image data, it is established in anatomy and evolutionary biology.

Shape analysis requires that different shapes are transformed in a common space, typically by a rigid transformation (translation and rotation). The parameters of this rigid transformation are determined in an optimization process that minimizes the distances of corresponding points. A major challenge is to determine these corresponding points that serve as landmarks. In particular for soft tissue structures there are not sufficient recognizable landmarks and therefore a parameterization is necessary to define these points. Without going into detail, we assume that this process is applied to many individual shapes S_i, say livers in a cohort study. Then, for each S_i an optimal non-rigid transformation to a reference shape R defines a deformation with displacement vectors for each landmark.

For use in epidemiology, a large set of displacement vectors is not the right level of granularity. Instead, a few dimensions are desirable that characterize major differences. Thus, typically a dimension reduction technique, such as PCA, is employed to characterize the directions that represent the major differences. This process may be adapted to specific analysis questions by assigning individual weights to the landmarks expressing a strong interest in particular displacements [21]. Thus, epidemiological hypotheses may be incorporated.

While the establishment of point correspondences is often a major challenge, recently alternative approaches were developed. The GAMES algorithm (Growing and adaptive meshes) [13] creates a data structure to represent the shape variance if no pairwise correspondence between points is given. However, it can be prone to errors since it requires a prior registration of segmentation masks.

Shape analysis, of course, may also be supported by appropriate visualizations that enable pairwise comparisons and emphasize differences. In this vein, [7] presented a system for shape space exploration based on carefully designed multiple coordinated views.

4.3 Analysis of Lumbar Spine Canal Variability

In Sect. 4.1, we introduced a case study related to the analysis of the lumbar spine in cohort data and explained how the spine and the vertebrae are detected in MRI data from the SHIP. Here, we extend this discussion by the analysis of the spinal canal and non-imaging attributes related to back pain. In the SHIP, attributes related to back pain history, e.g., working habits, physical activities, size and weight, are available to identify and analyze potential correlations with findings from the MRI data. After careful discussions, we selected 77 attributes (60 are ordinal or nominal and 17 scalar) to investigate back pain [26]. Ordinal data are primarily results of multiple choice questions. The epidemiologists suggested to focus on the overall shape and curvature of the spine in that region instead of individual vertebrae. This overall shape is well characterized by the lumbar canal. Thus, correlations between the shape of the lumbar spine, attributes of back pain history and activities both in leisure and working time may be analyzed.

Klemm and colleagues [25] extracted, clustered and visualized spine canal centerlines. Image segmentation of 493 MRI data sets was carried out automatically using tetrahedron-based finite element models of vertebrae and spinal canal [35]. Using barycentric coordinates of the tetrahedrons, a centerline consisting of 93 discrete points was extracted for each segmentation, as seen in Fig. 6 (left, middle). They served as input for an agglomerative hierarchical clustering which created groups of subjects based on differences in shape. This special clustering technique was chosen, since it produces meaningful results in the clustering of similar structures, such as fiber tracts derived from MR-Diffusion Tensor Imaging data [25].

The cluster visualization in Fig. 6 (right) displays each cluster representative as ribbon in a sagittal plane. The representative is the centerline with the smallest sum

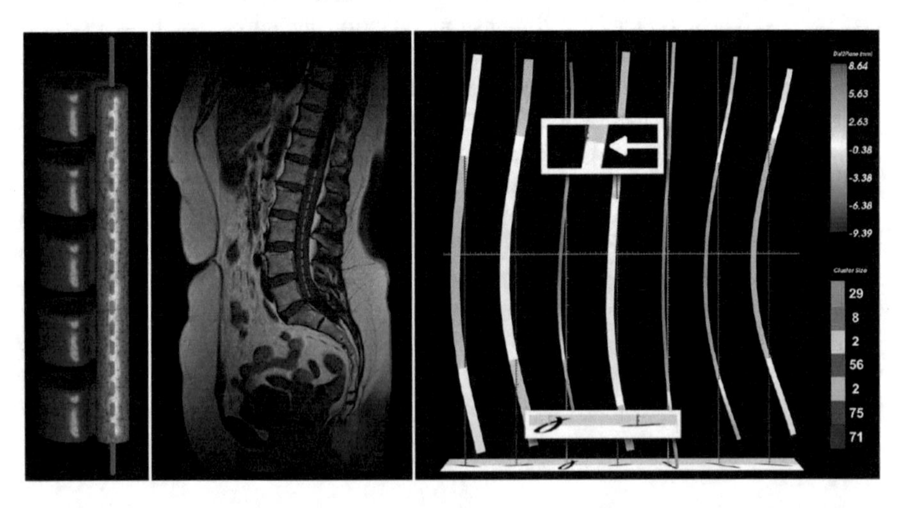

Fig. 6 Lumbar spine visualization of 243 female subjects. *Left:* Tetrahedron-based finite element model from Rak and colleagues [35]. The *dashed purple line* indicates the lumbar spine canal centerline *Middle:* Model used to detect lumbar spine canal in an MRI scan *Right:* Agglomerative hierarchical clustering of 243 centerlines yields seven clusters. Their representatives are visualized as ribbons mapping cluster size to width. The *ribbon color* encodes the distance to the semi-transparent plane orthogonal to the view direction (*lower inset*). Shadow projections (*upper inset*) provide an additional visual hint on the curvature extent [25]

of distances to all other centerlines, i.e., the centroid line of the cluster. The width of the ribbon encodes the cluster count and the color encodes the distance to the sagittal plane. Shadow projections also (redundantly) convey the distance to the sagittal plane. This allows to assess the 3D shape in a 2D projection. The results of the clustering can be used in different ways.

- **Outlier detection**: Extraordinary shapes yield clusters of small size that differ strongly from the global mean shape. This can point to pathologies or errors in the segmentation process.
- **Hypothesis generation**: Shape groups serve as a starting point for an exploratory analysis to analyze disease-related correlations. The usual workflow requires epidemiologists to define groups, e.g., age ranges. Groups calculated solely using shape information can be analyzed to detect statistically relevant associations in other expositions which can lead to new hypotheses.
- **Hypothesis validation**: Clustering based on non-image related features can be used to analyze if these clusters are correlated to characteristic shapes, e.g., a strongly bent spine canal representative.

Calculating curvature on groups created according to body height starting from 150 cm in 10 cm steps was performed. Klemm and colleagues found that taller people have a more straight spine compared to small people. They also found multiple clusters of people 10 years above average age across all groups that exhibit a strong “S” shape of the spine, which was the starting point for new investigations using expert chosen spine-related attributes. This method was extended to integrate the relevant information for identifying correlations. Thus, for a selected cluster, information related to the distribution of attributes, such as the back pain history (frequency and intensity of back pain), may be displayed as a tool tip. The initial observations show, that a box plot summarizing the distribution is more suitable than the full histogram. For routine use in epidemiology, the lumbar spine visualization (Fig. 6) has to be complemented with at least simple statistics to answer questions, such as: Is there a statistically significant difference between the curvature of the lumbar spine canal and back pain frequency? If so, what is the effect size?

4.4 Cluster Analysis and Information Space Reduction

A crucial task in epidemiology is the definition of groups of subjects. Differences and similarities among groups are investigated and control groups are defined to detect and assess the impact and interaction of risk factors to define the relative risk. A straightforward approach is the manual definition based on study variables and ranges of interest. A data-driven extension is the automatic detection of potentially relevant subgroups in the often high-dimensional data by means of clustering algorithms [26]. In particular, the generation of new hypotheses, which may be tightly connected to the identification of new groups, benefits from the latter. In clustering, subjects, being similar with respect to a certain similarity metric, are grouped in clusters with a low intra-cluster and a high inter-cluster variance. Particular challenges in the cluster analysis of cohort study data are:

- Missing data, e.g., denied answers to inconvenient questions [10]
- Mixture of scala and categorical study variables [2]
- Time-varying variables in longitudinal studies [15]

As a consequence of the first problem, it should be reported to the user how many datasets were actually used for clustering. Depending on the chosen attributes, this may be a subset of the overall amount of data. Incomplete datasets may and should be used when the relevant data is available. It is essential to use similarity metrics that consider also ordinal or categorical data. Usually, the following convention is used: The distance between datasets equals 1 if their categorical data is different and 0 if it is the same. With ordinal data, more care is necessary. The difference between "strongly agree" and "strongly disagree" on a Likert scale is larger than the difference between "agree" and "disagree". However, the precise quantification is not straightforward. As a first step to explore the SHIP data, a parallel coordinate view is combined with scatterplots and clustering (Fig. 7).

With respect to clustering algorithms, it is essential that the number of resulting groups has not to be specified in advance. Moreover, algorithms are preferred that allow outliers instead of forcing all elements to be part of a cluster. Outliers may be particularly interesting and thus serve as a starting point for further investigation. Of course, they may also indicate a bad quality of some data. In our experiments, density-based clustering with DBSCAN [12] produced plausible results when applied to non-image data of the SHIP study. The DBSCAN result is sensitive to the *minPoints* parameter that determines the minimum size of a cluster. Some cycles of clustering are necessary to make a suitable choice. In this process, an appropriate visualization is essential to easily understand the results. A visualization that conveys the location, size and shape of clusters is difficult in case that clustering is applied to more than two-dimensional data. A recently developed

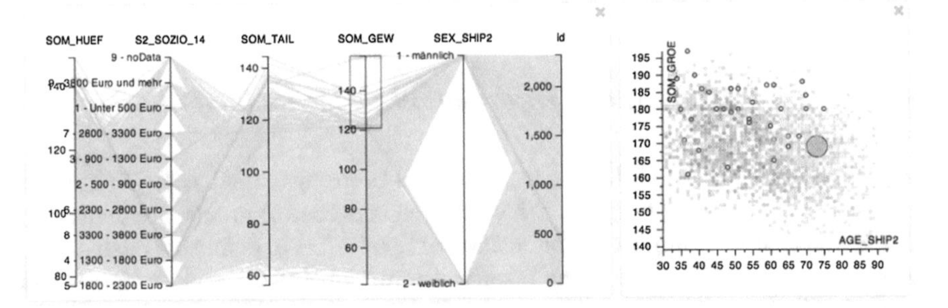

Fig. 7 A parallel coordinate view enables the selection of a relevant subset (here persons with a weight larger than 120 kg). The *scatterplots* represent correlations between age, size and weight. The elements are color-coded according to a clustering result that yields three clusters. The *encircled elements* correspond to the selection in the parallel coordinates view

approach enables 3D visualizations of clustering results with very low levels of occlusion [16].

The majority of cohort-studies are restricted to non-image data, i.e., categorical and scalar data, which may directly serve as input for a clustering algorithm. In [3], 176 patients with lower back pain have been monitored over six months via text messages describing their bothersomeness. All patients received chiropractic treatment. A hierarchical clustering of the individual temporal courses of bothersomeness revealed groups of patients who responded differently to the therapy, which may improve the optimal individual treatment selection. Hypotheses about the differences between paternal age-related schizophrenia (PARS) and other cases of schizophrenia were generated in [27]. A k-means clustering of demographic variables, symptoms, cognitive tests and olfaction for 136 subjects (34 with PARS) delivered clusters containing a high concentration of PARS cases. Significant characteristics of these clusters may give a hint on features of PARS improving its dissociation of other cases of schizophrenia.

In analyzing image-centric cohort study data, besides categorical and scalar data, images may be clustered for group definition. Image intensities, segmentation results, e.g., the surface of the segmented liver or the centerline of the spinal canal (recall Sect. 4.3) and derived information, e.g., liver tissue texture, liver volume and spinal canal centerline geometry, may serve as input. In [37], the anatomical variation of the mandibles is assessed across a population. For treating mandible fractures, subjects with similar characteristics are grouped in clusters and a suitable implant is designed per cluster. The clustering algorithm k-means is applied to transformation parameters of a locally affine registration between all mandible surfaces segmented in CT data. A cohort of 50 patients with suspicious breast lesions was investigated by means of dynamic contrast enhanced MRI (DCE-MRI) in [34]. Each lesion was clustered according to its perfusion characteristics by means of a region merging approach. Perfusion is represented by the temporal course of the DCE-MRI signal intensities. The clustering itself did not generate groups of patients here, but based on each individual number of clusters and their perfusion characteristics, two groups of lesions could be defined: benign and malignant. These groups were then compared to histological results from core needle biopsy.

Investigating high-dimensional non-image cohort study data often benefits from an information space reduction, e.g., by means of PCA. Plotting the data for inspection in a lower dimensional space while capturing the greatest level of variation, e.g., a scatterplot of the first two principal components, as well as detecting trends in the data and ordering them according to the variance they describe are important applications of PCA. In [36], symptom data gathered in interviews of 410 people with Turret syndrome was investigated in order to specify homogeneous symptom categories for a better characterization of the disease's phenotype. First, clusters of symptom variables were generated using *agglomerative hierarchical clustering*. Then, for each cluster and each participant a score was computed equal to the sum of present symptom variables in the cluster. These scores were the input for PCA, which produced homogeneous symptom categories, sorted according to their percentage of represented symptomatic variance. In [38], scatterplots of variables from cohort study data are extended by superimposing PCA ellipses. The ellipses are computed per group of subjects and illustrate its global distribution with respect to the two opposed variables. They are spanned by the principal component axes of a groups data points and centered at their mean. Optionally, their transparency is adjusted with respect to the groups confidence, i.e. the number of contained subjects.

4.5 Categorical Data

Cohort study data sets comprise many categorical data such as answers to questionnaires or categorizations that may result from binning continuous data, such as intervals of income or age groups. These data are *discrete* and exhibit a *low range*. While data resulting from binning or from answers marked at a Likert scale have an inherent order (ordinal data), often no inherent order exists. Standard information visualization methods like scatterplots and parallel coordinates are designed for continuous data and thus not ideal for displaying categorical data, since many data points occlude each other.

Categorical data are often visualized using boxes where the width is scaled to frequency [45]. These approaches use much space and are also not well suited for encoding multidimensional relationships. *Parallel sets*, introduced by Bendix and colleagues, comprise the same layout as parallel coordinates, "but the continuous axes were replaced with sets of boxes ... scaled to the frequency of the category" [5] (see Fig. 8). Selecting an attribute will map each category to a distinct color so that they can be traced through all visualized dimensions. Highlighting a category may be realized by drawing the selected category in a higher saturation leading to a pop-out effect. It is also useful to display a histogram for the selected category annotated with statistical information. Selecting a category will only display the particular box on an axis and make more room for connections to other axes. This is especially helpful if the number of categories for one dimension is large. In case of data without inherent order, reordering of axes is possible and supports the exploration by providing more comprehensible layouts.

5 Concluding Remarks

A variety of large prospective cohort studies are established and ongoing. They generate a wealth of potentially relevant information for assessing risks, for estimating costs involved in treatment and thus inform health policy makers with respect to potentially preventive measures or cost-limiting initiatives. Despite the great potential of data mining and interactive visualization, none of these studies included such activities in their original planning. The role of computer science, so far, was limited to database management and data security. Visual analytics has a great potential for exploring complex health-related data, as recently shown by Zhang and colleagues [50] for clinical applications, such as treatment planning. Similar techniques may be employed to address epidemiology research. In contrast to clinical applications, where a severe time pressure leads to strongly guided workflows, epidemiology research benefits from powerful and flexible tools that enable and support *exploration*. Currently, existing and widespread information visualization and analytics techniques are employed and adapted to epidemiologic data. The high-dimensional nature of these data, however, also requires to develop new techniques.

The specific and new aspect discussed in this paper was the integration of image data, information derived from image data, such as spine curvature-related measures, and more traditional socio-demographics data.

Future Work So far, visual analytics research and software was rarely focused on epidemiology. Thus, to adapt visual analytics to epidemiology and to integrate the solutions with tools familiar to epidemiologists is necessary. In epidemiology, national studies are prevailing, which is due to the large amount of legislative conditions to be considered. International studies would enable to explore diseases with lower prevalence, subtypes of diseases that occur rarely, e.g., cancer in early age, and specific questions of spatial epidemiology. The SHIP Brazil study will provide such information. It was recently initiated to perform a study in Brazil

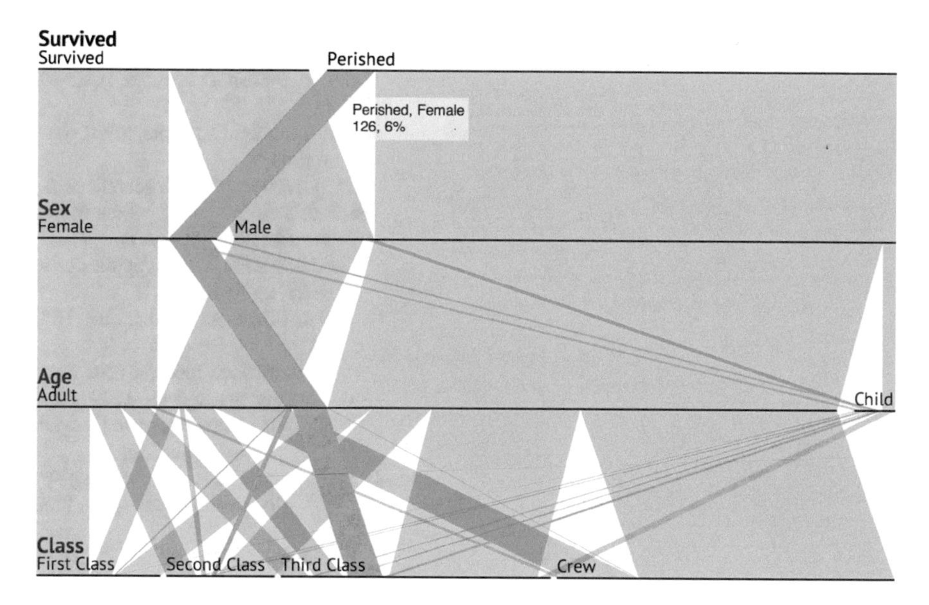

Fig. 8 Parallel sets are designed to explore categorical data. Parallel sets are based on the parallel coordinates layout, but it maps frequency of the data instead of rendering them just as data points. The user can interactively remap the data to new categorizations as well as highlight entries to examine their distribution along other mapped dimensions. The displayed data set shows the distribution of passengers of the RMS Titanic and whether they survived the sinking of the ship [9]

according to the standards and experiences gained in the German SHIP study. A nasty but essential problem of image-based epidemiologic study is quality control. Research efforts are necessary to automatically check whether image data fully cover the target region, whether the alignment of slices, e.g., in heart imaging, is correct and whether severe artifacts appear, e.g., motion artifacts. This chapter discussed exciting developments related to the combined use of radiologic image and more classical epidemiological data. The next wave is already clearly recognizable: genetics information will be integrated in the search for early markers associated with risks for diseases.

Acknowledgements We want to thank Lisa Fraunstein, David Kilias and David Perlich who supported our analysis of the SHIP data as student workers as well as Marko Rak who provided the detection algorithm for the vertabrae and Myra Spilopoulou for fruitful discussions on clustering and data mining (all University of Magdeburg). We thank Martijn Steenwijk for providing images from his work and Charl Botha for fruitful discussions. Matthias Günther (Fraunhofer MEVIS) explained us quality aspects of MR imaging in epidemiologic studies. This work was supported by the DFG Priority Program 1335: Scalable Visual Analytics. SHIP is part of the Community Medicine Research net of the University of Greifswald, Germany, which is funded by the Federal Ministry of Education and Research (grant no. 03ZIK012), the Ministry of Cultural Affairs as well as the Social Ministry of the Federal State of Mecklenburg-West Pomerania.

References

1. Ahlberg, C.: Spotfire: an information exploration environment. SIGMOD Rec. **25**(4), 25–29 (1996)
2. Ahmad, A., Dey, L.: A k-mean clustering algorithm for mixed numeric and categorical data. Data Knowl. Eng. **63**(2), 503–527 (2007)
3. Axén, I., Bodin, L., Bergström, G., Halasz, L., Lange, F., Lövgren, P.W., Rosenbaum, A., Leboeuf-Yde, C., Jensen, I.: Clustering patients on the basis of their individual course of low back pain over a six month period. BMC Musculoskelet. Disord. **12**, 99–108 (2011)
4. Beale, L.L., Abellan, J.J., Hodgson, S.S., Jarup, L.L.: Methodologic issues and approaches to spatial epidemiology. Environ. Health Perspect. **116**(8), 1105–1110 (2008)
5. Bendix, F., Kosara, R., Hauser, H.: Parallel sets: visual analysis of categorical data. In: IEEE Symposium on Information Visualization, pp. 133–140 (2005)
6. Blaas, J., Botha, C.P., Post, F.H.: Interactive visualization of multi-field medical data using linked physical and feature-space views. In: Proceedings of EuroVis, pp. 123–130 (2007)
7. Busking, S., Botha, C.P., Post, F.H.: Dynamic multi-view exploration of shape spaces. Comput. Graph. Forum **29**(3), 973–982 (2010)
8. Chui, K.K., Wenger, J.B., Cohen, S.A., Naumova, E.N.: Visual analytics for epidemiologists: understanding the interactions between age, time, and disease with multi-panel graphs. PloS One **6**(2), e14,683 (2011)
9. Davies, J.: Parallel set of the titanic data set. http://www.jasondavies.com/parallel-sets/ (2012). Accessed 30 Jan 2014
10. Donders, A.R., van der Heijden, G.J., Stijnen, T., Moons, K.G.: Review: a gentle introduction to imputation of missing values. J. Clin. Epidemiol. **59**(10), 1087–1091 (2006)
11. Engel, K., Toennies, K.D.: Hierarchical vibrations for part-based recognition of complex objects. Pattern Recogn. **43**(8), 2681–2691 (2010)

12. Ester, M., Kriegel, H.P., Sander, J., Xu, X.: A density-based algorithm for discovering clusters in large spatial databases with noise. In: Proceedings of the second international conference on knowledge discovery and data mining (KDD), pp. 226–231 (1996)
13. Ferrarini, L., Olofsson, H., Palm, W., Vanbuchem, M., Reiber, J., Admiraalbehloul, F.: GAMEs: growing and adaptive meshes for fully automatic shape modeling and analysis. Med. Image Anal. **11**(3), 302–314 (2007)
14. Fletcher, R.H., Fletcher, S.W.: Clinical Epidemiology. Lippincott Williams & Wilkins, Philadelphia (2011)
15. Genolini, C., Falissard, B.: KmL: k-means for longitudinal data. Comput. Stat. **25**(2), 317–328 (2010)
16. Glaßer, S., Lawonn, K., Preim, B.: Visualization of 3D cluster results for medical tomographic image data. In: Proceedings of Conference on Computer Graphics Theory and Applications (VISIGRAPP/GRAPP), pp. 169–176 (2014)
17. Gloger, O., Kuhn, J., Stanski, A., Völzke, H., Puls, R.: A fully automatic three-step liver segmentation method on LDA-based probability maps for multiple contrast MR images. Magn. Reson. Imaging **28**(6), 882–897 (2010)
18. Gloger, O., Toennies, K.D., Liebscher, V., Kugelmann, B., Laqua, R., Völzke, H.: Prior shape level set segmentation on multistep generated probability maps of MR datasets for fully automatic kidney parenchyma volumetry. IEEE Trans. Med. Imaging **31**(2), 312–325 (2012)
19. Gresh, D.L., Rogowitz, B.E., Winslow, R.L., Scollan, D.F., Yung, C.K.: Weave: a system for visually linking 3-D and statistical visualizations, applied to cardiac simulation and measurement data. In: Proceedings of IEEE Visualization, pp. 489–492 (2000)
20. Hegenscheid, K., Kühn, J.P., Völzke, H., Biffar, R., Hosten, N., Puls, R.: Whole-body magnetic resonance imaging of healthy volunteers: pilot study results from the population-based SHIP study. Fortschritte auf dem Gebiet der Röntgenstrahlen und der bildgebenden Verfahren (Röfo) **181**(8), 748–759 (2009)
21. Hermann, M., Schunke, A.C., Klein, R.: Semantically steered visual analysis of highly detailed morphometric shape spaces. In: Proceedings of IEEE Symposium on Biological Data Visualization (BioVis), pp. 151–158 (2011)
22. Hofman, A., Breteler, M.M.B., van Duijn, C.M., Janssen, H.L.A., Krestin, G.P., Kuipers, E.J., Stricker, B.H.C., Tiemeier, H., Uitterlinden, A.G., Vingerling, J.R., Witteman, J.C.M.: The Rotterdam Study: 2010 objectives and design update. Eur. J. Epidemiol. **24**, 553–572 (2009)
23. Hofman, A., van Duijn, C.M., Franco, O.H., et al.: The Rotterdam Study: 2012 objectives and design update. Eur. J. Epidemiol. **26**, 657–686 (2011)
24. Jerrett, M., Gale, S., Kontgis, C.: Spatial modeling in environmental and public health research. Int. J. Environ. Res. Public Health **7**(16), 1302–1329 (2010)
25. Klemm, P., Lawonn, K., Rak, M., Preim, B., Tönnies, K., Hegenscheid, K., Völzke, H., Oeltze, S.: Visualization and analysis of lumbar spine canal variability in cohort study data. In: Proceedings of Vision, Modeling, Visualization (VMV), pp. 121–128 (2013)
26. Klemm, P., Frauenstein, L., Perlich, D., Hegenscheid, K., Völzke, H., Preim, B.: Clustering Socio-demographic and medical attribute data in cohort studies. In: Proceedings of Bildverarbeitung für die Medizin (BVM) (2014)
27. Lee, H., Malaspina, D., Ahn, H., Perrin, M., Opler, M.G., Kleinhaus, K., Harlap, S., Goetz, R., Antonius, D.: Paternal age related schizophrenia (PARS): latent subgroups detected by k-means clustering analysis. Schizophr. Res. **128**(1–3), 143–149 (2011)
28. Mackinlay, J., Hanrahan, P., Stolte, C.: Show me: automatic presentation for visual analysis. IEEE Trans. Vis. Comput. Graph. **13**(6), 1137–1144 (2007)
29. Marathe, M., Vullikanti, A.K.S.: Computational epidemiology. Commun. ACM **56**(7), 88–96 (2013)
30. McInerney, T., Terzopoulos, D.: Deformable models in medical image analysis: a survey. Med. Image Anal. **1**(2), 91–108 (1996)
31. Pearce, N., Merletti, F.: Complexity, simplicity, and epidemiology. Int. J. Epidemiol. **35**(3), 515–519 (2006)

32. Petersen, S.E., Matthews, P.M., Bamberg, F., et al.: Imaging in population science: cardiovascular magnetic resonance in 100,000 participants of UK Biobank - rationale, challenges and approaches. J. Cardiovasc. Magn. Reson. **28**, 15–46 (2013)
33. Petyt, M.: Introduction to Finite Element Vibration Analysis. Cambridge University Press, Cambridge (1998)
34. Preim, U., Glaßer, S., Preim, B., Fischbach, F., Ricke, J.: Computer-aided diagnosis in breast DCE-MRI-quantification of the heterogeneity of breast lesions. Eur. J. Radiol. **81**(7), 1532–1538 (2012)
35. Rak, M., Engel, K., Tönnies, K.D.: Closed-form hierarchical finite element models for part-based object detection. In: Proceedings of Vision, Modeling, Visualization (VMV), pp. 137–144 (2013)
36. Robertson, M.M., Althoff, R.R., Hafez, A., Pauls, D.L.: Principal components analysis of a large cohort with tourette syndrome. Br. J. Psychiatry **193**(1), 31–36 (2008)
37. Seiler, C., Pennec, X., Reyes, M.: Capturing the multiscale anatomical shape variability with polyaffine transformation trees. Med. Image Anal. **16**(7), 1371–1384 (2012)
38. Steenwijk, M.D., Milles, J., van Buchem, M.A., Reiber, J.H.C., Botha, C.P.: Integrated visual analysis for heterogeneous datasets in cohort studies. In: Proceedings of IEEE VisWeek Workshop on Visual Analytics in Health Care (2010)
39. Stolte, C., Tang, D., Hanrahan, P.: Polaris: a system for query, analysis, and visualization of multidimensional relational databases. IEEE Trans. Vis. Comput. Graph **8**(1), 52–65 (2002)
40. Thew, S., Sutcliffe, A., Procter, R., de Bruijn, O., McNaught, J., Venters, C.C., Buchan, I.: Requirements engineering for e-Science: experiences in epidemiology. IEEE Softw. **26**(1), 80–87 (2009)
41. Turkay, C., Lundervold, A., Lundervold, A.J., Hauser, H.: Hypothesis generation by interactive visual exploration of heterogeneous medical data. In: Proceedings of Human-Computer Interaction and Knowledge Discovery in Complex, Unstructured, Big Data, pp. 1–12 (2013)
42. Völzke, H., Baumeister, S.E., Alte, D., Hoffmann, W., Schwahn, C., Simon, P., John, U., Lerch, M.M.: Independent risk factors for gallstone formation in a region with high cholelithiasis prevalence. Digestion **71**, 97–105 (2005)
43. Völzke, H., Alte, D., Schmidt, C., et al.: Cohort profile: the study of health in pomerania. Int. J. Epidemiol. **40**(2), 294–307 (2011)
44. Weaver, C.: Cross-filtered views for multidimensional visual analysis. IEEE Trans. Vis. Comput. Graph. **16**(2), 192–204 (2010)
45. Wittenburg, K., Lanning, T., Heinrichs, M., Stanton, M.: Parallel bargrams for consumer-based information exploration and choice. In: Proceedings of the ACM Symposium on User Interface Software and Technology (UIST), pp. 51–60 (2001)
46. Ystad, M.: Quantitative structural and functional brain imaging in cognitive aging. Ph.D. thesis, University of Bergen (2010)
47. Ystad, M., Lundervold, A.J., Wehling, E., Espeseth, T., Rootwelt, H., Westlye, L., Andersson, M., Adolfsdottir, S., Geitung, J., Fjell, A., Reinvang, I., Lundervold, A.: Hippocampal volumes are important predictors for memory function in elderly women. BMC Med. Imaging **9**(1), 1–15 (2009)
48. Ystad, M., Eichele, T., Lundervold, A.J., Lundervold, A.: Subcortical functional connectivity and verbal episodic memory in healthy elderly—resting state fmri study. NeuroImage **52**(1), 379–388 (2010)
49. Zhang, Z., Gotz, D., Perer, A.: Interactive visual patient cohort analysis. In: Proceedings of IEEE VisWeek Workshop on Visual Analytics in Healthcare (2012)
50. Zhang, Z., Wang, B., Ahmed, F., Ramakrishnan, I., Viccellio, A., Zhao, R., Mueller, K.: The five W's for information visualization with application to healthcare informatics. IEEE Trans. Vis. Comput. Graph. **19**(11), 379–388 (2013)

Three Dimensional Visualisation of Microscope Imaging to Improve Understanding of Human Embryo Development

Anna Leida Mölder, Sarah Drury, Nicholas Costen, Geraldine Hartshorne, and Silvester Czanner

Abstract The analysis of processes on a cellular and sub-cellular level plays a crucial role in life sciences. Commonly microscopic assays make use of stains and cellular markers in order to enhance image contrast, but in many cases, cell imaging requires the sample to be undisturbed during the imaging process, making staining, dying and fixing impractical. Non-destructive techniques are especially useful in long term imaging or in the study of sensitive cell types, such as stem cells, embryos or nerve cells. Novel advances in computation, imaging and incubator technology have recently made it possible to prolong the imaging time, reduced the cost of storing data and opened a door to the development of new computer aided analytical tools based on microscopic image data.

Here we illustrate how Hoffman Modulation Contrast imaging and Confocal Microscopy can be combined with visual computing and present results from determination of cell number, volume, spatial location and blastomere connectivity, using examples from embryos grown for in vitro fertilisation. We give examples of how knowledge of the imaging technique can be used to further improve the computer analysis and also how visually guided tools may aid in the diagnostic interpretation of image data and improve the result. Finally we discuss how the use of microscopic data as a basis for embryo modelling may help in both research and educational purposes. The aim of this chapter is to give an example of how microscopic imaging can be combined with standard computer vision techniques to aid in the interpretation of microscopic data, and demonstrate how visual computing techniques can make an essential difference in terms of scientific output and understanding.

A.L. Mölder (✉) • N. Costen • S. Czanner
School of Computing, Mathematics and Digital Technology, Manchester Metropolitan University, Manchester, UK
e-mail: mail@annaleida.com

S. Drury • G. Hartshorne
Warwick Medical School, University of Warwick, and University Hospitals Coventry and Warwickshire NHS Trust, Coventry, UK

L. Linsen et al. (eds.), *Visualization in Medicine and Life Sciences III*, Mathematics and Visualization, DOI 10.1007/978-3-319-24523-2_11

1 Introduction

Microscopic methodologies can roughly be divided into destructive or non-destructive techniques. Non-destructive techniques are preferable in many cases, where there is a need to keep interference with the sample at a minimum. In in vitro fertilisation (IVF), the embryo under observation cannot normally be manipulated or disturbed in any way, but must be observed "as is", if it is to be used for implantation. For research purposes, destructive techniques can sometimes be necessary in order to perform a particular measurement. Two destructive techniques commonly used in the study of embryos are fluorescence microscopy [17] and confocal microscopy [18]. Microscopic techniques which can be counted as non-destructive include bright- and dark-field microscopy, phase-contrast microscopy (PC) [4], Hoffman Modulation Contrast (HMC) microscopy [13], Differential Interference Contrast (DIC) [20] microscopy and Digital Holography (DH) [2]. Of these, bright- and dark-field microscopy produce an image of the amplitude of the transmitted (or reflected) light. However, cellular material is usually highly transparent, and for such objects, a better sample-to-background contrast can be obtained by recording the phase of light instead of the amplitude. PC microscopy and HMC imaging are techniques where the phase information of diffracted light is optically converted to amplitude information. Techniques such as these are very good for visualisation, but the image grey scale cannot be directly translated to quantitative data. DIC and DH are techniques where the sample phase-shift is imaged directly, yielding information on sample thickness and refractive index.

An increasing amount of image material available due to lower cost of hardware and increased ease of data storage has made it more and more cumbersome to perform image analysis manually. An increasing amount of work has been done in both open source and commercial projects, developing computational solutions for bioimaging problems. Generalised software, with a large number of tools as well as plugins for specific tasks include ImageJ [1], Icy [5], BioImage XD [15], CellProfiler [16] and Fiji [25]. Each imaging modality often requires its own computational approach, demanding high technical skills from the user, in order to choose the correct algorithm for the analytical problem at hand. In addition to the multi-purpose solutions, there exists a variety of software solutions, particularly designed for a narrower purpose [24, 31]. A majority of the available computational tools for microscopic imaging focus on fluorescent microscopy imaging - a reflection of its common use in research. However, for stem cell research, when working with tissues intended for transplantation or in IVF, non-destructive imaging modalities are the only option. Interesting work has been done in modelling fixed embryos [3, 12, 29], but little has been done so far on modelling and analysing growing embryos in the early stages after fertilisation using non-destructive imaging.

1.1 The Embryo Selection Process

When selecting an embryo suitable for implantation, the embryologist may look at a number of criteria, such as pronuclear appearance and orientation [6, 27], number, size, shape of blastomeres, degree of fragmentation [10], degree of blastocoelic expansion, cellular composition and compactness of the inner cell mass and trophectoderm [26]. Discussions concerning the relevance of embryo morphology in quality assessment exist [11], but it is likely that such evaluations will continue to play a large part in IVF embryo evaluation also in the future. Traditionally, embryos have been studied using a microscope (commonly HMC) only at certain time points during the course of their development. It has been shown in time-lapse studies that the timing of key occurrences can vary greatly between individual embryos with similar morphologic appearance at the conclusion of the recording period, and correlation has been shown between the timing of key developmental events and embryo quality [21]. Such indications, in combination with new possibilities for time-lapse imaging of human embryos for an extended period of time with fewer negative effects to their health, make it likely that the use of time-lapse recordings will continue to increase in the future.

1.2 Embryo Imaging Using Hoffman Modulation Contrast

HMC imaging is a popular optical set-up for non-destructive microscopic imaging, routinely used in embryology. Here, light is passed through a pair of off-axis slits, converting gradients in sample optical path to bands of light and dark appearance, depending on the spatial sample direction (Fig. 1). This effect is most apparent when performing any kind of non-symmetrical image operations. Compare for instance the output of a derivative of the raw image taken along the horizontal and vertical axis, respectively (Fig. 2). In computer vision, a common approach is the application of an edge finding filter. Here too, the anti-symmetry of the HMC must be taken into account, because the angle of incident light produces a shift in edges in the upper half of the image compared to the lower half. In Fig. 2c), a Canny edge detection filter has been applied to the raw image. It is clear that the filter may find edges on both sides of the lighter and darker bands, resulting in an uncertainty when trying to determine the location of the border of the embryo, or of a single blastomere. However, since the direction of light depends on the azimuthal angle between splits and sample, this effect can be reduced by rotating the sample around a vertical axis, and combining information from several images along the rotation.

Long-term imaging of sensitive material requires the process of capture to be non-destructive, but also requires that the sample can be kept undisturbed in a favourable atmosphere for an extended period of time. Novel construction of incubators and cultivation chambers has recently made it possible to monitor embryos over the course of several days, without any known consequences to their health.

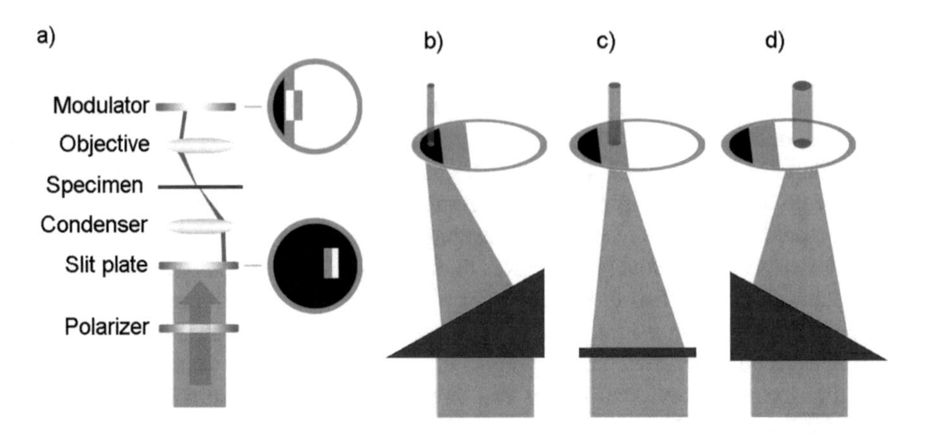

Fig. 1 (**a**) Layout of Hoffman Modulation Contrast optical set-up. Undiffracted light falls on the **grey** portion of the modulator (**c**), and the background of the image appears light *grey*. Refractive index gradients in the sample result in deflection of the light to either the *black* (**b**) or the *white* (**d**) section of the modulator, so that variations are imaged darker on one side, and brighter on the other, producing a pseudo three dimensional shadowing effect

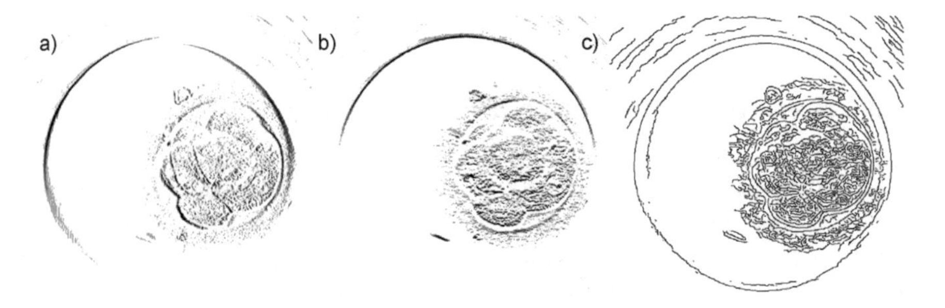

Fig. 2 Gradient of Hoffman Modulation Contrast image in X-direction (**a**) and Y-direction (**b**), respectively, and the direction invariant Canny Edge transformation (**c**). Note that the contrast of these images have been reversed for display purposes

However, there are difficulties other than the purely technical when combining automatic long term time-lapse imaging and microscopy. When examining embryos under the microscope, the three dimensional structure is very much of interest. In a traditional, manually handled microscope, much information can be gained by making proper use of the microscope controls: moving the sample around, scanning the focus, adjusting strength of illumination or making use of various filters and apertures in order to scan the three dimensional object in real time. In an automated time-lapse set-up, the possibility to manipulate optics is reduced when the optical set-up must incorporate a climate chamber to accommodate the living cellular material. If the microscope is instead meant to sit inside an incubator or other external chamber, the possibility of manipulating the optics is equally reduced, either because its operation requires the doors of the chamber to be opened, or because the optics are shielded to protect them from the high humidity of the chamber. In many time-lapse set-ups, the possibility of adjusting image quality in real time has vanished, and the user is now limited to studying the images some time after they are captured. This calls for new techniques to visualise this previously captured data in creative ways, possibly of regaining some of the interactivity lost to the user.

2 Materials and Methods

The anonymised embryos used in these studies were donated by consenting patients and the study of them was approved by Coventry Research Ethics Committee (04/Q2802/26) and the Human Fertilisation and Embryology Authority (R0155). Fresh embryos unsuitable for transfer or cryopreservation and frozen stored embryos no longer required for treatment were cultured using Medicult media (Origio, Redhill, UK) for up to 7 days and incubated in 37 °C CO_2 in air. Embryos were otherwise untreated or undisturbed during culture, while some were fixed and stained at the end of culture in preparation for the imaging process.

For confocal imaging, embryo nuclei were marked with DAPI, fixed on microscope slides and images were captured using the LSM510 confocal system (Zeiss, Hertfordshire, UK), using 400× magnification, with 1 μm between scans. Confocal images were segmented using a combination of region growing segmentation filters (Neighbourhood Connected Thresholding and Confidence Connected Thresholding) and Watershed segmentation [23]. A Delauney triangulation was used to compute the surfaces from the segmented outlines, and the body of each nucleus was then put together in a three dimensional representation of the complete embryo.

For HMC imaging, time-lapse series were captured using the Embryoscope® system (Fertilitech, Copenhagen, Denmark), with a 20 min interval between images. The embryos were mounted in wells in an EmbryoSlide® (Fertilitech, Copenhagen, Denmark), one embryo per well, and the imaging used a 635 nm LED, and a Hoffman Modulation Contrast Microscope. Up to seven focal planes, 20 μm apart, were captured simultaneously, resulting in an image stack, where the image in optimal focus was selected for each embryo using a variation of contrast detection adapted to Hoffman Modulation Contrast microscopy. Blastomere outlines were carefully selected manually using the software EmbryoSegmenter, as shown in Fig. 3, and then further processed using Python scripting, where a spherical model was adjusted to the segmentation outline as described in [8], assuming the segmentation to be at the waist, i.e. on the widest part of the blastomere. The spherical shape represents a first order simplification of the true blastomere shape, and can readily be extended to a more complex model, if information from several focal planes is considered. However, due to the longer field of view in HMC imaging, compared to confocal microscopy, in combination with a short depth of field, there is often bleed-through from objects out-of-focus and a higher uncertainty in the extension of objects in the focal (z) direction, compared to the xy-direction.

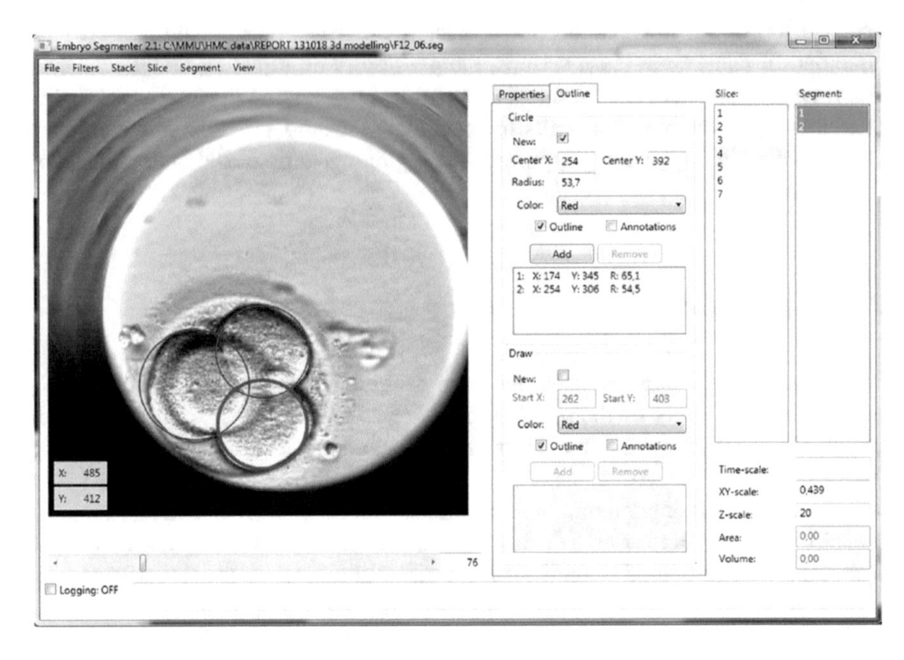

Fig. 3 EmbryoSegmenter for annotation, selection of focus level and blastomere outlines

The image at each focal distance can be described by the two dimensional function *I(x,y)*. Prior to handling, captured images were filtered with a Gaussian filter to remove speckle noise. To detect image sharpness, *i. e.* the focus, a Laplace filter *L(I)* was used,

$$L(I) = \frac{\nabla^2 I}{4} = \frac{1}{4} \cdot \left(\frac{d^2 I}{dx^2} + \frac{d^2 I}{dy^2}\right), \tag{1}$$

where the Laplacian has been applied to a 4-connected neighbourhood,

$$l_{i,j} = \frac{1}{4}(I_{i+1,j} + Ii - 1, j + Ii, j + 1 + Ii, j - 1) - I_{i,j}. \tag{2}$$

Note that due to the asymmetrical nature of Hoffman Modulation Contrast Imaging (Fig. 2), the symmetrical Laplace was chosen, rather than the direction-dependent gradient.

To detect areas of high sharpness, an H-maxima transform, *H*, was applied to the result *L(I)*, using an 8-connected neighbourhood. The H-maxima transform consists of a morphological grey-scale reconstruction of a marker image using the original image as mask, followed by a subtraction of the result from the original image

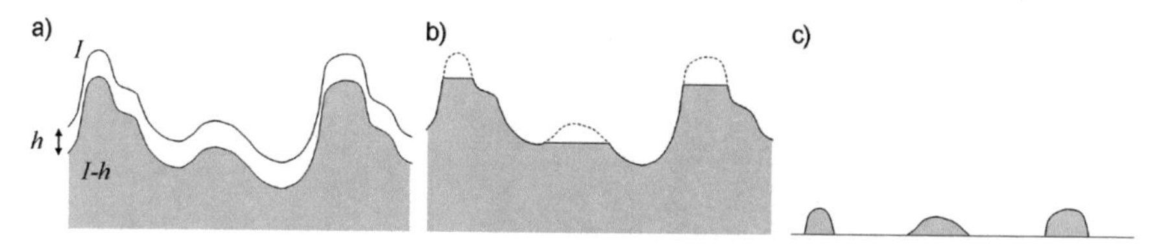

Fig. 4 The H-maxima transform extracts peaks higher than h. (**a**) Original image I, subtracted by a scalar h. (**b**) *Greyscale* reconstruction of I using I-h as marker. (**c**) Subtraction of *grey* scale reconstruction from original image

(Fig. 4). The marker image is obtained from the original image, subtracted with a constant value h. The result is a set of regional maximas lower than h. For this study, $h = 15\%$ of the image max was chosen. The detected maxima were extended using a Close transform with a 7 pixel diameter circular structure element, and the resulting image was thresholded at 99% of maximum image amplitude and converted to a binary mask. Holes were removed from the mask using a filling function, and the mask was then used to extract the corresponding region from the original image, I.

3 Modelling Using Confocal Microscopy

The short field of view of a confocal microscope makes it possible to section a sample volume into image slices and several scans may be used to compile a complete three dimensional visualisation of embryo structure. In Fig. 5, semi-automatically segmented outlines of nuclei from several slices have been combined to show the complete nuclear shape. Figure 6 shows the three dimensional layout of a human blastocyst, consisting of 121 blastomere nuclei. The nuclei along the perimeter form the trophectoderm. A denser pack of nuclei to the right of the top view image form the inner cell mass. A flattening effect due to the mounting of the embryo between microslides is clearly visible in the side view image. The accuracy of the 3D model depends on the amount of available data, the xy-resolution and the number of scans in the z-direction. Confocal microscopy has been chosen here because the low depth of field allows us to separate the signal between images in the stack, thus obtaining cleaner data. Optical sectioning is to some extent also possible in some non-destructive techniques, in particular in HMC, which has a relatively low depth of field compared to other types of light microscopy.

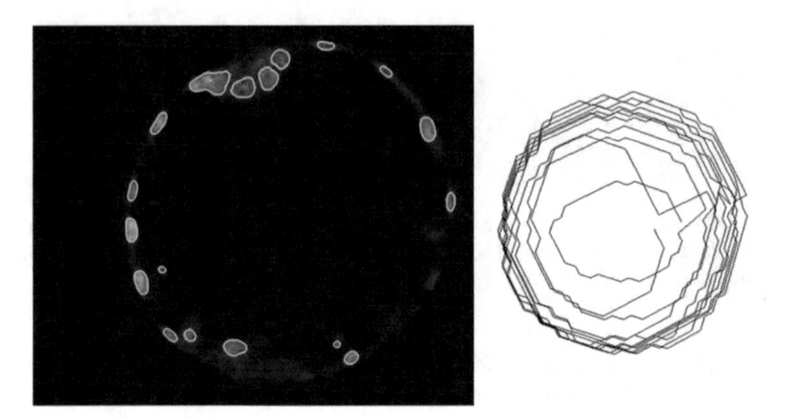

Fig. 5 Confocal image of blastocyst stage human embryo. Nuclei stained with DAPI have been segmented. Line plot of eleven segmented outlines, forming the body of a single blastomere nucleus; *top* view

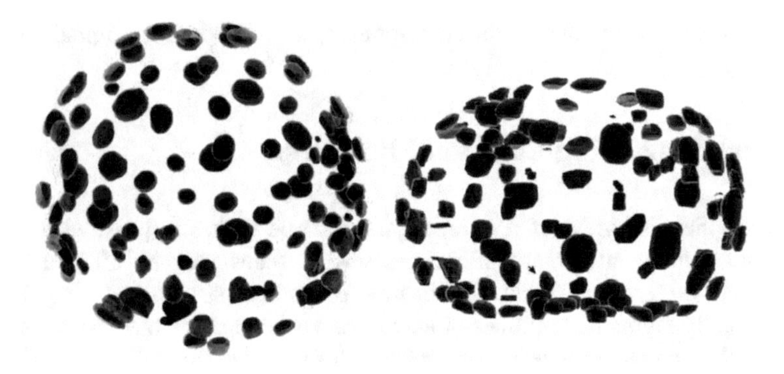

Fig. 6 Nuclei of blastocyst stage embryo, modelled after confocal microscope image scans. Front view and side view, respectively. A flattening effect is visible, where the embryo has been deformed between the microscope slide and cover slip [19]

4 Three Dimensional Modelling of Hoffman Modulation Contrast Images

When images are unsuited for automatic segmentation, or when the clinical nature of the material calls for manual methods to be applied, creative visualisation can still add value to the analysis. Figure 7 shows a basic spherical model of a 4-cell stage embryo, using 4 selected outlines, and a more detailed model, including the zona pellucida, and internal structures of the embryo. Blastomeres may be located at different positions along the focal axis and the selection of the correct position must be performed separately for each individual blastomere (Fig. 8). The crucial step in the modelling is the selection of the appropriate focal plane from the stack of available images.

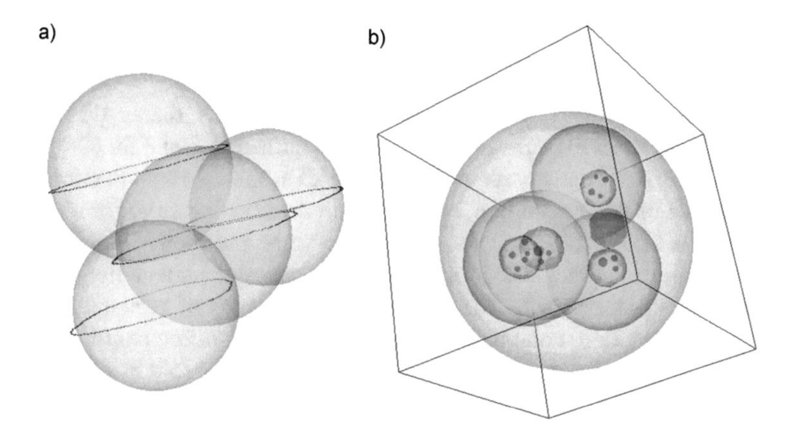

Fig. 7 (**a**) Segmentation outlines at blastomere waist are used to guide spherical cell models. (**b**) All structures present in the original image can be visualised. Here showing nuclei and nucleoli, as well as a large fragment (visible to the right of image centre). The different structures have been artificially coloured for clarity

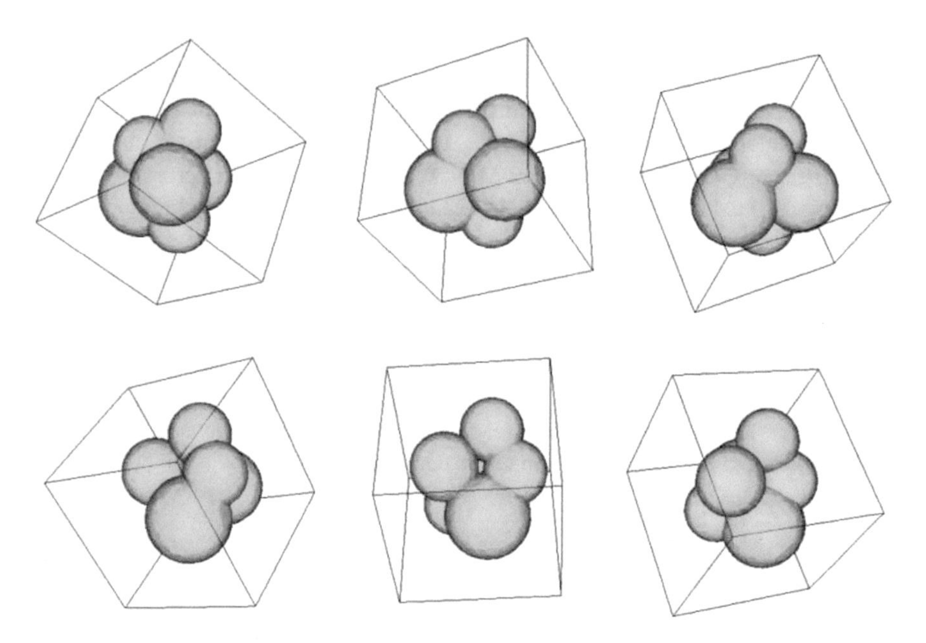

Fig. 8 6 cell stage embryo, with blastomeres modelled as spheres, rotated about its central axis. Bounding box added for clarity. Connected regions between blastomeres differ depending on blastomere size and location

4.1 Focal Selection

The most critical property of quality for an image is its focus. Without good sharpness, the image will be meaningless for both human eyes and for computerised analysis. As the use of camera recordings increases around us, both in our daily life, for the purpose of documentation, communication, surveillance or for recreational purposes, so it also increases in the laboratory. More and more often, images are captured as part of the scientific process, as a method of documentation or as a part of the analytical process itself. The number and complexity of available algorithms for image segmentation, computer vision or pattern recognition continues to grow and is likely to play a large role in how we handle data in the future. When we hand over more of medical surveillance and diagnostic tasks to automation, it is crucial that we can rely on the accuracy of these automatic procedures. One way to assure a level of quality, and to make sure we do not waste time trying to analyse material of poor standard, is to make sure that the images introduced to an analytical pipeline are captured at an optimal focus.

Many algorithms for automatic focus rely on computation of the power spectrum [30]. In cameras with moving lenses, it is also possible to adjust the focus based on the image contrast, a method referred to as contrast detection autofocus. Several images are then captured in sequence, while searching for local maxima of the image contrast or the gradient of the image contrast. This is usually not done for the entire image but for a selected area of interest. Several autofocus algorithms have also been evaluated for microscopy [7, 9, 22, 28]. In optical microscopy, where the depth of field is usually very short, the entire embryo is rarely in focus at one optical setting.

Figures 9 and 10 illustrates the extraction of regions in focus, based on the contrast detection algorithm, from the image stack of a human embryo. The embryo is at the blastocyst stage and has the shape of a hollow sphere with cells covering its walls. Due to the varying distribution of cells, the spherical shell of the embryo may vary in tissue content. The stack moves from slightly above the horizontal embryo central plane and downwards until it reaches the embryo base, where also structures on the bottom of the embryo container are encountered.

4.2 Calculation of Blastomere Connectivity

During the cleavage stage, as the blastomeres undergo mitosis, contact regions between them often increase, as cellular communications are initiated. At the onset of the compaction stage and at the following cavitation, the blastomeres adhere together more closely and start to form a single interacting multicellular organism. In this way, the developing embryo eventually transforms into a complete entity, rather than a package of loosely aggregated cells.

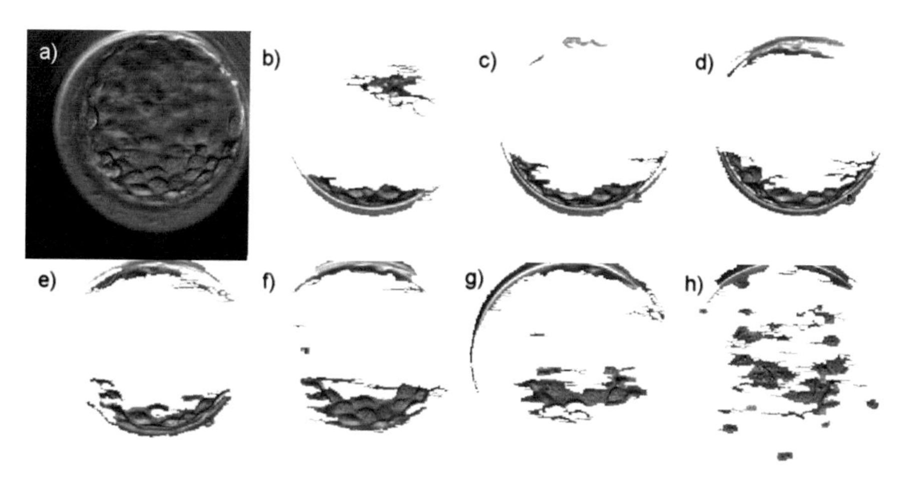

Fig. 9 Hoffman Modulation Contrast image of blastocyst stage embryo at 136.2 h, showing one original image (**a**) and seven images, captured at separate focal planes, with extracted regions in optimal focus (**b–h**). Focal planes cover the embryo from slightly above the embryo waist (**b**), down to bottom of the containing well (**h**). The inner cell mass is in focus on the first slide (**b**). The original image corresponds to the fifth image in the series (**f**). Images (**b**)–(**h**) have reversed contrast for display purposes

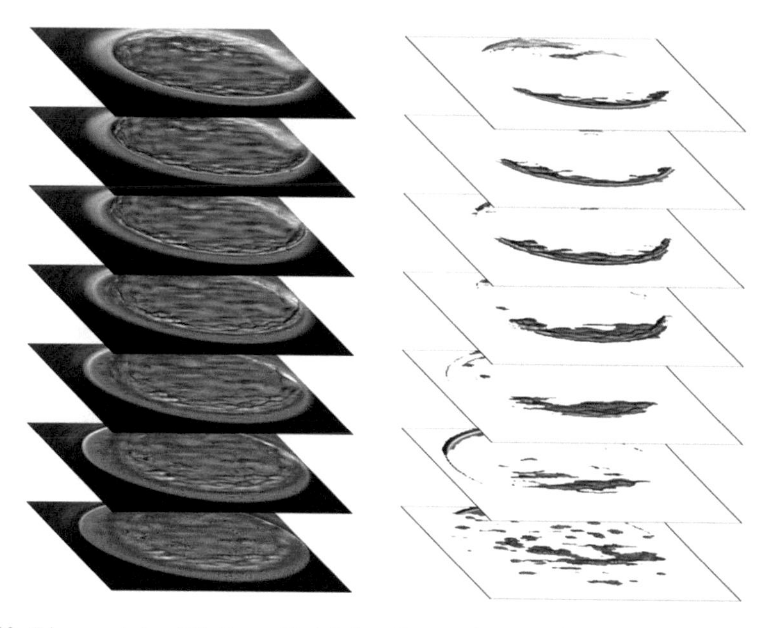

Fig. 10 7-image stack of Hoffman Modulation Contrast images of blastocyst stage embryo at 136.2 h, and the extracted focused regions. The extracted regions have reversed contrast for display purposes

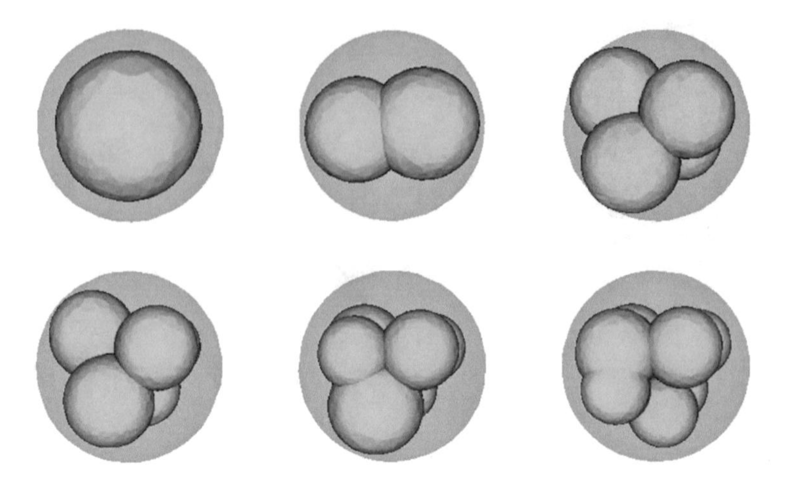

Fig. 11 Cleavage stage embryo observed during development from 0 to 34 h, showing 1, 2, 4, 5, 6 and 7 cell embryos, respectively. The boundary of the zona pellucida is indicated in *grey*

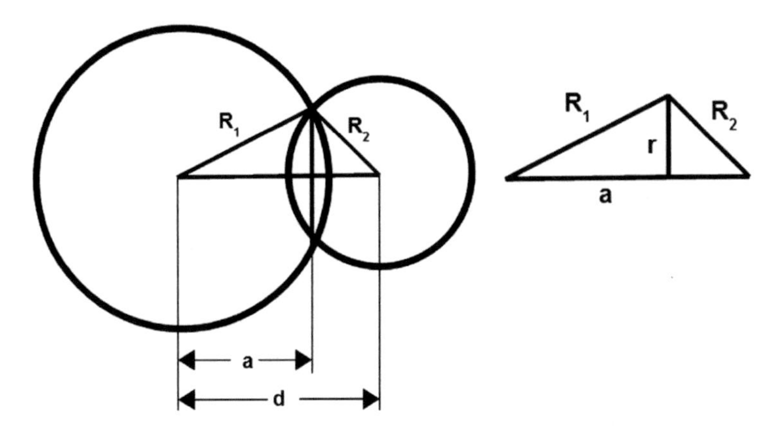

Fig. 12 The intersection plane between two spheres is described by a *circle*, definitions

In Fig. 11, six steps of development from one to seven blastomere stages of a cleavage stage embryo have been modelled.

When modelling each blastomere as a sphere, it is possible to compute an estimate of the blastomere connectivity, C. Let R_1 and R_2 be the radius of two intersecting model spheres, located a distance $d \leq R_1 + R_2$ from each other (Fig. 12). We assume the two spheres follow the physics of a double bubble [14]. Let a be the distance from the centre of the sphere with radius R_1 to the central point of intersection between the two bodies. The intersection will take the shape of a circle, with radius r, giving the contact area,

$$A_{con} = \pi r^2. \tag{3}$$

The connectivity

$$C = \frac{A_{con}}{(A_{con} + A_{uncon})} \tag{4}$$

where the part of the spherical surface outside the intersection, A_{uncon} can be calculated numerically from the model;

$$A_{uncon} = 4p\pi R^2 \tag{5}$$

where p is the part of the spherical surface of the model not contained within the neighbouring sphere. The value p is given by integrating over the surface of the sphere across an angle φ and dividing with the total area of the sphere,

$$p = \frac{1}{4\pi r^2} \cdot \int_{-\pi}^{\pi} \int_{0}^{\varphi} r^2 sin(\theta)\delta\theta \cdot \delta\varphi, \tag{6}$$

where

$$sin(\varphi) = \frac{r}{R_{1,2}}. \tag{7}$$

In Fig. 13, the blastomere volume and connectivity have been calculated for the embryo in Fig. 11.

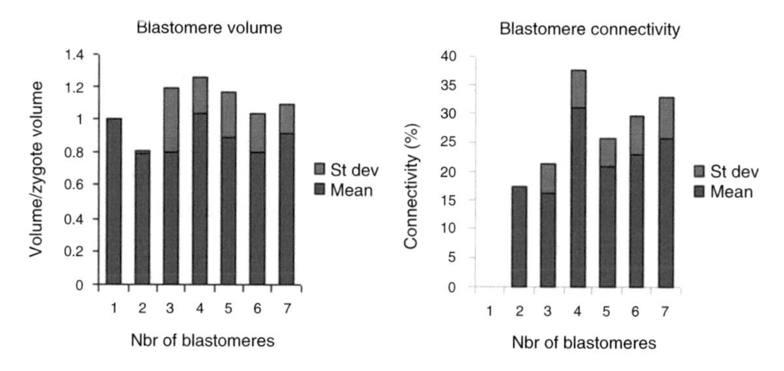

Fig. 13 Blastomere volume and connectivity for the embryo in Fig. 11. Blastomere volume has been normalised to the volume of the zygote and to number of cells in the embryo. The connectivity is displayed as percentage of surface area. Bars show both mean and standard deviation, calculated for each number of blastomeres. Apart from the 4 cell stage, which differs in both volume and connectivity, the total volume is constant, whereas the connectivity increases

5 Conclusion

We have shown how two dimensional image data can be used to construct a simple three dimensional model, which in turn can be used to extract additional data from embryological data sets and also how computer analysis may be used to supplement two dimensional imaging. With a three dimensional display, it is possible to view the sample from different directions, thus getting a clearer view of its spatial layout, and improving interaction with the sample. Using 3D plotting, not only can we get a much more intuitive understanding of the embryo structure and the positioning of blastomeres relative to each other, it is also possible to take measurements of cellular or nuclear volume and blastomere connectivity, which are not possible with a single scan.

When optically sectioning microscopy data, it is not always possible to rely on a short illumination path along the focus axis, as is possible in confocal microscopy, and the image will always suffer bleed-through from out-of-focus areas. We illustrate here how this problem to some extent may be overcome using pattern recognition and image treatment. A reliable and robust algorithm for focus level detection may be useful not only as a pre-processing step for image analysis, but also as a method for automatic focus of microscope hardware, when the images are being captured during long periods of time, when the capturing process is automatic, or when the microscope is placed in a climate chamber, as is often the case in time-lapse sequencing.

It is a challenging task for future developers, researchers and entrepreneurs to emulate real life events and actions by visualisation tools used to carry out analytical medical tasks. Generally speaking, the more technical, detailed and descriptive a function is, the longer, harder and more time consuming it is to replicate. Microscopic images are among the most detailed to be found, and the implications of making wrong decisions in a clinical environment are potentially severe. Hence, any software solution must be implemented with great care. However, it is not always necessary to simulate every aspect of reality, and a healthy scepticism for technology must not prevent us from embracing it in areas where it can increase knowledge and help reduce human suffering. The task of designing a model which simulates all the relevant features of a growing human embryo in a realistic manner will without doubt require a high degree of future cooperation between software developers and clinical scientists active in the field.

Acknowledgements The authors are very grateful for the raw image material needed for this survey, of which some was graciously provided to us by Fertilitech, Copenhagen, Denmark. We also thank everyone at Warwick Medical School who has helped us along the way.

References

1. Abramoff, M.D., Magalhaes, P.J., Ram, S.J.: Image processing with ImageJ. Biophoton. Int. **11**(7), 36–42 (2004)
2. Alm, K., Cirenajwis, H., Gisselsson, L., Gjorloff, A., Janicke, B., Molder, A., Oredsson, S., Persso, J.: Digital holography and cell studies. In: Rosen, J. (ed.) Holography, Research and Technologies. InTech (2011)
3. Baldock, R.A., Bard, J.B., Burger, A., Burton, N., Christiansen, J., Feng, G., Hill, B., Houghton, D., Kaufman, M., Rao, J., et al.: EMAP and EMAGE: a framework for understanding spatially organized data. Neuroinformatics **1**, 309–325 (2003)
4. Davidson, M.W., Abramowitz, M.: Optical Microscopy. Encyclopedia of Imaging Science and Technology. Wiley, New York (2002)
5. de Chaumont, F., Dallongeville, S., Chenouard, N., Hervé, N., Pop, S., Provoost, T., Meas-Yedid, V., Pankajakshan, P., Lecomte, T., Le Montagner, Y., Lagache, T., Dufour, A., Olivo-Marin, J.-C.: Icy: an open bioimage informatics platform for extended reproducible research. Nat. Methods **9**, 690–696 (2012)
6. Garello, C., Baker, H., Rai, J., Montgomery, S., Wilson, P., Kennedy, C.R., Hartshorne, G.M.: Pronuclear orientation, polar body placement, and embryo quality after intracytoplasmic sperm injection and in-vitro fertilization: further evidence for polarity in human oocytes? Hum. Reprod. **14**, 2588–2595 (1999)
7. Geusebroek, J.M., Cornelissen, F.: Robust autofocusing in microscopy. Cytometry **39**, 1–9 (2000)
8. Giusti, A., Corani, G., Gambardella, L.M., Magli, M.C., Gianaroli, L.: Blastomere segmentation and 3D morphology measurements of early embryos from Hoffman Modulation Contrast image stacks. IEEE Int. Symp. Biomed. Imaging Nano Macro 1261–1264 (2010)
9. Groen, F.C.A., Young, I.T., Ligthart, G.: A comparison of different focus functions for use in autofocus algorithms. Cytometry **6**, 81–91 (1985)
10. Guerif, F., Le Gouge, A., Giraudeau, B., Poindron, J., Bidault, R., Gasnier, O., Royere, D.: Limited value of morphological assessment at days 1 and 2 to predict blastocyst development potential: a prospective study based on 4042 embryos. Hum. Reprod. Oxf. Engl. **22**, 1973–1981 (2007)
11. Hardarson, T., Caisander, G., Sjögren, A., Hanson, C., Hamberger, L., Lundin, K.: A morphological and chromosomal study of blastocysts developing from morphologically suboptimal human pre-embryos compared with control blastocysts. Hum. Reprod. Oxf. Engl. **18**, 399–407 (2003)
12. Heid, P.J., Voss, E., Soll, D.R.: 3D-DIASemb: a computer-assisted system for reconstructing and motion analyzing in 4D every cell and nucleus in a developing embryo. Dev. Biol. **245**, 329–347 (2002)
13. Hoffman, R.: The modulation contrast microscope: principles and performance. J. Microsc. **110**, 205–222 (1977)
14. Isenberg, C.: The Science of Soap Films and Soap Bubbles. Dover Publications, New York (1992)
15. Kankaanpää, P., Paavolainen, L., Tiitta, S., Karjalainen, M., Päivärinne, J., Nieminen, J., Marjomäki, V., Heino, J., White, D.J.: BioImageXD: an open, general-purpose and high-throughput image-processing platform. Nat. Methods **9**, 683–689 (2012)
16. Lamprecht, M.R., Sabatini, D.M., Carpenter, A.E.: CellProfiler: free, versatile software for automated biological image analysis. BioTechniques **42**(1), 71–75 (2007)
17. Lichtman, J.W., Conchello, J.-A.: Fluorescence microscopy. Nat. Methods **2**, 910–919 (2005)
18. Liu, Y.-C., Chiang, A.-S.: High-resolution confocal imaging and three-dimensional rendering. Methods **30**, 86–93 (2003)
19. Mölder, A., Czanner, S., Costen, N.: Multidimensional visualisation to improve the understanding of biological data sets. In: Central European Seminar on Computer Graphics, vol. 17 (2013)

20. Nomarski, G.: Microinterferometrie differentiel a ondes polarisees. J. Phys. Radium **16**, 9S–11S (1955)
21. Payne, D., Flaherty, S.P., Barry, M.F., Matthews, C.D.: Preliminary observations on polar body extrusion and pronuclear formation in human oocytes using time-lapse video cinematography. Hum. Reprod. Oxf. Engl. **12**, 532–541 (1997)
22. Price, J.H., Gough, D.A.: Comparison of phase-contrast and fluorescence digital autofocus for scanning microscopy. Cytometry **16**, 283–297 (1994)
23. Rafferty, K., Drury, S., Hartshorne, G., Czanner, S.: Use of concave corners in the segmentation of embryological datasets. Rev. Bioinforma Biom **1**, 1–8 (2012)
24. Rajaram, S., Pavie, B., Wu, L.F., Altschuler, S.J.: PhenoRipper: software for rapidly profiling microscopy images. Nat. Methods **9**, 635–637 (2012)
25. Schindelin, J., Arganda-Carreras, I., Frise, E., Kaynig, V., Longair, M., Pietzsch, T., Preibisch, S., Rueden, C., Saalfeld, S., Schmid, B., Tinevez, J.-Y., White, D.J., Hartenstein, V., Eliceiri, K., Tomancak, P., Cardona, A.: Fiji: an open-source platform for biological-image analysis. Nat. Methods **9**, 676–682 (2012)
26. Schoolcraft, W.B., Gardner, D.K., Lane, M., Schlenker, T., Hamilton, F., Meldrum, D.R.: Blastocyst culture and transfer: analysis of results and parameters affecting outcome in two in vitro fertilization programs. Fertil. Steril. **72**, 604–609 (1999)
27. Scott, L.A., Smith, S.: The successful use of pronuclear embryo transfers the day following oocyte retrieval. Hum. Reprod. Oxf. Engl. **13**, 1003–1013 (1998)
28. Sun, Y., Duthaler, S.: Autofocusing in computer microscopy: selecting the optimal focus algorithm. Microsc. Res. Tech. **65**, 139–149 (2004)
29. Tassy, O., Daian, F., Hudson, C., Bertrand, V., Lemaire, P.: A quantitative approach to the study of cell shapes and interactions during early chordate embryogenesis. Curr. Biol. **16**, 345–358 (2006)
30. Welch, P.: The use of fast Fourier transform for the estimation of power spectra: a method based on time averaging over short, modified periodograms. IEEE Trans. Audio Electroacoust. **15**, 70–73 (1967)
31. Zhong, Q., Busetto, A.G., Fededa, J.P., Buhmann, J.M., Gerlich, D.W.: Unsupervised modeling of cell morphology dynamics for time-lapse microscopy. Nat. Methods **9**, 711–713 (2012)

Quantitative Analysis of Knee Movement Patterns Through Comparative Visualization

Khoa Tan Nguyen, Håkan Gauffin, Anders Ynnerman, and Timo Ropinski

Abstract In this paper, we present a novel visualization approach for the quantitative analysis of knee movement patterns in time-varying data sets. The presented approach has been developed for the analysis of patellofemoral instability, which is a common knee problem, caused by the abnormal movement of the patella (kneecap). Manual kinematic parameter calculations across time steps in a dynamic volumetric data set are time-consuming and prone to errors as well as inconsistencies. To overcome these limitations, the proposed approach supports automatic tracking of identified features of interest (FOIs) in the time domain and, thus, facilitates quantitative analysis processes in a semiautomatic manner. Moreover, it allows us to visualize the movement of the patella in the femoral groove during an active flexion and extension movement, which is essential to assess kinematics with respect to knee flexions. To further support quantitative analysis, we propose kinematic plots and time-angle profiles, which enable comparative dynamics visualization. As a result, our proposed visualization approach facilitates better understanding of the effects of surgical interventions by quantifying and comparing the dynamics before and after the operations. We demonstrate our approach using clinical time-varying patellofemoral data, discuss its benefits with respect to quantification as well as medical reporting, and describe how to generalize it to other complex joint movements.

1 Introduction

Recent advances in medical imaging technologies enable the acquisition of organ functions. This leads to many possible clinical applications, such as studies of activities in the brain using functional magnetic resonance imaging (fMRI), or studies of the dynamics of a beating heart through dynamic computed tomography (CT)

K.T. Nguyen (✉) • A. Ynnerman • T. Ropinski
Scientific Visualization Group, Linköping University, Linköping, Sweden
e-mail: tankhoa@gmail.com

H. Gauffin
Orthopedic Department, Linköping University Hospital, Linköping, Sweden
e-mail: hakan.gauffin@lio.se

L. Linsen et al. (eds.), *Visualization in Medicine and Life Sciences III*, Mathematics and Visualization, DOI 10.1007/978-3-319-24523-2_12

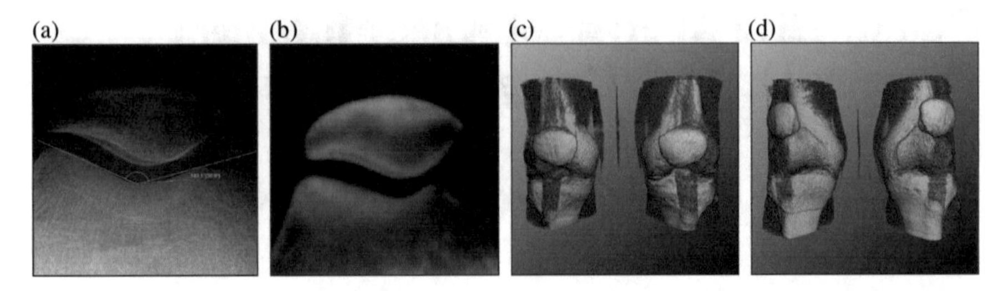

Fig. 1 2D and 3D views of a knee joint. While the conventionally used Skyline view captured at 30–45° of knee flexion looks normal based on X-ray (**a**) as well as 3D CT (**b**), only the joint movement pattern imaged through 4D CT [(**c**) and (**d**)] reveals the patella dislocation issue at the full extension. (**a**) Skyline view. (**b**) Rendering of 3D CT. (**c**) Rendering of a knee at active flexion. (**d**) Rendering of a knee at full extension

scans. Another potential clinical application is dynamic image-based orthopedics. Although, static 3D CT and magnetic resonance (MR) imaging data have been used to support the analysis of joints, they cannot capture the dynamics of joint movement patterns. As illustrated in Fig. 1, while the renderings from the static data indicate no patellofemoral instability (see Fig. 1a, b), the renderings of data captured in the dynamic CT scan reveal patellofemoral instability at extreme positions (see Fig. 1c, d).

There are three main challenges that arise from 4D CT imaging-based orthopedics. The first challenge is the quantification of complex joint movements based on dynamic CT data. While inspecting an animation of the dynamic data might help the medical doctor to get a first impression about underlying conditions, quantification is necessary to support a proper categorization. Since dynamic CT scans contain a vast amount of information, manual calculation of kinematic parameters is time-consuming and can lead to inconsistent results. The second challenge lies in the comparative analysis of joint movement patterns. When detected conditions require a surgical intervention, a follow up scan is often performed to measure the success of the operation. However, measuring surgical success requires a comparison of the status before and after the intervention, which makes a comparative analysis necessary. Finally, as medical reporting is required in all medical disciplines, sufficient techniques are required to be able to report on detected conditions.

To meet the challenges in 4D CT image-based orthopedics, we propose novel visualization metaphors, which enable quantification, comparison, and reporting of joint movements in 4D CT data. We propose *radial plot*, which not only conveys the relations between angles but also helps to emphasize the changes and implicitly incorporates spatial information into the rendering of the plot. The *time-angle profile* presents different kinematic parameters and the joint movement patterns in a single image in such a way, that both quantitative and comparative visual analysis are supported. The proposed visual metaphors have been developed in close collaboration with medical experts who wish to investigate patellofemoral instabilities using time-varying CT acquisitions. In addition, we propose a semiautomatic GPU-based feature tracking technique that allows the manually selected features of interest (FOIs) to be tracked automatically across volumes in dynamic CT data. This enables automatic and consistent quantitative measurements of kinematic parameters from the selected FOIs. The main contributions of the paper can be summarized as follows:

- Visualization metaphors for the quantitative assessment and comparison of joint movement patterns.
- Component separation and visualization to enable medical reporting of joint movement patterns.
- Interactive visual inspection of joint movement data based on an enhanced GPU-based feature identification and tracking technique based on the SIFT algorithm.

2 Medical Background

The patella (kneecap) is a bone that is incorporated into the tendon of the quadriceps muscles of the thigh and moves within a groove at the lower end of the femur (thigh bone). When the knee bends, the patella engages the groove and is fixed to the center of the groove by muscle forces. On the other hand, the patella moves to the upper shallower end of the groove when the knee is straightened. Consequently, the patella will be more loose, and may move somewhat more lateral (to the outside) [1]. The medial patellofemoral ligament (MPFL) is the primary medial restraining structure against lateralization of the patella when the knee is straight or mildly flexed, and contributes up to 80 % of the medial restraining forces to the patella [2].

A dislocation of the patella occurs when the patella comes completely out of its groove, gets fixed to the outside of the knee joint, which causes significant pains. This problem commonly happens during sporting activities to young and physically active people. The annual incidence of primary patella dislocations has been estimated to 43 per 100,000 in children under the age of 15 [3]. The first dislocation usually occurs as a significant injury with the knee in near full extension, but the patella may dislocate much more easily thereafter. Recurrent patella dislocations eventuate in 15–45 % of primary dislocation cases and can cause significant problems that not only prevent sport activity but also be a hindrance in daily activity [4].

There are more than hundred surgical options for patients with repeated dislocations [5]. It is possible to restore the normal anatomy by repairing the torn ligaments on the inner aspect of the knee, to deepen the groove, or to realign the patella tendon to stabilize the patella in a more medial (inner) position [1]. In the last decade, it has been increasingly popular to reconstruct the MPFL, often as an isolated procedure [6]. However, the lack of accurate follow-up methods to assess the movement of the patella in relation to the groove after rehabilitation or surgical intervention has led to the multitude of operations.

Ordinary radiographic methods are inadequate for the assessment of patellofemoral malalignment. The most commonly used technique for visualization of the patellofemoral joint requires 30–45° of knee flexion (see Fig. 1a) [5, 7, 8]. This is a static examination and the images are not obtained near full extension where the groove is most shallow and the patella is unstable. While conventional CT and MRI scans can image a straight knee, they are static and will not display patella tracking under functional movements [5]. With the use of 4D CT, it becomes possible to get a 3D visualization of the patella's tracking in the femoral groove during a functional active flexion and extension movement with normal muscular forces. However, it is difficult to describe 3D motion in a distinct way. Fixed body axes in either the patella or femur can give confusing results [5]. Thus, there is a great need for quantitative analysis of joint movements through apparent visualization methods to give a better understanding and also to get quantifiable results.

3 Related Work

Over past years, many visual representations have been proposed to convey the relation between angles, which is the major question in the analysis of joint movements [9]. The angle-angle plot, in which each axis represents a kinematic or kinetic signal, is commonly used. By plotting one signal against the others, an angle-angle plot illustrates the coordinated motion of the two segments or joints. Manal et al. [10] used an advanced color-coding scheme to incorporate additional parameters into the standard angle-angle plot. Later, the authors proposed the use of an additional dimension within the plot so that additional parameters can be visualized [11]. Côté et al. [12] presented 2D projected stick figures to show the kinematic results of continuous hammering performed by different patient groups. While the proposed approach was sufficient when looking at joint height, it failed to reveal the kinematic relationships. Keefe et al. [13] proposed a visualization system based on multiple views to show relationship within sequential kinematic data. Krekel et al. [14] focused on the visualization of the relationships between multiple connected joints by combining interactive filtering with visualization. The approach probably most relevant to our research is that of Krekel et al. [15], in which a technique for visualizing the range of motion of the shoulder joint is presented.

Scale-invariant feature transform (SIFT) algorithm was first proposed to match features between 2D images [16, 17]. Later, SIFT algorithm has been extended to handle n-dimensional data [18, 19] and successfully applied to different application scenarios [20–23]. Yu et al. [24] have shown that SIFT algorithm provides high performance and stable result in comparison to other techniques. Recently, Nguyen and Ropinski proposed an enhanced GPU-based implementation of SIFT algorithm that improves both performance and feature matching accuracy when dealing with dynamic volumetric datasets [25].

While other research has been conducted on the visualization of complex joints [14, 26], to our knowledge, the work in this paper is the first to tackle time-varying volumetric CT data in order to support the quantitative analysis of joint movement patterns. By directly deriving the kinematic parameters from the volumetric data in a semiautomatic manner, we can avoid patient instrumentation of markers.

4 Method

Figure 2 presents the workflow of the proposed approach. First, the dynamic CT scan is pre-processed. Particularly, stable feature locations are identified, extracted and matched between scans. In the interactive visual analysis stage, users can interactively select FOIs and the system automatically tracks these features between scans using the information from the pre-processing stage. Depending on the application in mind, different quantitative measurements of kinematic parameters based on the selected features are performed. The results are then visualized using the proposed visual metaphors, which we describe in detail in Sect. 5, to support quantitative and comparative analysis of the underlying kinematic information.

In the following subsections, we first describe the proposed feature identification and tracking technique. We then describe in detail a geometry setup that facilitates robust calculations of different kinematic parameters applied to the investigation of patellofemoral instability.

4.1 *Enhanced GPU-Based Feature Identification and Tracking*

The proposed feature tracking technique is based on the GPU-based implementation of the SIFT algorithm proposed by Nguyen and Ropinski [25]. To further improve the accuracy of the feature matching process, we propose a novel feature descriptor construction technique called *ring descriptor*. The algorithm is performed in three successive stages: *feature location detection*, *feature descriptor construction*, and *feature matching*.

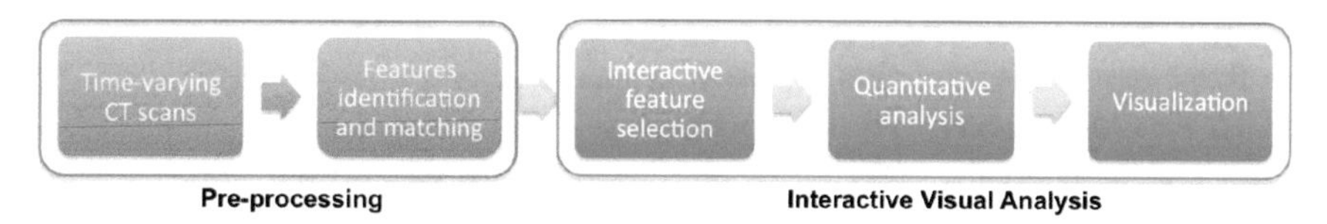

Fig. 2 The workflow of the proposed approach. In the pre-processing stage, features in a dynamic CT data set are extracted and matched between scans. In the second stage, user can interactively select FOIs. Based on the information from the pre-processing stage, selected FOIs are tracked automatically in the time domain. Depending on the application in mind, different quantitative measurements can be performed and visualized for assessment

Feature Location Detection In the first stage, stable features in the input image are identified through a scale-based analysis approach. The input volumetric data is convoluted with variable-scale Gaussian functions to generate a scale space. The local extrema of the difference-of-Gaussian functions applied to the constructed scale space represent the potential local features in the original data. Lindeberg [27] shows that these local extrema are close approximations to the scale normalized Laplacian-of-Gaussian, which are the most stable features in the input image [28].

The construction of the scale space from the input data and the detection of local extrema are computation-demanding processes. Fortunately, the calculations at each voxel are independent of the calculations at the others. As a result, by parallelizing the computation using the GPUs, a higher performance can be achieved.

Feature Descriptor Construction The aim of the second stage is to construct a unique descriptor to represent the detected feature location in such a way that is most invariant with respect to rotation, scaling, and translation. The construction of such a descriptor is based on the gradient orientations in the neighborhood of the detected feature location and presented as a histogram of gradient orientations. In 3D, a gradient orientation comprises three components: azimuth, elevation, and tilt angle. While the elevation and azimuth components can be derived directly from the gradient itself, the tilt angle can only be derived through a more complex analysis of the neighborhood [29].

In the first step of constructing a descriptor, the dominant orientation in the neighborhood around the detected feature location is identified. The other gradients are then re-oriented to the identified dominant orientation to achieve rotation invariant property. In addition, the neighborhood is subdivided into sub-regions to avoid disruptive changes of the gradient orientations and, thus, improve the uniqueness of the constructed descriptor. As a result, a descriptor is basically a concatenation of histograms of gradient orientations from sub-regions.

Previous research has shown that the size of the neighborhood can have a large impact on the uniqueness of the constructed descriptors, which affects the accuracy of the feature matching process [29, 30]. This is due to the fact that the size of the neighborhood is a global setting and is dependent on the input data. While a large size neighborhood (e.g., greater than $16 \times 16 \times 16$ voxels) might fail to capture the local characteristic of the detected feature, a small size neighborhood (e.g., less than or equal to $8 \times 8 \times 8$ voxels) might put too much emphasis on the local property.

To overcome this limitation, we propose a novel descriptor construction approach called *ring descriptor*. The neighborhood is divided into non-overlapping sub-regions centered at the detected feature location. For each non-overlapping sub-region, the dominant gradient orientation is identified. This allows us to capture not only the local characteristic of the detected features but also reduce the impact of the neighborhood size setting on the uniqueness of the final descriptor. The gradients in the neighborhood are re-oriented to the identified dominant gradient orientation of each non-overlapping ring before the construction of the descriptor.

One obvious advantage of the proposed technique is the ability to introduce a weighting factor into the construction process, which helps to emphasize or de-emphasize the local property of the detected feature depending on the input data. For instance, even with large neighborhood size settings, users can increase the weighting factor for the dominant orientation identified in the inner non-overlapping ring thus emphasize the local property of the feature and regard the property of the whole neighborhood at the same time. As a result, the size of the neighborhood has less impact on the result, which makes the proposed approach more flexible in handling different type of dynamic volumetric data.

It is a challenge to transform a standard CPU-based implementation of SIFT to a GPU-based one. Due to the limitation of the memory architecture on the GPU, i.e. a large histogram cannot be fit into local/share memory, the size of the histogram has a big impact on the performance. Since the proposed technique allows us to reduce the size of the constructed descriptors, and improve the uniqueness at the same time by subdividing the neighborhood into non-overlapping ring, we can fully exploit the computing power of the GPU to achieve high performance and, thus, support interactive visual analysis.

It is worth mentioning that a smoothing operator is usually applied to the constructed descriptor to reduce the impact of disruptive changes of gradient orientations in a standard implementation of SIFT [17]. As the type and the bounds of the smoothing operator are abstracted from the original input data, it is difficult to identify a good smoothing operator to improve the quality of the constructed descriptor. In the proposed technique, by moving the computing from CPU to GPU, we do not only achieve a higher performance but also implicitly achieve the interpolation supported by the hardware. Particularly, the GPU hardware enables us to achieve a bi-linear interpolation without any cost. Moreover, instead of re-orienting gradient orientations in the neighborhood around a detected feature location, the whole neighborhood is re-oriented to the dominant orientation. Hence, gradient orientations are updated accordingly. This allows us to minimize the effect of different interpolation scheme applied to the constructed descriptors. More importantly, the uniqueness of the constructed descriptors are improved as the interpolation in our approach is directly based on the underlying information from the input data.

Feature Matching Once the features descriptors have been constructed for two data sets to be compared, they can be used to identify matching features using different techniques such as RANSAC [31], Best-Bin-First (BBF) [32]. The minimum Euclidean distance between descriptors is commonly used as an indicator for a possible match. Consequently, to identify possible matches between two data sets, the minimum Euclidean distance between each descriptor in the first data set and all descriptors in the second data set must be identified. We refer to the work of Nguyen and Ropinski [25] for further detail on the GPU-based implementation of SIFT algorithm.

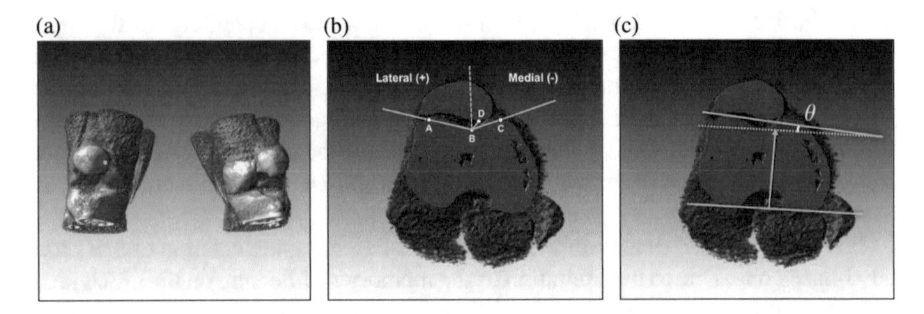

Fig. 3 Congruence and tilt angle measurement. (**a**) is the visualization of the knee overlain with a FOI (*green*), which marks the widest part of the on the left knee. (**b**) and (**c**) are the illustrations of the congruence and tilt angle measurement through a clip plane setup. (**a**) Knee overlain with a FOI (*green*). (**b**) Congruence angle measurement. (**c**) Tilt angle measurement

4.2 Kinematic Parameters Measurement

Over past years, many researches focusing on the methods to describe patello-femoral joint motion have been made. The results can be classified into two groups: those that describe motion of the patella in relation to the femoral groove and those that describe the motion of the patella in relation to a fixed-body axis [5]. The former are primarily image-based studies [33–35]. One significant drawback of previous results is the inability to describe the patella motion in three-dimensional space.

The congruence and tilt angles between the patella and the femoral groove are two commonly used kinematic parameters in quantitative analysis of the patello-femoral issue. As the congruence and tilt angles reflect the relation between the patella and the femoral groove, we propose a geometry setup, in which the patella and the clip plane cut through its widest part serve as the reference, for quantitative kinematic measurement (see Fig. 3). The congruence angle is then defined by the angle between the bisected femoral trochlea (dashed line in Fig. 3b) and the vector connecting the apex of sulcus angle, ABC, to the lowest point of the articular ridge of the patella (BD in Fig. 3b). By using the same geometry setup, the tilt angle, θ, is defined as the angle subtended by a line joining the lateral edge and the lowest point of the articular ridge of the patella and the posterior condylar line as illustrated in Fig. 3c.

Besides the ability to describe the patella motion in three-dimensional space, the combination of the proposed kinematic parameters measurement approach and the feature identification and tracking allows us to have a semi-automatic measurement of kinematic parameters. Through an interactive visualization setup, users can manually select FOIs as illustrated in Fig. 3a. These selected FOIs are then automatically tracked across scans in the input dynamic volumetric data set based on the information derived in the pre-processing stage. Particularly, based on matched features in the pre-processing stage, affine transformations can be derived and applied to the selected FOIs to identify the corresponding positions across scans. The results can be further refined by taking a small neighborhood, i.e., 5×5×5 neighborhood, around the calculated locations into consideration. Then, for each voxel in the neighborhood, a descriptor is constructed and matched with the corresponding element in the selected FOIs to refine the tracking result.

It is worth pointing out that when the patella moves out of the femoral groove, there is no intersection between the clip plane and the femoral groove, thus the congruence angle cannot be calculated. Consequently, the proposed approach reflects the fact that there is no congruence between the patella and the femoral groove, which the previous research results failed to capture and present.

In addition to the measurement of the congruence and tilt angle, we propose the measurement of the distance, d_G, from the lowest point of the articular ridge of the patella, R, to the femoral groove to convey the translation of the patella in relation to the femoral groove as follows

$$d_G = \min_{1 \leq i \leq N} d(R, p_i) \tag{1}$$

where N is the number of point on the femoral groove, and d is the Euclidean distance function. In this approach, the femoral groove is approximated by a set of discrete points in order to reduce the computation demand. The presented feature tracking technique allows us to track the identified points, which approximate the femoral groove, in the time domain to facilitate the calculation and reduce the time-consuming process of manually select features through out all the scans in a dynamic data set. By applying the same approach, the distance between the drill holes on the patella and the drill hole on the femoral can also be calculated. The quantitative results enable doctors to assess the success of the surgical intervention in comparison to the pre-surgical intervention as well as the effect after the surgical intervention.

5 Comparative Visualization

In Sect. 4, we describe the proposed feature identification and tracking algorithm and the geometry setup that facilitates robust calculations of different kinematic parameters for the quantitative analysis of the patellofemoral instability. In this section, we describe in detail the proposed visual metaphors to visualize the derived kinematic information.

5.1 Radial Angle Plot

The angle-angle plot is one of the most commonly used visual metaphor for depicting kinematic parameters [10]. In a traditional set up, each axis of the plot

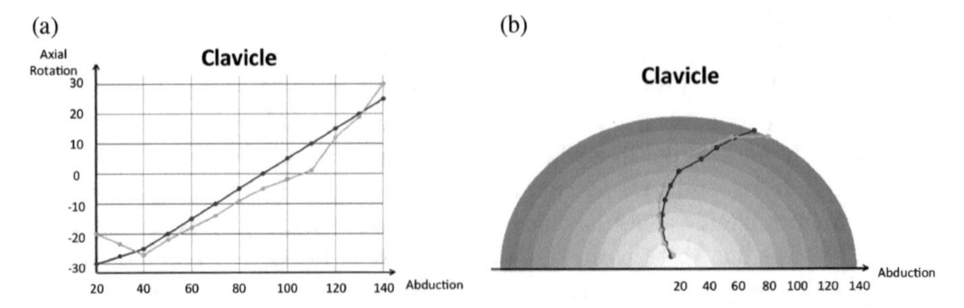

Fig. 4 The relations between the abduction angle and the axial rotation of two subjects. The standard angle-angle plot allows quantification of the relation but makes mental registration with the spatial context difficult (**a**). The proposed radial angle plot represents the same data in a joint centered way, while at the same time exaggerating the differences for critical abduction angles (**b**). (**a**) Angle-angle plot. (**b**) Radial angle plot

represents a kinematic parameter, whereby the plot reveals the relation of the individual parameters to each other. Figure 4a shows an example of the angle-angle plot depicting the relations between the abduction angle and the axial rotation of two subjects. Although the angle-angle plots help to reveal the relations between the two kinematic parameters, its representation does not reflect the visual mapping of the presented angular values within the context of the joint. Consequently, the interpretation as well as the mental linking of the plot to the actual movement patterns requires a lot of experience and time. This becomes especially apparent when observing the comparative qualities of the angle-angle plot. When comparing the plotted data for Patient 1 (*orange*) and Patient 2 (*blue*), it becomes clear that the movement of Patient 1 includes a lower degree of axial rotation for abduction levels 40–130. However, as the spatial reference of the joint is missing, the rotation difference with respect to the joint is not immediately visible.

To address these shortcomings, we present *radial angle plots* as a joint-centered visualization metaphor to depict kinematic parameter relations. The radial angle plots have been designed in such a way that they can be easily applied to various joint types. Instead of using linear mappings for the x and the y-axis as illustrated in Fig. 4a, we combine linear and polar mappings of the angular values. While the x-axis and, thus, the corresponding concentric circles, is the linear representation of one kinematic parameter, the y-axis is converted into a polar representation as shown in Fig. 4b. This enables the visual mapping of the plotted values to the actual angular changes, as given within the frame of reference of the joint under investigation. The concentric layout of the angular relations between kinematic parameters provides an implicit exaggeration of the most interesting kinematic parameter range, i.e., an angle representing full abduction or flexion. When dealing with joint issues, these extreme angles are the most problematic, even small changes need to be considered and should thus be emphasized by the used visualization metaphor.

Figure 4 compares the traditional angle-angle plot to the proposed radial angle plot. In Fig. 4b, the x-axis represents the abduction angle while the axial rotations

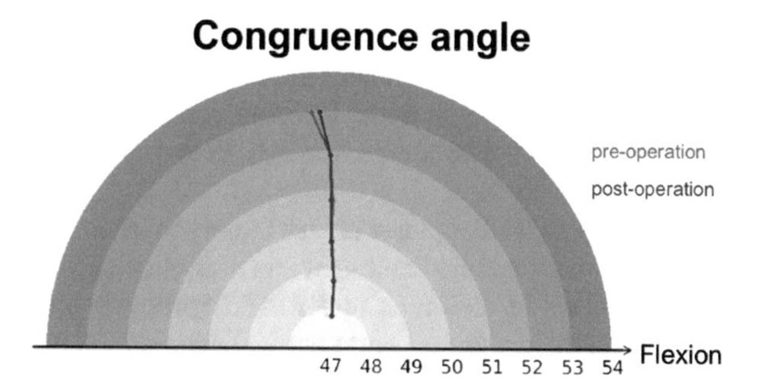

Fig. 5 A radial angle plot of the congruence angles from pre-operation (*red*) and post-operation (*blue*)

are placed on the corresponding shaded concentric half-circle (from −90 degrees to +90 degrees, from left to right). The gray scale gradient has been chosen to allow an intuitive quantification of the polar angle coordinates, as well as an emphasis of the joint center, which is located in the middle at 0 degrees. While the angle-angle plot also depicts the relations between the two kinematic parameters for both patients, the radial angle plot reflects the actual angular changes directly through its visual mapping.

When dealing with the patellofemoral analysis, which is addressed in Sect. 6, an additional benefit of radial angle plots is the fact, that they can be directly co-registered with the 2D Skyline view representation. As illustrated in Fig. 5, the concentric layout and the polar representation of the congruence values not only enables us to emphasize the difference between pre- and post-operation result at extreme position (53 degrees flexion) but also implicitly co-registers the rotation of the kneecap to the femoral groove (center of the radial angle plot). Moreover, the radial angle plot also reflects the medial and lateral patella angles corresponding to the negative and positive halves of the rendering. It is worth pointing out that in the design of the radial angle plot, the range of the angular value can be normalized accordingly to emphasize the changes.

5.2 *Time-Angle Profile*

Although the 3D visualization of a full scan in a time-varying data set can provide both the visual context and the detail information to support visual analysis, there is a large amount of data presented to the user, especially in the case of 4D CT. In addition, the occlusions in the 3D visualization of volumetric data can change drastically in the time domain, and affect the visual access to areas of interest. Consequently, this can cause distractions when a user changes from one time frame to another. As the kinematic analysis process usually focuses on a defined organ and derive its movement patterns over time, an abstract visualization focusing on the organ and its kinematic characteristic can help to improve the visual analysis process. We propose an abstract visualization of a defined organ in the time domain called *time-angle profiles*. The proposed visual metaphor is designed in such a way that it helps to filter out unnecessary information while maintaining the kinematic parameters of the organ under investigation. At the same time, it captures the whole movement within a single picture and, thus, supports medical reporting and pattern comparison.

To generate this representation, the organ must be identified in all time frames before all instances are combined into a single visualization. In a standard approach, the extraction of the organ under investigation in all scans of a dynamic data set is very time-consuming. Fortunately, based on the proposed feature identification and matching technique, affine transformations can be derived using the matched features contained in the identified organ. As a result, a single extraction of the organ is enough for the generation of the proposed visual representation. The combination of all instances of the organ in all time frames is then visualized through a 3D sweep structure. This sweep representation provides an overview of the whole kinematic changes in the time domain. However, it does not facilitate visual assessment of kinematic parameters due to the fact that a complex movement pattern is usually a combination of different components such as rotations, translations, and tilting. Inspired by the magic mirror metaphor introduced by König et al. [36], we exploit the use of projection techniques to decompose the movement patterns for the extracted organ. By projecting the whole movement pattern onto a multi-planar plane geometry, we can reveal the underlying components of the kinetic parameters. For instance, a projection from the top view can reflect the rotation, and a projection from the side view can shows the translation over time. In addition, the semi-transparent sweep structure is color coded to depict the various degrees of kinematic parameter changes.

Figure 6 illustrates the proposed visual metaphor applied to the visualization of the movement pattern of the patella. At the center of the image is the extracted movement pattern of the patella over time. The transition from a lighter to a darker color depicts the degree of kinematic parameter changes. For instance, as the patella rotates, tilts and translates more at the extreme position in comparison to the initial relaxed one, the color goes from a dark to a light shade of blue. By projecting the movement pattern using different viewing angles, the proposed visualization approach can achieve the decomposition of the complex kinematic characteristics. The orthographic projection of the trajectory onto the side plane reveals the shifting of the patella out of its groove. The projection onto the bottom plane reflects the changes in congruence angles, while the projection onto the back plane shows the degree of tilting. Although the proposed technique does not provide the contextual information by focusing only on the movement pattern, it can be used as an adjunct to the 3D visualization of the time-varying data in a multiple linked view setup. As a result, contextual information and an overview of the kinematic changes over time can be visualized in an integrated manner.

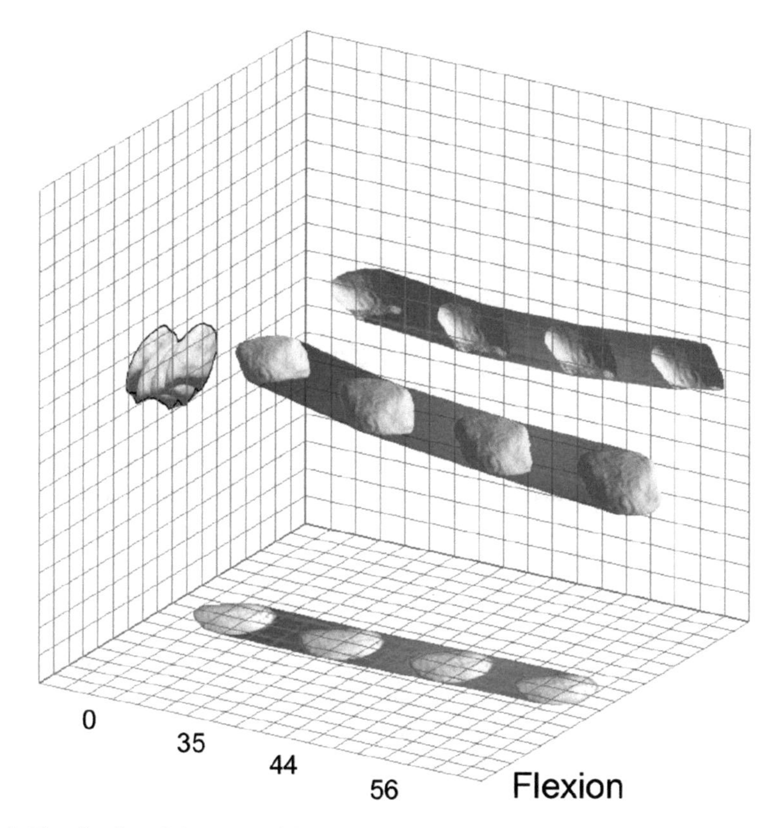

Fig. 6 Visualization of the proposed time-angle profile metaphor. The extracted movement pattern of the patella over time is rendered at the center of the image. Kinematic parameters are incorporated through projection technique onto the *bottom*, *side*, and *back plane*

6 Results and Discussion

We applied to proposed approach to two real clinical data sets to verify the ability to facilitate quantitative analysis of the patella movements. The first subject, patient A, is a 19 years old female with bilateral pronounced instability problems without any significant injury since she was 13 years old. The left and right side of patient A were operated in 2011 and 2012 respectively. The left side has been stable since the operations and is now asymptomatic. However, patient A had an injury on the right side after the operation. The situation makes it hard to assess the reconstructed tendon with CT scan on the right side on which patient A still has some instability symptoms. As a result, a re-rupture or an elongation of the reconstruction is under suspicion. The second subject, patient B, is a 23 years old male with distinct problems after an injury on the left side. Patient B also has minor instability problems on the right side. In 2013, patient B had a successful operation on the left side. Figure 7a illustrates an overview of the patellofemoral issue in both cases,

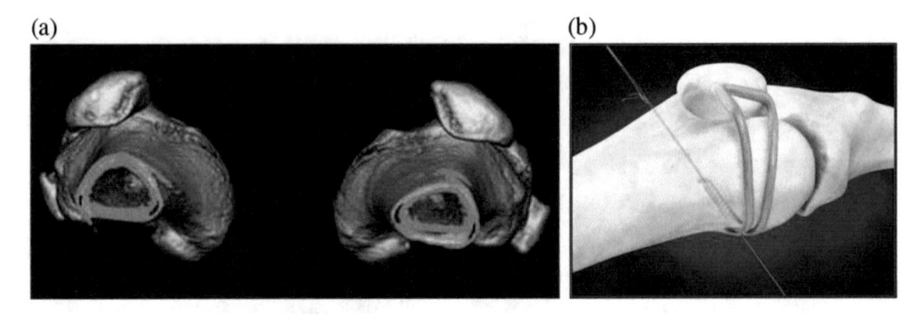

Fig. 7 Visualization of straight knees before the operation (**a**), and the reconstruction of MPFL with a gracilis tendon (**b**) (permitted by Storz)

in which the kneecaps are far out and tilted on the lateral side. The operations were performed with a reconstruction of the MPFL with a gracilis tendon as illustrated in Fig. 7b.

Data Preparation About a month before the operation, a dynamic CT scan is performed. The patient is supine with both legs resting on a radiolucent knee-support. The maximal knee angle is about 40–50 ° when the heels are touching the bed. The patient performs bilateral active knee extensions until full extension, referenced as 0 degrees, and then flexions. The acquired dynamic CT scans were pre-processed using the GPU-based enhanced feature identification and tracking approach proposed in Sect. 4.1. In the interactive visual analysis stage, the domain experts interactively select FOIs (see Fig. 3a) in one time frame. Based on the information from the pre-processing stage, selected FOIs are tracked automatically through the whole dynamic CT scan. Kinematic parameters are then calculated in an automatic manner. The results are then presented using the proposed visualization metaphors to convey the underlying kinematic parameters of interest.

Quantitative Visual Analysis In the following paragraphs, we report quantitative measurement results as well as the comparative visualization facilitated by the proposed approach. Besides the commonly used kinematic parameters such as the congruence and tilt angles, we also calculate the distance from the lowest point on the articulate ridge to the femoral groove, the distance between the drill holes on the kneecap to the drill hole on the femur, which are useful for the investigation of the patellofemoral instability.

Figure 8 presents the quantitative measurements of the distance from the lowest point of the articular ridge of the patella to the femoral groove. Patient A had subluxations of both kneecaps when straightening the knees before the operation. On the postoperatively asymptomatic left side, the distance to the groove has decreased from about 60–25 mm with straight knees. On the right side, patient A still has some instability symptoms; however, it is less than before the operation, the distance to the groove has decreased from about 53–28 mm. The result of the facilitated the measurement of the distance from the two drill holes on the kneecap

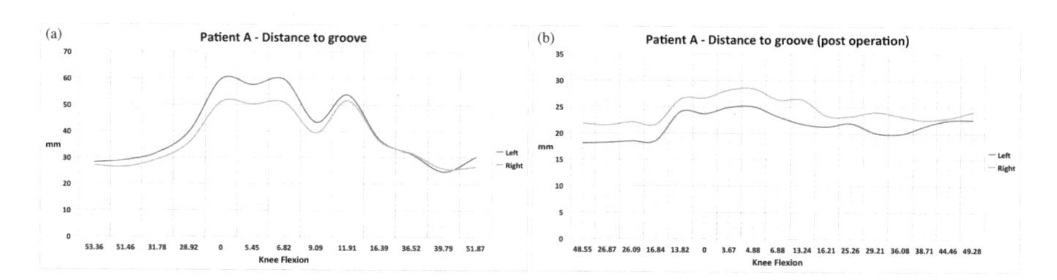

Fig. 8 Quantitative measurement of the distance from the lowest point on the articulate ridge to the femoral groove before operation (**a**), and post operation (**b**) from patient A

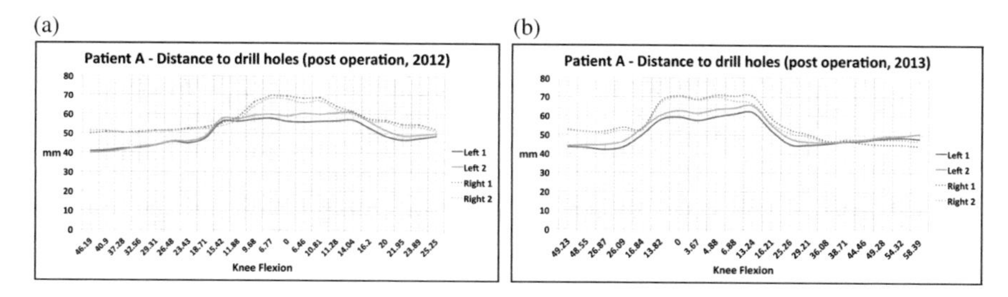

Fig. 9 Distance between the drill holes on the kneecap to the drill hole on the femur 3 months after the operation (**a**), and 1 year after the operation (**b**) for patient A. The measurements showed that the distances are stable between 3 and 12 months postoperatively

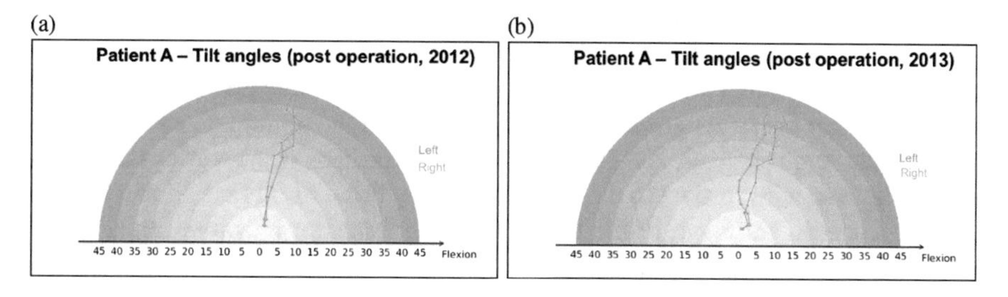

Fig. 10 Quantitative measurement of tilt angles of the kneecap 3 months (**a**) and 12 months after operation (**b**) for patient A

to the one on the femur for patient A is presented in Fig. 9. The measurement shows that the distances are stable between the 3 months post-operation and 12 months post-operation.

Figure 10 reports the measurements of tilt angles using the technique proposed in Sect. 4.2. The proposed radial angle plot presents not only the movement patterns of the kneecap but also present them with respect to the reference frame, which is the femoral groove. Figure 10a shows that the tilting pattern of the left and the right kneecap are similar. Due to the injury happened after the operation in 2012, patient A still has instability issue on the right side. Although the tilt angles are similar between the left and right side at full extension, the patterns are different

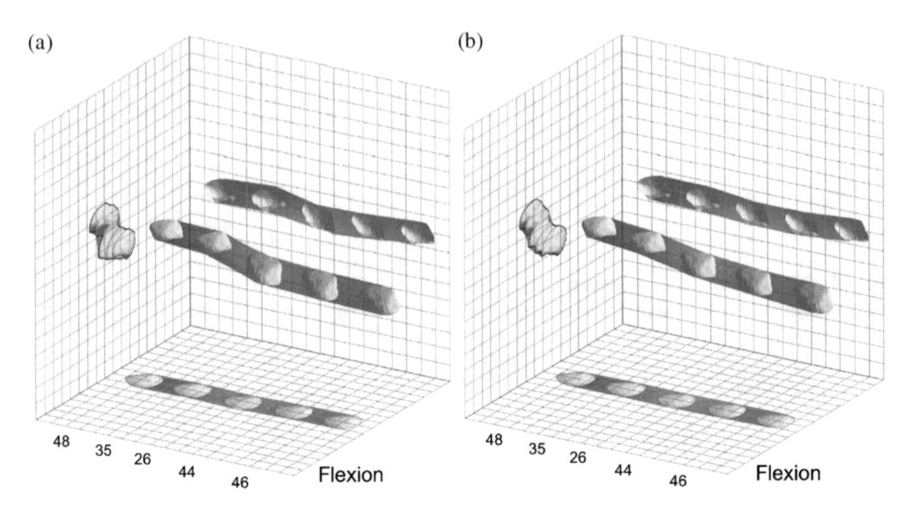

Fig. 11 Visualization of the movement pattern of the left kneecap of patient B using the time-angle profile visual metaphor. (**a**) and (**b**) present the measurements of kinematic parameters from pre- and post-operation stage respectively

when the functional movement comes to 45° flexion, which is emphasized through the radial angle plot in Fig. 10b.

The movement patterns of the left kneecap of patient B are reported in Fig. 11. The decomposition of the complex kinematic properties are done through the projection technique. Particularly, the projection onto the side presents the shifting of the kneecap out of its groove. The projection onto the bottom plane reflects the congruence angle changes, and the projection onto the back plane conveys the tilting of the kneecap with respect to the femoral groove. From Fig. 11a, b, it can be seen that the kneecap shifts out of the femoral groove less after operation. In addition, the changes of tilt angles as well as the congruence angles are also less than before. In addition to the detail rendering of the transformation of the kneecap, the color-coding scheme also quickly provides the information about the kinematic changes. For instance, the color transition from dark to light between the 48° and 35° flexions in the rendering on the back plane indicates that the tilt angle reduces rapidly. On the other hand, the small color transition from 44 to 46° flexion in the rendering on the bottom plane indicates that the congruence angle stays almost unchanged. This reflects the fact that the operation on the left side of patient B was successful.

It is worth pointing out that the proposed approach not only helps to significantly reduce the time required for the calculation of different kinematic parameters but also produces consistent results. According to the feedback from domain experts, the quantitative measurements facilitated by our proposed approach support the suspicion of the doctor about the instability of the right knee of patient A. For instance, patient A had a more pronounced decrease of the groove distance for the asymptomatic left knee postoperatively compared to the right one, but both legs of the ligament reconstruction were unchanged according to the distances between the drill holes between 3 months and 12 months postoperatively. This indicates that the injury happened after the operation could have elongated the reconstructed ligaments on the right side, but there were no further elongation the rest of the first postoperative year. Moreover, we also receive positive feedback from domain experts about the time-angle profiles. This visual metaphor is very useful in providing an overview of the movement pattern of the patella under investigation. Besides the ability to decompose a complex movement pattern into several kinematic parameters through projection technique, the combination with the facilitated automatic kinematic parameters measurement enables the transition from an overview into a detail quantitative analysis. In addition, the design of the proposed time-angle profiles and radial angle plots can facilitate the communication between domain experts and patients about the result of the surgical intervention.

7 Generalization

Although the data analyzed and visualized in this work has been extracted from the patellofemoral analysis scans, the presented techniques can be extended and applied to different kind of joint movement analysis. The proposed GPU-based feature identification and tracking can be applied to dynamic CT scans in general. Depending on the application in mind, different criteria for FOIs selection and different kinematic parameter calculations can be implemented based on the tracking results.

The time-angle profile visual metaphor was designed in such a way that it can be easily extended through the use of different projecting criteria. By deploying different projection criteria, the time-angle profile visual metaphor can be used for visual analysis of different underlying kinematic parameters. Moreover, these projections can be co-related through a multiple linked-views setup to facilitate both the overview of the whole movement pattern as well as detail decomposition of the kinematic behavior under investigation.

Figure 4 shows the application of the radial angle plot to visualize the kinematics of a ball joint. The joint-centered nature of the radial angle plot is clearly an advantage, as it allows a quantified interpretation of the rotational angle without hindering the embedding into the spatial context. To incorporate the additional degrees of freedom, either the plot can convey several kinematic properties, or multiple radial angle plots can be combined. As an alternative to these two variants, also parallel coordinate views could be integrated to show the correlation between different angles [14]. Finally, while the current radial angle plot supports a angle domain of $[-90; +90]$, increasing the arc length can be used to represent larger angle domains.

8 Conclusions and Future Work

In this paper, we have presented new visual metaphors for the interpretation kinematic information derived from dynamic CT data sets. In addition, we proposed an enhanced GPU-based feature identification and tracking based on the SIFT algorithm. Besides the performance improvement, thus allows interactive feature identification and tracking, the proposed technique also achieve a higher accuracy in the process of feature matching based on the novel descriptor construction approach. As a result, the proposed approach facilitate the ability to automatically track FOIs across scans in a dynamic CT data set which helps to significantly reduce the time required for quantitative measurements of kinematic parameters and provide consistent results. We have applied the introduced techniques to the analysis of the patellofemoral issue. Integrating the presented techniques into the analysis process of 4D CT scans has shown great potential as the quantitative measurement results from the system reflect the real clinical situation of the patients. Moreover, the presented visual metaphor enables medical doctors to have not only a quick overview of the movement pattern of feature under investigation in the dynamic CT scans but also the underlying kinematic information.

In the future, we would like to acquire 4D CT data of the shoulder joint and verify the application of the proposed visualizations. Furthermore, we plan to use our approach to derive a normal collective, which would allow us to automatically emphasize abnormal movement patterns. When dealing with a system of multiple joints for each patient, we would also like to combine multiple radial angle plots into a single view. Finally, we would like to investigate how time-angle profiles could be extended to support an embedding of several individuals.

References

1. Rhee, S.-J., Pavlou, G., Oakley, J., Barlow, D., Haddad, F.: Modern management of patellar instability. Int. Orthop. **36**(12), 2447–2456 (2012)
2. Desio, S.M., Burks, R.T., Bachus, K.N.: Soft tissue restraints to lateral patellar translation in the human knee. Am. J. Sports Med. **26**(1), 59–65 (1998)
3. Nietosvaara, Y., Aalto, K., Kallio, P.E.: Acute patellar dislocation in children: incidence and associated osteochondral fractures. J. Pediatr. Orthop. **14**(4), 513–515 (1994)
4. Hing, C.B., Smith, T.O., Donell, S.: Surgical versus non-surgical interventions for treating patellar dislocation. Cochrane Database Syst. Rev. **11**, (2011). http://onlinelibrary.wiley.com/doi/10.1002/14651858.CD008106.pub2/abstract
5. Bull, A.M.J., Katchburian, M.V., Shih, Y.-F., Amis, A.A.: Standardisation of the description of patellofemoral motion and comparison between different techniques. Knee Surg. Sports Traumatol. Arthrosc. **10**(3), 184–193 (2002)
6. Christiansen, S.E., Jacobsen, B.W., Lund, B., Lind, M.: Reconstruction of the medial patellofemoral ligament with gracilis tendon autograft in transverse patellar drill holes. Arthrosc. J. Arthrosc. Relat. Surgery **24**(1), 82–87 (2008)
7. Smith, T.O., Davies, L., Toms, A.P., Hing, C.B., Donell, S.T.: The reliability and validity of radiological assessment for patellar instability. A systematic review and meta-analysis. Skelet. Radiol. **40**(4), 399–414 (2010)

8. Davis, D.K., Fithian, D.C.: Techniques of medial retinacular repair and reconstruction. Clin. Orthop. Relat. Res. **402**, 38–52 (2002)
9. Whittle, M.: Gait Analysis: An Introduction. Butterworth-Heinemann Medical, Edinburgh (2007)
10. Manal, K., Stanhope, S.J.: A novel method for displaying gait and clinical movement analysis data. Gait Posture **20**(2), 222–226 (2004)
11. Manal, K., Chang, C.-C., Hamill, J., Stanhope, S.J.: A three-dimensional data visualization technique for reporting movement pattern deviations. J. Biomech. **38**(11), 2151–2156 (2005)
12. Côté, J.N., Raymond, D., Mathieu, P.A., Feldman, A.G., Levin, M.F.: Differences in multi-joint kinematic patterns of repetitive hammering in healthy, fatigued and shoulder-injured individuals. Clin. Biomech. **20**(6), 581–590 (2005)
13. Keefe, D.F., Ewert, M., Ribarsky, W., Chang, R.: Interactive coordinated multiple-view visualization of biomechanical motion data. IEEE Trans. Vis. Comput. Graph. **15**(6), 1383–1390 (2009)
14. Krekel, P.R., Valstar, E.R., De Groot, J., Post, F.H., Nelissen, R.G.H.H, Botha, C.P.: Visual analysis of multi-joint kinematic data. Comput. Graph. Forum **29**(3), 1123–1132 (2010)
15. Krekel, P.R., Botha, C.P., Valstar, E.R., de Bruin, P.W., Rozing, P.M., Post, F.H.: Interactive simulation and comparative visualisation of the bone-determined range of motion of the human shoulder. In: Proceedings of Simulation and Visualization (SimVis), pp. 275–288 (2006)
16. Lowe, D.G.: Object recognition from local scale-invariant features. In: IEEE International Conference on Computer Vision, vol. 2, pp. 1150–1157 (1999)
17. Lowe, D.G.: Distinctive image features from scale-invariant keypoints. Int. J. Comput. Vis. **60**(2), 91–110 (2004)
18. Cheung, W., Hamarneh, G.: n-SIFT: n-dimensional scale invariant feature transform. IEEE Trans. Image Process. **18**(9), 2012–2021 (2007)
19. Scovanner, P., Ali, S., Shah, M.: A 3-dimensional SIFT descriptor and its application to action recognition. In: International Conference On Multimedia, pp. 357–360 (2007)
20. Toews, M., Wells III, W.M.: Efficient and robust model-to-image alignment using 3D scale-invariant features. Med. Image Anal. **17**(3), 271–82 (2012)
21. Ni, D., Qu, Y., Yang, X., Chui, Y.P., Wong, T.-T., Ho, S.S., Heng, P.A.: Volumetric ultrasound panorama based on 3D SIFT. In: Conference on Medical Image Computing and Computer-Assisted Intervention, Part II, pp. 52–60 (2008)
22. Flitton, G., Breckon, T., Bouallagu, N.M.: Object recognition using 3D SIFT in complex CT volumes. In: British Machine Vision Conference, pp. 11.1–12 (2010)
23. Flitton, G., Breckon, T.P., Megherbi, N.: A comparison of 3D interest point descriptors with application to airport baggage object detection in complex CT imagery. Pattern Recogn. **46**(9), 2420–2436 (2013)
24. Yu, T.-H., Woodford, O.J., Cipolla, R.: A performance evaluation of volumetric 3D interest point detectors. Int. J. Comput. Vis. **102**(1–3), 180–197 (2012)
25. Nguyen, K.T., Ropinski, T.: Feature tracking in time-varying volumetric data through scale invariant feature transform. In: SIGRAD Conference on Visual Computing, pp. 11–16 (2013)
26. Pronost, N., Sandholm, A., Thalmann, D.: A visualization framework for the analysis of neuromuscular simulations. Vis. Comput. Int. J. Comput. Graph. **27**(2), 109–119 (2011)
27. Lindeberg, T.: Scale-space theory: a basic tool for analysing structures at different scales. J. Appl. Stat. **21**, 225–270 (1994)
28. Mikolajczyk, K., Tuytelaars, T., Schmid, C., Zisserman, A., Matas, J., Schaffalitzky, F., Kadir, T., Gool, L.V.: A comparison of affine region detectors. Int. J. Comput. Vis. **65**(1–2), 43–72 (2005)
29. Allaire, S., Kim, J.J., Breen, S.L., Jaffray, D.A., Pekar, V.: Full orientation invariance and improved feature selectivity of 3D SIFT with application to medical image analysis. In: IEEE Conference on Computer Vision and Pattern Recognition Workshops, pp. 1–8 (2008)
30. Paganelli, C., Peroni, M., Pennati, F., Baroni, G., Summers, P.: Scale invariant feature transform as feature tracking method in 4D imaging: a feasibility study. In: IEEE Conference on Engineering in Medicine and Biology Society (EMBC), pp. 6543–6546 (2012)

31. Fischler, M.A., Bolles, R.C.: Random sample consensus: a paradigm for model fitting with applications to image analysis and automated cartography. Commun. ACM **24**(6), 381–395 (1981)
32. Beis, J.S., Lowe, D.G.: Shape indexing using approximate nearest-neighbour search in high-dimensional spaces. In: Conference on Computer Vision and Pattern Recognition (CVPR), pp. 1000–1006 (1997)
33. Brossmann, J., Muhle, C., Schröder, C., Melchert, U.H., Büll, C.C., Spielmann, R.P., Heller, M.: Patellar tracking patterns during active and passive knee extension: evaluation with motion-triggered cine MR imaging. Radiother. Oncol. **187**, 205–212 (1993)
34. Powers, C.M., Shellock, F.G., Pfaff, M.: Quantification of patellar tracking using kinematic MRI. J. Magn. Reson. Imaging **8**(3), 724–732 (1998)
35. Shellock, F.G., Mink, J.H., Deutsch, A.L., Foo, T.K., Sullenberger, P.: Patellofemoral joint: identification of abnormalities with active-movement, "unloaded" versus "loaded" kinematic MR imaging techniques. Radiother. Oncol. **188**, 575–578 (1993)
36. König, A.H., Doleisch, H., Gröller, E.: Multiple views and magic mirrors - fMRI visualization of the human brain. Technical report, Institute of Computer Graphics and Algorithms, Vienna University of Technology (1999)

Part V
Visualization in Life Sciences

Interactive Similarity Analysis and Error Detection in Large Tree Collections

Jens Fangerau, Burkhard Höckendorf, Bastian Rieck, Christian Heine, Joachim Wittbrodt, and Heike Leitte

Abstract Automatic feature tracking is widely used for the analysis of time-dependent data. If the features exhibit splitting behavior, it is best characterized by tree-like tracks. For a large number of time steps, each with numerous features, these data become increasingly difficult to analyze. In this paper, we focus on the problem of comparing and contrasting hundreds to thousands of trees to support developmental biologists in their study of cell division patterns in embryos. To this end, we propose a new visual analytics method called *structure map*. This two-dimensional, color-coded map arranges trees into tiles along a Hilbert curve, preserving a tree similarity measure, which we define via graph Laplacians. The structure map supports both global and local analysis based on user-selected tree descriptors. It helps analysts identify similar trees, observe clustering and sizes of clusters within the forest, and detect outliers in a compact and uniform representation. We apply the structure map for analyzing 3D cell tracking from two periods of zebrafish embryogenesis: blastulation to early epiboly and tailbud extension. In both cases, we show how the structure map supported biologists to find systematic differences in the data set as well as detect erroneous cell behaviors.

1 Introduction

Developmental biologists analyze the process of how embryos develop from single cells into complete organisms. During this *embryogenesis*, patterns in cell movements and divisions are believed to play a crucial role to determine cell orga-

J. Fangerau (✉) • B. Rieck • H. Leitte
Computer Graphics and Visualization, Heidelberg University, Heidelberg, Germany
e-mail: jens.fangerau@iwr.uni-heidelberg.de; bastian.rieck@iwr.uni-heidelberg.de; heike.leitte@iwr.uni-heidelberg.de

B. Höckendorf • J. Wittbrodt
Centre for Organismal Studies, Heidelberg University, Heidelberg, Germany
e-mail: burkhard.hoeckendorf@cos.uni-heidelberg.de; jochen.wittbrodt@cos.uni-heidelberg.de

C. Heine
Scientific Visualization Group, ETH Zürich, Zürich, Switzerland
e-mail: cheine@inf.ethz.ch

L. Linsen et al. (eds.), *Visualization in Medicine and Life Sciences III*, Mathematics and Visualization, DOI 10.1007/978-3-319-24523-2_13

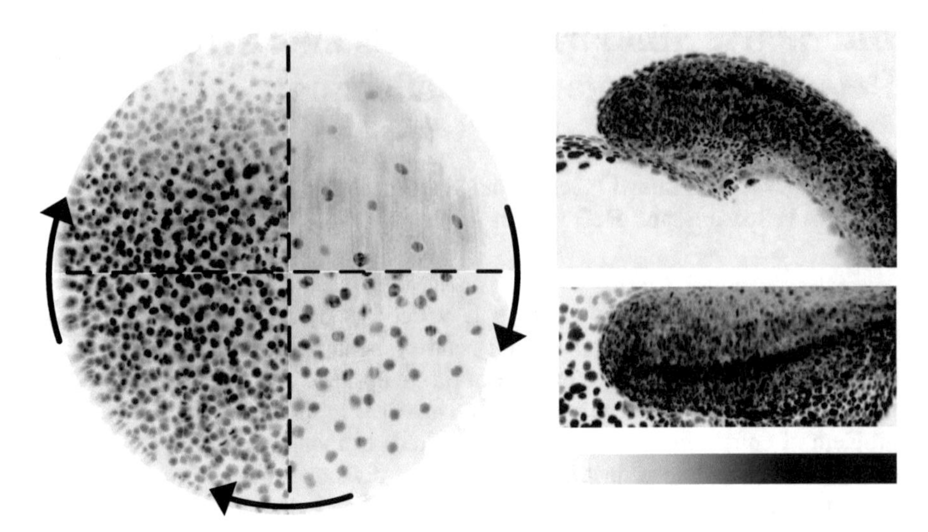

Fig. 1 Maximum intensity projections (MIPs) of zebrafish data sets. *Left*: Each quadrant shows a different time step (0, 25, 75, and 99) of blastulation to early epiboly. Note the increase in cells. *Right*: MIP of one time step showing tailbud extension

nization into tissue and organs. Light sheet microscopy enables the non-destructive observation of organisms at subcellular resolution. Its use for developmental biology was demonstrated by Keller et al. [11], who used digitally scanned laser light sheet microscopy to observe zebrafish (*danio rerio*) embryos. A survey of selective plane illumination microscopy for developmental biology is given by Huisken and Stainier [9]. Figure 1 shows maximum intensity projections of selected time steps in two different phases of zebrafish development.

Light sheet microscopy is a technique for producing the 3D intensity image sequences that can be processed by tracking algorithms (e.g. Lou et al. [14]) to determine so-called *cell lineages*. Each lineage encodes, in a tree, the developmental history of a cell such as its movements and divisions. Analyzing local tree structures is believed to yield insights into the corresponding cell division patterns and thus help unveil the blueprint of how organisms develop. Lineage trees have a simple structure: Nodes either have two or one descendants. In the former case, a cell divides, and in the latter case, a cell goes through its *cell cycle*, i.e. the time period between divisions. Stem cell divisions are commonly classified as *symmetric*, resulting in two identical stem cells, or *asymmetric*, resulting in one stem cell and one progenitor cell. Progenitor cells divide faster than stem cells, and their descendants stop dividing after a short period and change behavior. Of particular interest is the detection of asymmetric cell division.

Unfortunately, the obtained cell lineage information is imperfect in three aspects. First, due to technical reasons [11], observations cannot start before embryos comprise multiple dozen to few hundred cells. Second, noise in the images makes some cells vanish and appear again, leaving the automatic segmentation and tracking algorithm no choice but to start one new tree for each later cell found. However, the cell division patterns we are interested in manifest themselves locally. Hence, we can detect them regardless of the global tracking quality. By analyzing a sufficient number of trees, our analysis thus results in a good approximation of the global tracking quality. Third, some trees imply cells dividing faster than it is biologically plausible. This is another processing error, but can be detected reliably using structural tree motifs. Such errors have only a local effect on the lineage, but hamper automatic analysis severely. Fortunately, the largest part of the trees remains available for visual inspection.

In this paper, we present a new visual analytics approach, called the *structure map*. It visualizes a large number of trees highlighting structural motifs. In the spirit of Ware, who claims that a core property of visualization is not only the finding of patterns in the data but also the ability to indicate problems of data acquisition [24, p. 3], our method serves two purposes: On the one hand, to spot processing errors so that developers of the cell tracking algorithm can improve it and on the other hand, to explore motifs in trees so that biologists can search for cell division patterns masking parts affected by processing errors or ascertaining confidence in the analysis of imperfect data.

This approach requires a global overview of trees as well as local details of individual tree substructures. The structure map employs *spectral analysis* to sort trees globally based on their structural similarity. It then uses this order to render trees into tiles along a Hilbert curve, positioning similar trees close together. To aid exploration, numeric descriptors that characterize the structure of trees and subtrees are used to highlight globally which trees are strong matches and locally where matches manifest. This enables comparing different tree structures as well as detecting outliers and anomalous structures. We demonstrate our approach on two real-world data sets from developmental biology and provide a supplemental video showing the results [6]. In an epiboly data set, homogeneous, largely symmetric cell divisions could be identified, while in the tailbud data set, longer cell cycles throughout the data could be recognized.

The contributions of this paper are:

- A new visualization method, the *structure map*, that supports the interactive analysis of thousands of trees, especially from developmental biology. The structure map arranges unique tree structures in a compact and uniform design, using a Hilbert curve and spectral differences of trees.
- Both a global analysis based on user-defined descriptors and a local detailed indication of these descriptors in each individual tree structure.
- A similarity analysis and application-based error detection by steering cell-based descriptor parameters with an immediate visual feedback.

2 Related Work

Schulz [19] and Landesberger et al. [23] illustrate multiple existing tree visualizations. These visualizations are commonly divided into three groups: *Space-filling*, *node-link-based*, and *hybrid* approaches. Space-filling techniques use the complete area of a display in order to illustrate hierarchies in a tree, e.g. *treemaps* [20] or *beamtrees* [21]. Node-link-based techniques use links between vertices to represent their relationships. Trees can be layered horizontally, radially, or in balloon layouts in 2D [8], for example. However, simple layouts often do not scale to larger sets of trees, making user interaction more complex. Hybrid approaches combine node-link-based techniques with treemaps and visualize subsets of the hierarchy information [27]. Wong et al. introduce *GreenCurve* [26] for the compact visualization of large graphs with small-world properties. Their method arranges nodes on a space-filling curve using the *Fiedler vector* of the graph Laplacian. The method results in a compact visualization with little node overlaps. *GreenCurve* is unsuitable for visualizing similarities between multiple graphs, though.

Graham and Kennedy [7] review visualizations for tree collections. For calculating similarities between trees, the most common technique is the *tree edit distance* [2]. The distance between two trees is defined as the minimum number of edit operations, i.e. insertions, deletions, and modifications, that are required to transform one tree into another. Calculating the tree edit distance requires *labeled ordered trees*, though, which makes this method inapplicable to cell lineage trees. The *maximum agreement subtree* (MAST) problem [13] is also often used to define a distance between trees. Here, the distance is defined by the largest subtree that two trees have in common. However, MAST only works for leaf-labeled trees, making it also unsuitable for cell lineage trees. Munzner et al. [16] presented *TreeJuxtaposer*, a method that is geared towards comparing phylogenetic trees. *TreeJuxtaposer* combines visual analysis with interactive leaf similarity highlighting, but does not scale to the comparison of larger amounts of trees. A different approach employs methods from *algebraic graph theory*. Here, *graph descriptors* and *graph invariants*, i.e. measures that aim to define graphs up to isomorphism, are used to quantify similarities. Navigli and Lapata [17] survey common descriptors for developing unsupervised algorithms for graph analysis. Landesberger et al. [22] use *graph motifs* to partition graphs into weakly-connected components. The components are subsequently clustered using a *self-organizing map* [12]. While this approach scales well to larger data sets, it does not indicate local descriptor properties for individual tree structures, which is required when analyzing erroneous cell division behavior, for example.

When working with larger data sets, interactivity is paramount for similarity analysis. Bremm et al. [3] work in a similar setting than this paper. They present a system for comparing phylogenetic trees and providing interactive visual analysis on multiple levels of detail. Using a set of similarity scores that are geared towards phylogenetic trees, their visualization employs a color-coded matrix for representing similarities. However, this matrix does not scale to data sets in developmental biology.

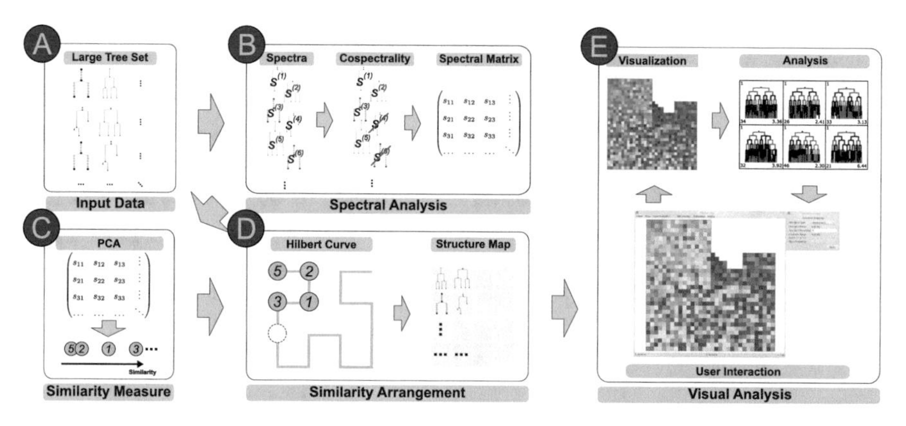

Fig. 2 Workflow chart. *A*: A large set of trees are selected as input data. *B*: *Spectral analysis* of the tree data is used to derive structural similarity and to merge isomorphic trees. *C*: To transform the spectral similarity matrix to a lower dimension, we apply *principal component analysis*. *D*: The intrinsically 1D data is laid out in the plane along a *Hilbert curve*. *E*: The visual analysis is a loop consisting of the visualization of color-coded tree descriptors in the structure map (Sect. 3.2), the analysis and interpretation of the result (Sect. 4.3)

3 The Structure Map

In the following, we describe our workflow for generating the structure map of a set of trees. The structure map is an interactive matrix-based visualization that groups trees according to their structural similarity. Additional information can be derived using interactive color-coding and user interaction. Figure 2 illustrates the steps of our analysis pipeline: (A) The input data consists of a large tree set. (B) We use *spectral analysis* to derive a structural similarity measure of these trees. (C) In order to obtain an ordering from the similarity matrix, we apply *principal component analysis*. (D) We then arrange the trees into tiles of the plane using a *Hilbert curve*. (E) We color tiles using a set of biologically relevant *tree descriptors*.

3.1 Layout

We aim to support the analyst's investigation of large tree collections by grouping trees. As a prerequisite, we need to quantifying their similarities.

Spectral Analysis A generic and powerful set of methods for analyzing graphs is given by *spectral analysis*, which analyzes a graph through the spectrum of an associated matrix such as the adjacency matrix. The *spectrum*, i.e. the set of eigenvalues, is a *graph invariant*. It only depends on the graph structure and not on any labels or layouts. In particular, isomorphic graphs have the same spectrum.

Furthermore, so-called interlacing theorems prove that small changes to a graph result in small differences in its spectrum.

The literature describes multiple matrices that may be associated to a graph. A popular choice of matrix is the *normalized graph Laplacian*. Arsić et al. [1] give an overview of the properties of its eigensystem. Although there are upper bounds for the uniqueness of a tree spectrum, it has proven to be robust with respect to *cospectrality* [25] in practice, i.e. it is unlikely that trees with the same spectrum are non-isomorphic.

The *normalized Laplacian matrix* $\overline{L}$ of a graph (N, E) is defined as

$$\overline{L_{ij}} = \begin{cases} 1 & \text{if } i = j \text{ and } d_i \neq 0 \\ -(d_i d_j)^{-\frac{1}{2}} & \text{if } \{i, j\} \in E \text{ and } d_i, d_j \neq 0 \;, \\ 0 & \text{otherwise} \end{cases} \tag{1}$$

where d_i and d_j denote the *degree* of node i and j. The set of eigenvalues, sorted in descending order, yields the *spectrum* s_k of the kth tree. The eigenvalues λ_i of $\overline{L}$ are known to satisfy $0 \leq \lambda_i \leq 2$ and to sum up to the number of non-isolated nodes.

If we encounter cospectral trees, i.e. trees i, j with $s_i = s_j$, we apply a tree isomorphism test [5]. We then partition isomorphic trees (i.e. trees that are identical up to symmetry) into equivalence classes. This amounts to a *lossless compression* of the input data set, in practice providing a significant reduction in tree structures that need to be displayed. For our data sets, we observed compression rates between 80 % and 95 %.

We define the similarity between two trees as the Euclidean distance of their spectra. Because spectra typically differ in their number of eigenvalues, we pad the smaller spectrum with zeros. Wilson and Zhu [25] showed that adding an isolated node in the graph gives an additional zero eigenvalue, but preserves the other eigenvalues. The connectivity information, which is relevant for similarity analysis, is encoded by all *nonzero* eigenvalues, and thus remains unchanged.

Principal Component Analysis Prior to assigning the set of trees to the structure map, we need to sort them such that they are grouped according to their similarity. We require a linear ordering of the lineage trees to have a high discriminatory power. More precisely, the notion of size needs to be respected such that larger trees (many nodes) are clearly separated from smaller trees (few nodes). For obtaining this ordering, we apply a 1-dimensional *principal component analysis* (PCA) [10] to the spectral matrix. We observed that this 1-dimensional embedding already accounts for more than 90 % of the variance, thereby indicating that the ordering is well-suited for sorting our lineages.

Hilbert Curve Layout The structure map is defined by a grid of tiles, each of which represents one isomorphism class of trees (i.e. trees that are identical up to symmetry). Having obtained a 1-dimensional embedding of the trees, we use a space-filling *Hilbert curve* to determine the assignment of trees to tiles of the structure map. The Hilbert curve is a fractal curve whose range is a 2-dimensional

Fig. 3 The first six iterations of the Hilbert curve

square. The curve preserves *locality* very well, i.e. points that are near when traversing the curve are also likely to be close in the space the curve is embedded in. This property makes the curve suitable for generating an arrangement of trees. The Hilbert curve is usually calculated iteratively (see Fig. 3 for several iterations). In practice, we choose the number of iterations such that the curve contains all trees in a data set. In contrast to a simple linear arrangement of trees according to their similarity values, the Hilbert curve layout yields a compact visualization of the complete data set while preserving locality. The trade-off, however, is that the structure map cannot use the complete screen space in general. For example, the layout of $2^{12} + 1 = 4097$ trees requires a Hilbert curve with $2^{13} = 8192$ nodes, resulting in almost 50 % empty tiles. The empty tiles do *not* impede the visual analysis of tree structures in any way, though.

To make screen space usage more efficient, we plan to employ *squarified tree maps* [4] in conjunction with the structure map. This will enable us to scale different parts of the map according to user-defined criteria, such as the prevalence of biologically-implausible cell tracking behavior.

Tile Design A tile is drawn as a colored square that represents a single isomorphism class of trees. We draw a representative tree on top of the tile using the Reingold-Tilford tree drawing algorithm [18]. We chose this layout algorithm because it draws isomorphic subtrees congruently, and detecting symmetric and asymmetric cell division patterns is imperative for the domain experts. For each tree, we apply an adaptive anisotropic scaling method such that the tree fits completely into the tile. Parts of the tree that match a given tree descriptor are highlighted. Furthermore, a label in each tile states the number of isomorphic trees this tile represents. The tiles are also colored based on that value.

3.2 Tree Descriptors

In discussions with domain experts from developmental biology, we arrived at a set of tree descriptors well-suited for the tasks of similarity analysis and error detection. Each tree descriptor yields quantitative information about a tree. The tiles of the structure map encode this information through different colors. Tree descriptors augment the spectral analysis by incorporating additional domain knowledge to uncover both similarities and errors in a data set.

Number of Nodes and Leaves The number of nodes is positively correlated with the overall size and depth of a tree. If a tree has n nodes, we know that it must have exactly $n-1$ edges. For biological data sets, the number of nodes corresponds to the number of tracked cells. Even without any further biological analysis, trees with a comparatively small number of nodes are more likely to contain erroneously tracked cells. Moreover, the number of leaves also measures the number of cell divisions.

Mean Path Length The lengths of different paths within a cell lineage tree turned out to be an essential quantity for similarity analysis. We define a path p of length l as a sequence of l connected nodes in a tree, i.e. $p = \{n_1, \ldots, n_l\}$. In addition, we require p to start at the root of the tree or at a division node, and end at such a division node or at a leaf (see Fig. 4). By this definition, the path length corresponds to a cell cycle phase during which the cell changes its position. We hence use this information as a descriptor to find outliers or incorrect cell behavior. Since a tree commonly consists of multiple paths of different lengths, we compute its mean path

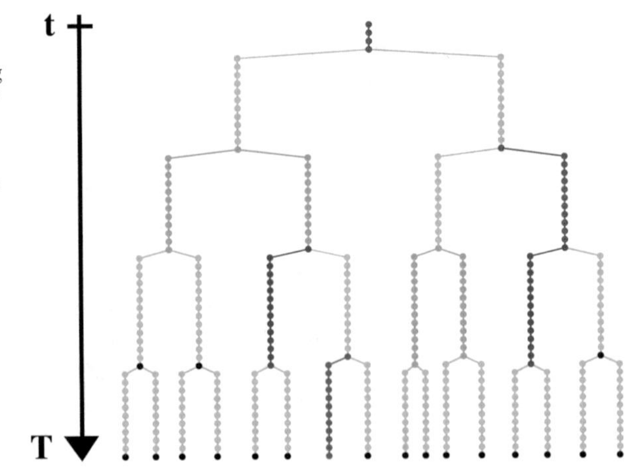

Fig. 4 Path and branch definitions in a cell lineage tree. Note that time is running from top to bottom instead of running from left to right. Paths start at the root or at a division node and end at such a node or at a leaf (examples marked in *red*). Inner paths neither start at the root nor end in a leaf node (examples marked in *dark blue*). Branches are defined by two paths starting at a common node (examples marked *orange*)

length $\overline{p}$ as

$$\overline{p} = \frac{1}{|p|} \sum_{i=1}^{|p|} p_i, \tag{2}$$

where $|p|$ is the number of paths in the tree k and p_i denotes the ith path in the tree.

Mean Inner Path Length By ignoring all paths that start at the root node or end at a leaf node (see Fig. 4), we obtain the mean inner path length $\overline{q}$, which is a variant of the previous measure $\overline{p}$. The mean inner path length only counts cell cycles that have been fully captured. Thus, this length can be used as a measure to identify two cell divisions that occur in quick succession, for example.

Mean Branch Asymmetry The length of a cell cycle also describes a biologically important substructure within a cell lineage tree. We define a branch b_n by a starting node n with two successors, i.e. a cell division node, and by the length of its left path l_n and its right path r_n. These paths contain all nodes between two cell division nodes (see Fig. 4). We call a branch *symmetric* if the lengths of l_n and r_n are equal. For each branch b_n, we calculate the *branch asymmetry* as the difference in lengths of its two paths: $\Delta b_n = |l_n - r_n|$. From this, we derive the mean branch asymmetry $\overline{b}$ of a tree as

$$\overline{b} = \frac{1}{|b|} \sum_{n} \Delta b_n, \tag{3}$$

where $|b|$ denotes the number of branches in the tree and the sum ranges over all cell division nodes. If a tree does not have any branches, we assign it $\overline{b} = -1$. Since Eq. (3) always results in positive numbers, we can easily identify trees without branches. We use red color to encode such trees in the structure map, in order to signify that they have maximum dissimilarity to all other trees.

4 Visual Analysis

The GUI used in our application is depicted in Fig. 5, left. The structure map is displayed in an interactive 2D window which supports standard camera navigations like panning and zooming (see Fig. 5A). A second widget controls the descriptor settings (see Fig. 5B).

4.1 Structure Map Widget

The structure map is the interface to the collection of trees. It supports the rapid analysis of large collections of trees in a compact representation. Each set of

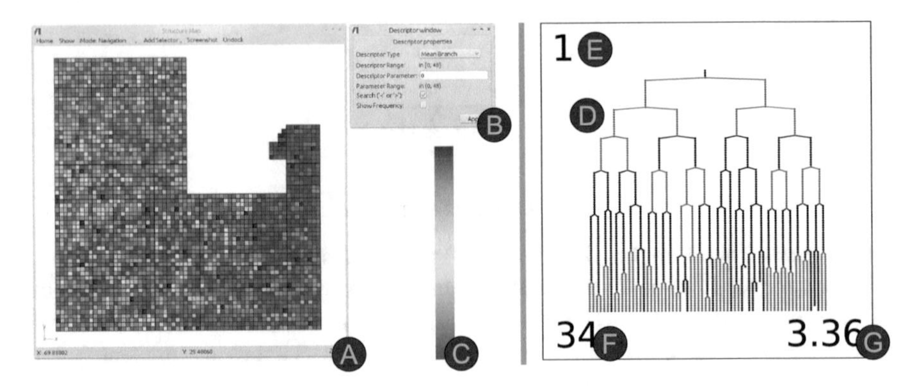

Fig. 5 *Left*: Workspace for visual analysis. The GUI consists of the structure map displayed in a 2D window (*A*) and a widget (*B*) for descriptor settings. We use a diverging color map described by Moreland [15] (*C*). *Right*: Magnified version of a common tile in the structure map. Local features in the tree are colored *red* (*D*) based on the selected descriptor type. The frequency of the unique tree structure in relation to the complete data set is shown in the *top left label* (*E*). The frequency of local features in the tree based on the selected descriptor (here: symmetric branches) is displayed at the *bottom left* (*F*). The descriptor value for the whole tree (here: mean branch asymmetry) is shown at the *bottom right* (*G*)

isomorphic trees is represented in a colored tile whose color represents the value of the currently selected tree descriptor. For coloring tiles, we use the color map of Moreland [15] that is well-suited for scientific visualization (see Fig. 5C). Each tile shows one representative tree structure colored in black to allow for rapid overview and orientation of the user. Figure 5, right, shows an example of such a tile with an assigned tree structure. Local features of the tree are highlighted in red (see Fig. 5D) based on user-selected parameters that are set in the descriptor widget (Fig. 5B). Here, a specific descriptor can be selected. For an explanation of its user-definable range parameters, please see Sect. 4.3.

4.2 Semantic Zooming

In order to improve the visual analysis as well as the performance for interaction, the structure map features *semantic zooming*. By this, relevant objects are displayed in different levels of detail depending on the zooming level. This affects both the tree structure and the tile colors. If the distance between the virtual camera and the structure map is large, no trees are rendered at all. The structure map focuses on the overall color of the tiles such that they can be compared to each other (see Fig. 6A). Upon zooming in, only parts of a tree structure (half the lines and no nodes) are rendered (see Fig. 6B). With decreasing distance, each tile color fades from its original color (set by a tree descriptor) to a white background, while all lines and no nodes are rendered for each tree (see Fig. 6C). At this stage, also the tile labels

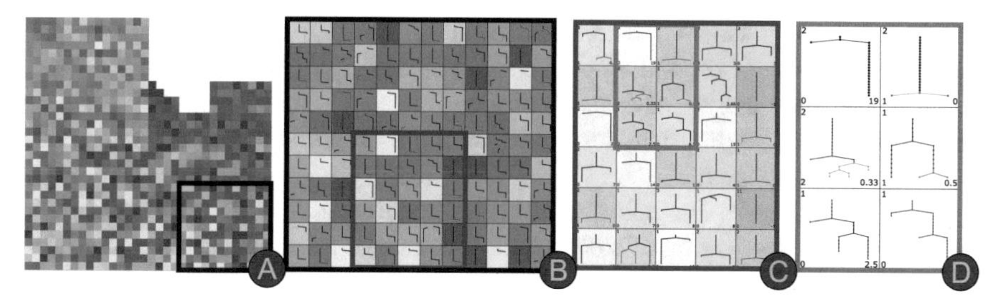

Fig. 6 Different levels of semantic zooming: The different levels emerge in the order from left to right when zooming in while same *rectangle colors* represent the same area. For large distances between camera and structure map, only the *tile colors* and no trees are rendered (*A*). In the next level, half the lines and no nodes of the tree structures are displayed (*B*). Further zooming starts fading from the original color to white and all lines and tile labels are drawn (*C*). The highest zoom level displays all details of the tree structure in a *white tile* (*D*)

as explained in Fig. 5 are displayed. Finally, for very small distances, all details of a tree are rendered while each tile becomes completely white (see Fig. 6D). The color change guarantees that the red color of the color map does not interfere with the red color used for highlighting local features. The functionality of the semantic zooming is also shown in our supplemental video [6].

4.3 Interactive Analysis

The structure map supports the following analysis strategies:

1. Users can investigate similarities and differences within the tree collection by investigating the spatial layout of the trees. This reveals structural differences encoded in the graph spectrum, such as the number of leaf nodes or division nodes, as well as the total number of nodes.
2. The tiles of the map can be colored according to the tree descriptors. Each tile is assigned a color depending on the value of the descriptor. The colors provide additional domain-specific information. By analyzing color patterns in the structure map, structural descriptions are combined with local tree features.
3. The color code can be adjusted to represent the difference to a predefined scalar descriptor parameter P that can be specified in the descriptor widget. For example, users can highlight tree structures with a specific mean branch asymmetry that are smaller or larger than P.
4. The last analysis method colors tiles according to the frequency of a certain feature. For example, users can search for cell cycles that are shorter than ten time steps. Each tile is now colored according to the number of such short cycles occurring in the tree. Simultaneously, the respective parts of the tree are highlighted in the small tree structure represented inside the tile (red color in Fig. 5D).

The descriptor widget (see Fig. 5B) can be used to choose a specific tree descriptor and its corresponding parameter. The descriptor type can be selected in a combo box. For analysis tasks such as descriptor depiction (see item 2), the user simply selects the appropriate descriptor from the drop down menu and the color map is instantly updated.

The analysis tasks for coloring the tiles require additional information and parameters. Upon selection of a descriptor, two ranges are updated. The *descriptor range* indicates the smallest and largest descriptor value for the current data set. The *element range* lists the smallest and largest value for the mean value computation of the descriptors. For example, if the *mean path length* descriptor is used, the descriptor range gives the shortest and longest mean path length over all trees. The element range gives the length of the shortest and longest path occurring in any of the trees.

As explained in item 3, users specify a descriptor parameter P to highlight all trees that have either a smaller or larger value than the given one. The color of the tiles is changed to represent distances to the given value, making outliers and patterns readily visible. Range limitation is realized with a check box that specifies that either features above or below the given threshold are to be highlighted in the tree (red coding in the tree) (see Fig. 8B, D). To simplify the identification of outliers, colors can be toggled between distance and frequency mode using the last check box. The distance mode was used for all previous tasks. In frequency mode (see item 4), the color indicates the number of occurrences of a given feature within the tree represented in the tile. Figure 8 gives an example of a frequency-based structure map. Here, the *mean branch asymmetry* is chosen with a descriptor parameter of $P = 0$. Thus, all symmetric (see Fig. 8A) or asymmetric (see Fig. 8C) branches are highlighted based on their number of appearances.

5 Results

We used the structure map to analyze two real-world data sets of segmented 3D videos showing the development of a zebrafish embryo. Our visual analysis approach works on a standard desktop computer, providing near-instantaneous visual feedback (see also the supplemental video [6]).

In the following, we focus on the visual analysis of similarities and errors in the tree structures. The workflow begins with the selection of a specific descriptor and the setting of the descriptor parameter P using the descriptor widget. The analysis is then further augmented by switching between the distance-based and frequency-based coloring of the structure map.

5.1 Lineage Tree Data

From now on, we will refer to the data sets as *epiboly* and *tailbud* data. The epiboly data set (Fig. 1A) shows early events from *blastula* to early *epiboly* stages ($\approx$ 3.5–4.5 hours post fertilization, hpf), while in the tailbud data set (Fig. 1B), the tail extension ($\approx$ 15–16 hpf) is recorded. For the epiboly data set, we thus expect homogeneous cell behavior and numerous stem cell divisions. Stem cell divisions are commonly classified as being symmetric or asymmetric. A symmetric division results in two identical stem cells whereas an asymmetric division generates one stem cell and one progenitor cell. Progenitor cells divide faster than stem cells. Furthermore, their daughter cells stop dividing after a short period and change their behavior. In contrast, cell division in the tailbud data set is much slower so that different division behaviors are expected to occur.

The epiboly data set starts with 90 cells and ends with 3253 cells covering 100 time steps in total. The complete data set includes 4896 lineage trees. The tailbud data has 9961 cells at the beginning and ends after 50 time steps with $10,173$ tracked cells. Here, $58,048$ lineage trees are generated. Spectral analysis uncovers that both data sets contain a large amount of isomorphic trees. Partitioning trees into equivalence classes enables us to significantly reduce the number of unique trees that need to be displayed by the structure map.

5.2 Preliminary Observations

Figure 7 shows the structure map for both data sets based on the number of nodes (A, C) and leaves (B, D). In the epiboly data set, merging isomorphic trees reduces their number from 4896 to 875. The remaining trees are arranged in a Hilbert curve of level 5, resulting in a structure map of $2^5 \times 2^5 = 1024$ tiles, 149 of which are empty. Since we are interested in analyzing the complete structure of all trees, we assign each tree to a tile regardless of the time step in which it was tracked. For the tailbud data set, 3221 of originally $58,048$ trees remain after merging. We thus generate a Hilbert curve of level 6, resulting in a structure map of $2^6 \times 2^6 = 4096$ tiles, 875 of which remain empty.

Both data sets contain numerous small trees with few nodes. However, the epiboly data set features several large trees with 2000 nodes on average, while the largest tree in the tailbud data set has a mere 190 nodes.

The number of nodes and the connectivity information of a tree is encoded by its Laplacian matrix. In Fig. 7A, C, we observe that the structure map thus tends to group trees according to the number of their nodes. A similar grouping according to the number of leaves can be identified in Fig. 7B. In contrast, the tailbud data set contains multiple smaller trees with approximately the same number of leaves. The structure map thus does not group them by the number of their leaves (see Fig. 7D).

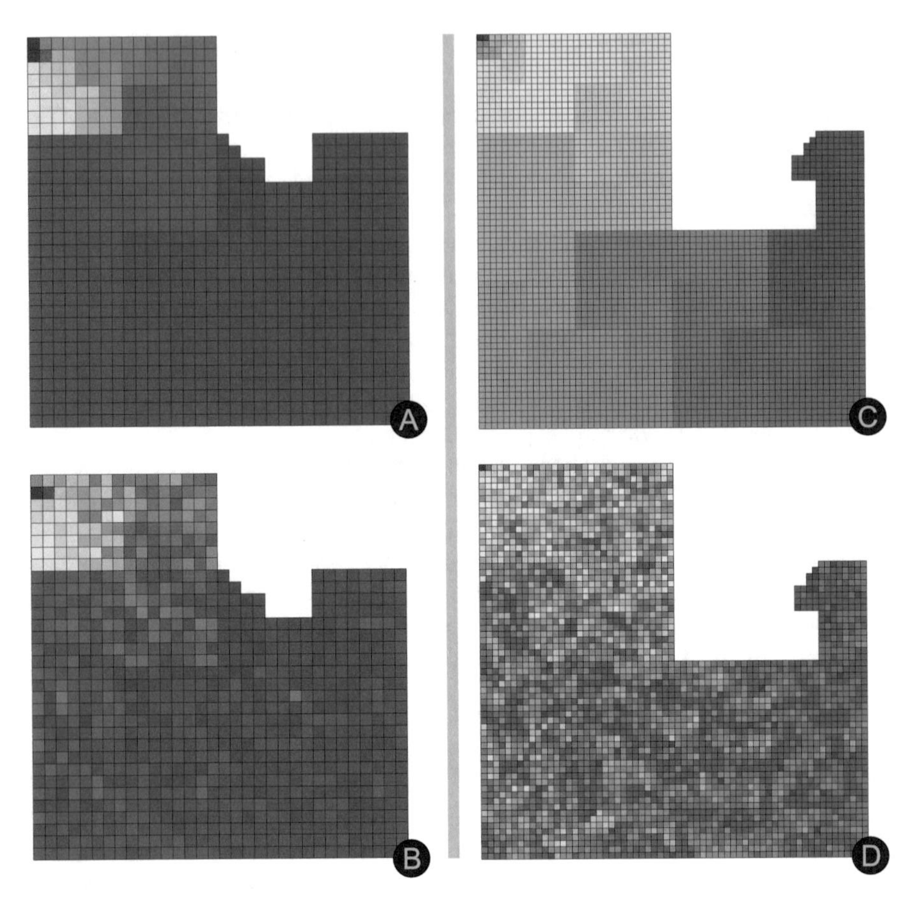

Fig. 7 Distance-based coloring of structure maps for both data sets: *Left*: Structure map of the epiboly data based on the number of nodes (*A*) and leaves (*B*). *Right*: Structure map of the tailbud data based on the number of nodes (*C*) and leaves (*D*)

5.3 Similarity Analysis and Error Detection

Since we expect the epiboly data set to contain homogeneous cell division patterns, we first analyze this data based on the *mean branch asymmetry* in order to identify symmetric and asymmetric cell division behavior. We set an initial parameter of $P = 0$ that represents symmetric branches. Figure 9A shows the resulting structure map with a distance-based coloring, while Fig. 8A uses a frequency-based coloring. The majority of blue tiles in Fig. 9A illustrate structures that have a small mean branch asymmetry. Red and orange tiles correspond to trees without branches or with large differences between their mean branch asymmetry value and P.

A further analysis of blue tiles with red substructures reveals several symmetric branches. For the most part, these are situated in the top left corner where the large trees are located (see Fig. 9B). These lineage trees exhibit the correct cell

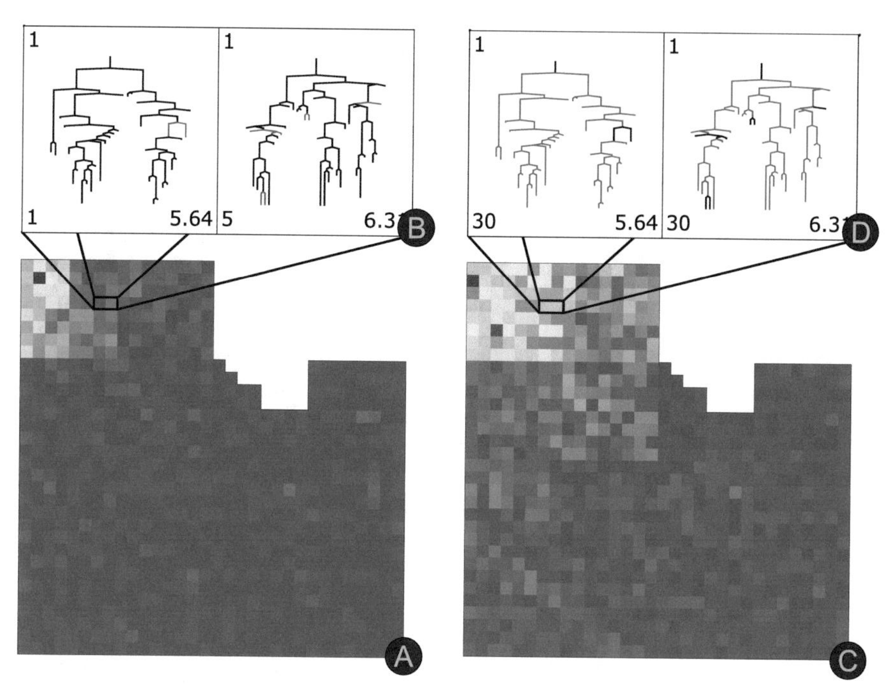

Fig. 8 Frequency-based coloring of the structure map applied to the epiboly data, based on mean branch asymmetry and $P = 0$. Structure map of symmetric (*A*) and asymmetric branches (*C*). A pair of trees with few symmetric (*B*) and many asymmetric branches (*D*) highlighted in *red*

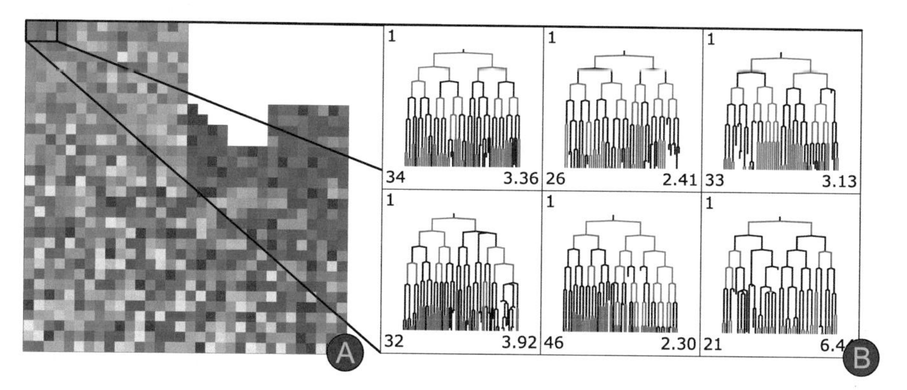

Fig. 9 Distance-based coloring of the structure map applied on the epiboly data based on mean branch asymmetry and $P = 0$: *A*: Structure map with focus on symmetric branches. *B*: Extraction of lineage trees with the most node numbers and their symmetric branches

division behavior and they are very similar with respect to the frequencies of symmetric branches. We also observe that large trees with many branches tend to have more symmetric branches than small trees with few branches. However, this does not apply to all large trees, as illustrated in Fig. 8B. Although the two

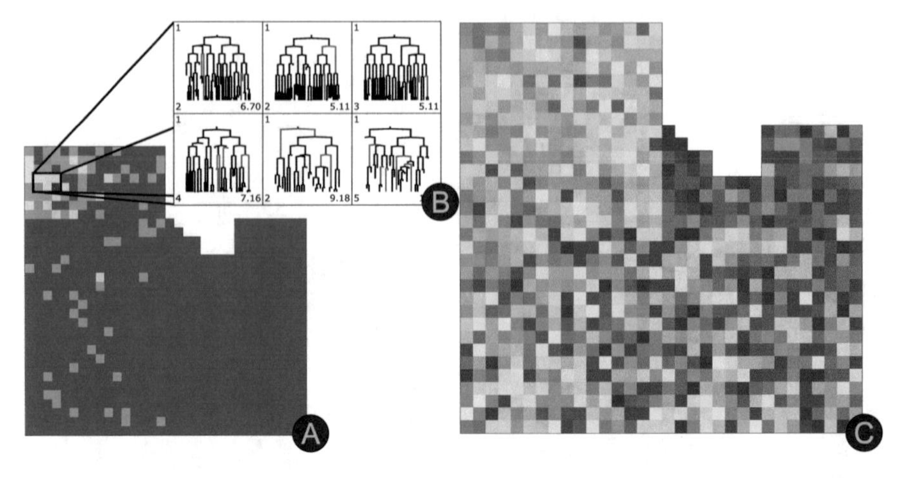

Fig. 10 Structure map of the epiboly data based on mean branch asymmetry and $P = 25$. *A*: Frequency-based structure map. *B*: Examples of trees with asymmetric branches. *C*: Distance-based structure map

lineage trees depicted in this figure have many branches, they both exhibit erroneous connectivities as well as biologically-incorrect cell division behavior. For these trees, we thus only identify a small number of symmetric branches (lower-left label: 1 and 5). We can easily switch between showing asymmetric and symmetric branches. Figure 8C shows the resulting structure map and all asymmetric branches. Using this visualization, we are able to identify the two lineage trees in Fig. 8D as outliers because their tile colors indicate a comparatively large number of asymmetric branches.

We now set the input parameter to $P = 25$ in order to detect large branch asymmetry values, i.e. outliers. For branch asymmetry values larger than 25, cell division has not been tracked correctly for the longer path in the branch. The occurrence of such branches is thus a certain indicator for errors. Figure 10A shows the structure map with frequency-based coloring using the mean branch asymmetry descriptor. The large amount of blue tiles indicates that the majority of trees have few occurrences of branches with large asymmetry values. Outliers can easily be identified by the tile colors, while the local highlighting of substructures serves to enable a more detailed investigation of the errors. Similar to the analysis of symmetric branches, we observe that large trees also tend to have large branch asymmetries (see Fig. 10B). Note that we cannot identify such a behavior in trees whose depth is smaller than P. Since the data set contains many of these small trees, the structure map consists primarily of blue tiles (see Fig. 10A).

The structure map in Fig. 10C shows a distance-based coloring for the mean branch asymmetry and $P = 25$. We observe numerous red tiles showing the large difference between their mean branch asymmetry and P. This also states that most of these tiles have a smaller mean branch asymmetry or no branches at all. The tree structures in the blue tiles indicate their small distance to the descriptor parameter

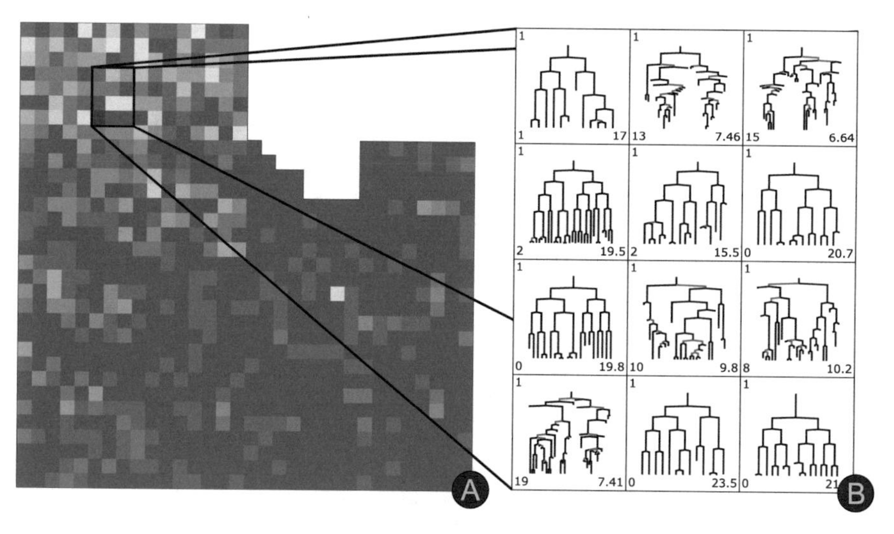

Fig. 11 Frequency-based coloring of structure map for the epiboly data based on mean inner path length and $P = 4$. *A*: Frequency-based structure map. *B*: Example of trees with short cell cycles

$P = 25$ and the highlighted branches have a branch length difference larger than 25. In both cases, tree structures in blue tiles represent biologically-incorrect cell division behavior that biologists consider as an outlier.

Another implausible biological behavior in the epiboly data set is the occurrence of two cell divisions within a small time frame. We thus select the *mean inner path length* descriptor and set $P = 4$ in order to identify inner paths with less than 4 nodes. Figure 11A shows the frequency-based coloring of the structure map. Again, we observe a majority of blue tiles (implying that these trees do not contain many short paths) but also some outlier tree structures highlighted in Fig. 11B. A detailed local investigation of these outliers reveals a multitude of apparent cell division errors. Moreover, we observe many small inner paths across the whole data set. The data set thus includes many outliers caused by erroneously-tracked cells.

We hypothesize that the tailbud data set contains fewer cell divisions, longer cell cycles, and more variety with respect to cell development. For the initial analysis, we select the *mean inner path* descriptor and a parameter of $P = 4$ in order to find implausible cell cycle phases. Figure 12A shows the frequency-based coloring of the tailbud data. We observe that blue and light-blue are the prevailing tile colors, meaning that there are few trees with very short inner paths. This substantiates the assumption that, by and large, the data set contains longer cell cycles. In contrast to Fig. 11A, tiles with light-blue colors are more evenly-distributed across the structure map. Furthermore, we can identify some red tiles that indicate trees with many short cell cycles (see Fig. 12B, C). In order to analyze the cell division behavior, we select the mean branch asymmetry descriptor and a parameter value of $P = 25$. We then proceed to highlight all branch asymmetries larger than 25. In the frequency-based structure map in Fig. 13A, we identify numerous asymmetric branches and thus a

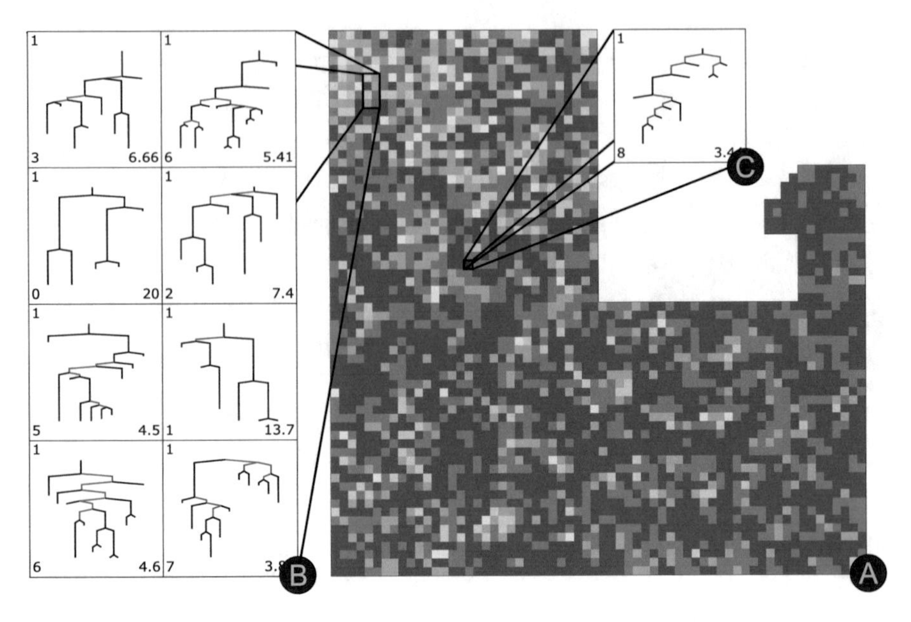

Fig. 12 Frequency-based coloring of structure map of the tailbud data based on mean inner path length and $P = 4$. *A*: Frequency-based structure map. *B, C*: Examples of trees with short cell cycles in relation to all short paths

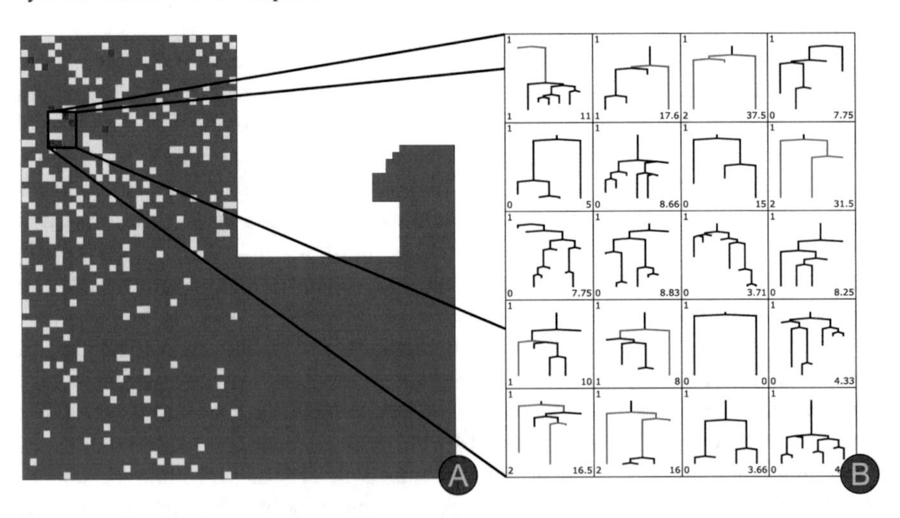

Fig. 13 Frequency-based coloring of structure map of the tailbud data based on mean branch asymmetry and $P = 25$. *A*: Frequency-based structure map. *B*: Example of trees with high branch asymmetries

variety of different cell division behaviors (Fig. 13B shows some examples of these trees). These large differences are most likely based on tracking errors missing the track for a cell movement in the smaller path.

In comparison, both structure maps differ significantly in the frequency of identified errors because of the different biological developments of the data sets. For example, consider the frequency of the local features in the tiles in Figs. 11A and 12A.

6 Conclusions and Future Work

We have shown that the structure map is well-suited for interactive visual analysis on large tree data, especially on data from developmental biology. The structure map yields a simple and compact overview of several thousands of cell developments. These developments are depicted in the form of cell lineage trees that are arranged such that local neighborhood similarity is maintained as much as possible. The structure map permits both global analysis based on color-coding, as well as local investigation of substructures in trees based on multiple cell descriptors. Moreover, it enables finding similarities and detecting both erroneous and outlying biological events. The structure map supports biological analysis in two ways: First, it yields immediate feedback about the overall tracking quality through the investigation of biological events such as cell division patterns and cell cycle lengths. Second, biologists may backtrack implausible cell behavior, thereby analyzing and improving both data processing and data acquisition.

Using the structure map for two different data sets, biologists were able to formulate and validate hypotheses about markedly different biological development processes. We also showed how to use the structure map as a tool for ascertaining the quality of cell lineages in different input data sets. Comparing epiboly and tailbud data sets uncovered significant differences in tracking quality.

A drawback of our method is the runtime required for the computation of eigensystems for spectral analysis (ca. 3 h for the epiboly data set and 92 s for the tailbud data set). Our current implementation precomputes them and stores them on the disk. In the future, we plan to use the multiscale approach of Wong et al. [26], who calculate converging approximations of the graph Laplacian, for speeding up eigensystem calculations.

The structure map has several potential extensions. As the similarity analysis presented in this paper does not require information about time steps, our current implementation ignores them. However, this information might be valuable for some analysis scenarios, so future implementations could incorporate it into the tree similarity measure in order to yield more biologically robust results. We furthermore note that the graph Laplacians we used for spectral analysis can be trivially extended to accommodate weighted edges. This could be used to integrate more information about cell movements, e.g. their velocity. We did not exhaust the possibilities for tree descriptors and finding local structural features in trees. Other applications of visualizing large tree collections may find different tree descriptors useful. Note that since the graph Laplacians and the associated similarity measure are not constrained to trees, our method can be extended to visualize large graph collections as well.

The next step is the integration of our method into a larger system where the structure map is used to mark interesting cell lineages. Prominent discoveries could then be highlighted and juxtaposed using, for instance, a maximum intensity projection or a volume rendering of the raw data. We expect that the focused observation of spatial cell migration and division will support the biologists in their investigation and interpretation of different cell development phenomena, e.g. learning more about the physical conditions that determine asymmetric cell divisions.

References

1. Arsić, B., Cvetković, D., Simić, S.K., Škarić, M.: Graph spectral techniques in computer sciences. Appl. Anal. Discrete Math. **6**, 1–30 (2012)
2. Bille, P.: A survey on tree edit distance and related problems. Theor. Comput. Sci. **337**(1), 217–239 (2005)
3. Bremm, S., von Landesberger, T., Hess, M., Schreck, T., Weil, P., Hamacherk, K.: Interactive visual comparison of multiple trees. In: IEEE VAST, pp. 31–40 (2011)
4. Bruls, M., Huizing, K., van Wijk, J.: Squarified treemaps. In: Proceedings of Joint Eurograph and IEEE TCVG Symposium on Visualization, pp. 33–42 (1999)
5. Buss, S.: Alogtime algorithms for tree isomorphism, comparison, and canonization. In: Computational Logic and Proof Theory, pp. 18–33. Springer, Heidelberg (1997)
6. Fangerau, J., Höckendorf, B., Rieck, B., Heine, C., Wittbrodt, J., Leitte, H.: Supplemental video. https://vimeo.com/86031849 (2014). Accessed 06 Feb 2014
7. Graham, M., Kennedy, J.: A survey of multiple tree visualisation. Inf. Vis. **9**, 235–252 (2010)
8. Herman, I., Melançon, G., Marshall, M.S.: Graph visualization and navigation in information visualization: a survey. IEEE Trans. Vis. Comput. Graph. **6**, 24–43 (2000)
9. Huisken, J., Stainier, D.Y.: Selective plane illumination microscopy techniques in developmental biology. Development **136**(12), 1963–1975 (2009)
10. Jolliffe, I.: Principal Component Analysis, 2nd edn. Springer, New York (2002)
11. Keller, P.J., Schmidt, A.D., Wittbrodt, J., Stelzer, E.H.: Reconstruction of zebrafish early embryonic development by scanned light sheet microscopy. Science **322**(5904), 1065–1069 (2008)
12. Kohonen, T.: Self-Organizing Maps. Springer, New York (2001)
13. Kubicka, E., Kubicki, G., McMorris, F.: An algorithm to find agreement subtrees. J. Classif. **12**, 91–99 (1995)
14. Lou, X., Kaster, F.O., Lindner, M.S., Kausler, B.X., Köthe, U., Hockendorf, B., Wittbrodt, J., Jänicke, H., Hamprecht, F.A.: Deltr: digital embryo lineage tree reconstructor. In: Biomedical Imaging: From Nano to Macro, pp. 1557–1560. IEEE, Chicago (2011)
15. Moreland, K.: Diverging color maps for scientific visualization. In: Proceedings of 5th International Symposium. Advances in Visual Computing: Part II, pp. 92–103. Springer, Heidelberg (2009)
16. Munzner, T., Guimbretière, F., Tasiran, S., Zhang, L., Zhou, Y.: TreeJuxtaposer: scalable tree comparison using focus+context with guaranteed visibility. ACM Trans. Graph. **22**(3), 453–462 (2003)
17. Navigli, R., Lapata, M.: Graph connectivity measures for unsupervised word sense disambiguation. In: Proceedings of 20th International Joint Conference on Artifical Intelligence, pp. 1683–1688 (2007)
18. Reingold, E.M., Tilford, J.S.: Tidier drawings of trees. IEEE Trans. Softw. Eng. **7**(2), 223–228 (1981)

19. Schulz, H.: Treevis.net: a tree visualization reference. IEEE Comput. Graph. **31**, 11–15 (2011)
20. Shneiderman, B.: Tree visualization with tree-maps: 2-D space-filling approach. ACM Trans. Graph. **11**, 92–99 (1992)
21. van Ham, F., van Wijk, J.J.: Beamtrees: compact visualization of large hierarchies. Inf. Vis. **2**, 31–39 (2003)
22. von Landesberger, T., Gorner, M., Schreck, T.: Visual analysis of graphs with multiple connected components. In: IEEE VAST, pp. 155–162. IEEE, Atlantic City (2009)
23. von Landesberger, T., Kuijper, A., Schreck, T., Kohlhammer, J., Wijk, J.V., Fekete, J.-D., Dieter, W. F.: Visual analysis of large graphs: state-of-the-art and future research challenges. Comput. Graph. Forum **30**, 1719–1749 (2011)
24. Ware, C.: Information Visualization: Perception for Design, 3rd edn. Morgan Kaufmann, San Francisco (2012)
25. Wilson, R.C., Zhu, P.: A study of graph spectra for comparing graphs and trees. Pattern Recogn. **41**(9), 2833–2841 (2008)
26. Wong, P.C., Foote, H., Mackey, P., Chin, G., Huang, Z., Thomas, J.: A space-filling visualization technique for multivariate small-world graphs. IEEE Trans. Vis. Comput. Graph. **18**, 797–809 (2012)
27. Zhao, S., McGuffin, M.J., Chignell, M.H.: Elastic hierarchies: combining treemaps and node-link diagrams. In: Proceedings of IEEE Symposium on Information Visualization (2005)

Efficient Reordering of Parallel Coordinates and Its Application to Multidimensional Biological Data Visualization

Tran Van Long and Lars Linsen

Abstract Multidimensional data visualization is a challenging research field with many applications in various fields of sciences. Parallel coordinate plots are one of the most common information visualization techniques for visualizing multidimensional data. Unfortunately, the effectiveness of parallel coordinates depends heavily on the order of the data dimensions and different orders exhibit different information about the structures in the multidimensional data. In this paper, we propose a method that supports an automatic dimension reordering and spacing of the axes in parallel coordinate plots. The underlying idea of our method is to find an asymptotic for the optimization of the permutation based on data dimension similarity. We present our method with two kinds of similarities, namely, Pearson's correlation similarity for unclassified data and class distance consistency for classified data. We present results on well-known multidimensional data sets to show how our method improves the parallel coordinate plots and to prove its efficiency. Finally, we demonstrate how our approach can be applied to the visualization of bivariate structures in biological data.

1 Introduction

Multidimensional data are acquired in many fields of research and applications such as transactions in business, numerical simulations in physics, and network security in computer sciences. In the context of biological data analysis, they have gained increasing importance, e.g., when considering gene expression data or when describing biological cells or structures with multiple attribute values. Data set sizes and data collections are constantly growing. The challenge is to detect valuable information in these complex data sets. Multidimensional data

T.V. Long (✉)
University of Transport and Communications, Hanoi, Vietnam
e-mail: vtran@utc.edu.vn

L. Linsen
Jacobs University, Bremen, Germany
e-mail: l.linsen@jacobs-university.de

L. Linsen et al. (eds.), *Visualization in Medicine and Life Sciences III*, Mathematics and Visualization, DOI 10.1007/978-3-319-24523-2_14

visualization techniques have proven to effectively support the exploration and analysis of the data in an intuitive manner.

The order of data dimensions is crucial for the effectiveness of a large number of multidimensional data visualization techniques [4] such as parallel coordinate plots [18], star coordinate plots [19], radial visualization (Radviz) [17], scatterplot matrices [13, 24], circle segments [3], and pixel recursive patterns [20]. The data dimensions have to be positioned in some one- or two- dimensional arrangement on the screen. This chosen arrangement of dimensions can have a major impact on the expressiveness of the visualization, as the relationships between adjacent dimensions are significantly easier to detect than relationships between dimensions positioned far from each other. Dimension ordering aims at improving the effectiveness of the visualization by giving suitable orders to the dimensions. Good orders are such that users can easily detect relationships or pay more attention to more important dimensions.

In many commonly used multidimensional data visualization techniques like most of the ones mentioned above (parallel coordinates, star coordinates, Radviz, and scatterplot matrix), uniform spacing and/or angles are placed evenly between two adjacent axes in the display. We presume that non-uniform spacing could help to convey information about dimensions, such as similarity between adjacent data dimensions or structure of a subspace.

In this paper, we propose a general approach to data dimension ordering for multidimensional visualization techniques and, exemplarily, demonstrate how to apply it to parallel coordinate plots. It can easily be applied in the context of other multidimensional visualization methods. Our approach is based on the ideas of Belkin and Niyogi [7] and Ding and He [12] that find positions of data dimensions according to their similarity. Inspired by Laplacian eigenmaps, we can find layouts of the data dimensions based on a similarity matrix. Our approach is applied to automatically find a dimension reordering and dimension spacing in parallel coordinates. The dimension reordering emphasizes the high similarity of adjacent axes. The dimension spacing emphasizes the pairwise relationships between adjacent data dimensions. Our approach supports data dimension ordering and data dimension spacing to enhance bivariate structures of multidimensional data sets.

After discussing previous related work on parallel coordinates and dimension reordering in multidimensional data visualization techniques in Sect. 2, we present the data dimensions similarity measurement in Sect. 3. In Sect. 4, we describe an asymptotic to finding the data dimension reordering and data dimension spacing in parallel coordinates. In Sect. 5, we show the effectiveness of our methods with well-known multidimensional data sets in the context of both classified and unclassified data, before we apply them to biological data. In Sect. 6, we further discuss and evaluate our approach, also in comparison to the literature.

2 Related Work

Dimension ordering is an important issue in multidimensional data visualization. Suitable dimension ordering supports the display of important structures such as relationships between dimensions. An effective way to improve the quality of multidimensional data visualizations is to reorder the dimension axes based on similarity of data attributes [4]. This is particularly evident in parallel coordinate plots.

The general problem of dimension ordering is presented by Ankerst et al. [4]. They describe the effectiveness of dimension ordering for many multidimensional visualization techniques, define the concept of similarity between data dimensions, and formulate the optimization problems. The optimization problems are presented based on the idea of dimension ordering such that the layout of data dimensions is optimized to place dimensions next to another based on their similarity. The task is to place axes of highly similar dimensions next to each other. The authors prove that the optimizations are NP-hard problems and equivalent to the traveling salesman problem. The authors also propose an automatic heuristic approach to generate a solution. We propose a method for dimension reordering based on their similarity that is not heuristics-based and, nevertheless, has polynomial time complexity.

Dimension ordering of matrices has been introduced by Bertin [8]. Bertin [8] proposed a display and an analysis approach for multidimensional data that is based on a matrix called the Bertin matrix. The Bertin matrix allows for reordering both rows and columns simultaneously to generate a homogeneous structure. Siirtola [26] proposed some algorithms for computing a reorderable matrix and integrating it with parallel coordinates. The reorderable matrix can be implemented with moving rows or columns, sorting the matrix, reorganizing it, and automatic permutation of the matrix. The linked views between the reorderable matrix and the parallel coordinates allow for operations such as selecting a single data item, selecting a subset of data items, selecting a dimension, permutation of dimensions, and forming a classification. These operators can help users to view the data structure on some interesting subspace. The spectral methods is introduced for seriation problem [6, 10]. The permutation is defined based on the Fiedler vector of the Laplacian matrix of the similarity matrix. The spectral method is also proposed for dimensionality reduction with many variation of principal component analysis [21].

Dimension reordering for the Radviz method was investigated by Di Caro et al. [11]. They proposed two methods for dimension arrangement on Radviz based on the optimization problem formulated using a similarity matrix and a neighborhood matrix between data dimensions on a unit circle [11]. Recently, Albuquerque et al. [2] used the cluster density measure to find good layouts of Radviz. They proposed a greedy incremental algorithm to successively add data dimensions to a Radviz layout to determine a suitable order.

Dimension reordering for star coordinates was introduced by Artero et al. [5]. They proposed the SBAA (Similarity Based Attribute Arrangement) algorithms for sorting data dimensions in the star coordinates. The SBAA is based on the nearest neighbor heuristic method of data dimensions.

In the context of dimension ordering for parallel coordinates, Dasgupta et al. [9] applied the branch-and-bound algorithm to optimize the dimension's order. They introduce many criteria for optimality such as optimal line crossings, optimal angles of crossing, and optimal parallelism, and also combine their optimizations with axes inversions of the parallel coordinates. Wang et al. [29] proposed an interactive hierarchical dimension reordering, spacing, and filtering in parallel coordinates. The similarities among data dimensions are determined as done by Ankerst et al. [4], but they construct a similarity hierarchy of data dimensions based on a bottom-up clustering. They also introduced the concept of automatic data dimension spacing. The default spacing between all adjacent axes is an equidistant spacing of the parallel coordinates. Their algorithm of adjusting the spacing of data dimensions is derived from the hierarchical structure of the dimensions. McDonnell and Mueller introduce illustrative parallel coordinates (IPC) [23]. The IPC method improve parallel coordinate visualization for large scale data set and clustering visualization.

Seo and Shneiderman [25] proposed the rank-by-feature framework for ranking with respect to a criterion of interest and sorting scatterplots according to that criterion. Dimension reordering is achieved by testing all permutations. Albuquerque et al. [1] proposed a method for ranking data dimensions by summing other data dimensions. Then, the permutation of data dimensions is obtained by sorting according to the quality of the data dimensions. Similarly, Tatu et al. [28] introduced some methods for measuring the quality of scatterplot matrix and parallel coordinates for both classified and unclassified data. The authors proposed the Rotating Variance Measure (RVM) for finding linear and nonlinear pairwise correlations between the data dimensions. The RVM is applied to unclassified data on scatterplot matrices. The class density measure is used to measure overlap between classes and is, then, applied for scatterplot matrices with unclassified data. For the measurements on parallel coordinates, they introduced a Hough-space measure for the number of accumulator cells that have a higher value than a threshold. The Hough-space measure is applied for unclassified data. They also proposed to use the overlap measure for classified data. They apply it to each class and compute the sum of the quality measures over all classes. The dimension order optimization, then, uses a heuristic algorithm. Recently, Ferdosi et al. [14] proposed a method for finding clusters structures in subspaces. The authors introduced a dimension ordering based on ranking sequential orthogonal subspaces. The subspace that has the highest quality is chosen and the full space provides the ordering of all dimensions. Lehmann et al. [22] introduced a method for sorting the data dimension in scatterplot matrices. For each scatterplot, they measure the quality and obtain the quality measurement of the scatterplot matrix. The authors use a hill-climbing algorithm to find the best permutation with three local cost functions for the matrix quality of scatterplot matrices.

3 Data Dimension Similarity

In this section, we introduce the similarity measures between dimensions of a multidimensional data set. The similarity measures are defined pairwise between any two dimensions. Given two-dimensional data $(x_i, y_i), i = 1, 2, \ldots, n$, from the random vector (X, Y). We present the data dimension similarity between X and Y for both cases unclassified data and classified data. We denote the similarity between X and Y by **similar**(X, Y). We normalize the similarity of data dimension to the interval $[0, 1]$. Thus, **similar**$(X, Y) \approx 1$ means that X and Y have high similarity, while **similar**$(X, Y) \approx 0$ means that X and Y are dissimilar.

3.1 *Unclassified Data*

For unclassified data, we base our similarity measure on Pearson's correlation. Pearson's correlation coefficient for a pair of variables (X, Y) is defined by

$$\rho = \rho_{XY} = \frac{E[(X - E(X))(Y - E(Y)))]}{\sqrt{Var(X)}\,\sqrt{Var(Y)}}.,$$

where $E(.)$ denotes the expected value and $Var(.)$ denotes the variance. Pearson's sample correlation coefficient of a sample $(x_i, y_i), i = 1, 2, \ldots, n$, can be determined by

$$r = r_{xy} = \frac{n\sum_{i=1}^{n} x_i y_i - \sum_{i=1}^{n} x_i \sum_{i=1}^{n} y_i}{\sqrt{n\sum_{i=1}^{n} x_i^2 - (\sum_{i=1}^{n} x_i)^2}\,\sqrt{n\sum_{i=1}^{n} y_i^2 - (\sum_{i=1}^{n} y_i)^2}}.$$

The range of values of Pearson's correlation coefficient is given by the interval $[-1, 1]$, which represents the linear relationship of the two variables. If the absolute value of Pearson's correlation coefficient is close to 1, it indicates a strong linear correlation, whereas, if it is close to 0, it indicates a weak relationship between the two variables.

For the application to parallel coordinates, we emphasize the linear relationship between two adjacent variables. Dimension spacing should also display the importance of the linear relationship. Therefore, we define the similarity between data dimensions X and Y in case of unclassified data as

$$\mathbf{similar}(X, Y) = |r|.$$

3.2 Classified Data

For classified data, we use the Class Distance Consistency (CDC) [27] to measure the similarity between data dimensions. Suppose we are given a data set $\{\mathbf{p}_i = (x_i, y_i) : 1 \leq i \leq n\}$, which is classified into m class, where the classes are labeled by $\{1, 2, \ldots, m\}$. We denote the class label of data point $\mathbf{p}$ by $label(\mathbf{p}) \in \{1, 2, \ldots, m\}$. For each class, we denote as $\mathbf{c}_i$ the centroid of the ith class. A data point $\mathbf{p}$ belongs to a class i, if the distance from the data point $\mathbf{p}$ to the centroid $\mathbf{c}_i$ of that class is smallest. Hence, we denote

$$class(\mathbf{p}) = \arg\min_{1 \leq i \leq m} distance(\mathbf{p}, \mathbf{c}_i).$$

A data point $\mathbf{p}$ is correctly classified, if its label matches the class with smallest distance to the centroid. Otherwise. the data point $\mathbf{p}$ is misclassified.

The CDC of the data set $\{\mathbf{p}_i = (x_i, y_i) : 1 \leq i \leq n\}$, is defined as the number of correctly classified data points. Using this criterion, we define

$$\mathbf{similar}(X, Y) = \frac{|\mathbf{p}_i = (x_i, y_i) : label(\mathbf{p}_i) = class(\mathbf{p}_i)|}{n}.$$

In parallel coordinates, two data dimensions shall be adjacent, if they are suitable for representing the class structures of the given data set. The dimension spacing shows the importance of the class structures in parallel coordinates.

4 Data Dimension Reordering

Given a p-dimensional data set X with dimensions $X_1, X_2, \ldots, X_p$. We compute the similarity matrix $S = (s_{ij})_{p \times p}$, where the entries are the pairwise similarities of the data dimensions as introduced in Sect. 3, i.e., $s_{ij} = \mathbf{similar}(X_i, X_j)$.

Let y_i denote the coordinates of the ith data dimension X_i, $i = 1, 2, \ldots, p$. The distance between two data dimensions is defined by $(y_i - y_j)^2$ and when weighted by its similarity is given as $(y_i - y_j)^2 s_{ij}$. We define our objective function as the summation of weighted distances of all pairwise data dimensions, i.e., the objective function is given by

$$\sum_{i \neq j} (y_i - y_j)^2 s_{ij}.$$

The task is to find the minimum of this objective functions. The minimization of the objective function ensures that, if s_{ij} is close to 1, then y_i is close to y_j. Hence, if two data dimensions have higher similarity, they are placed closer to each other. If all y_i

are equal, the objective function reaches its minimum. To avoid the trivial solution, we postulate $\sum_{i=1}^{p} y_i = 0$ and $\sum_{i=1}^{p} y_i^2 = 1$.

In parallel coordinates, the positions of the data dimensions $X_1, X_2, \ldots, X_p$ are fixed to locations $1, 2, \ldots, p$ on a line, e.g., $y_i = i$. The dimension reordering can be determined by computing the index permutation $\pi = \{\pi_1, \pi_2, \ldots, \pi_p\}$ that minimizes the objective function

$$J(\pi) = \sum_{i \neq j} (i - j)^2 s_{\pi_i, \pi_j}.$$

For a vector $\mathbf{y} = (y_1, y_2, \ldots, y_p)$, the vector permutation of $\mathbf{y}$ is denoted by $\pi(\mathbf{y}) = (y_{\pi_1}, y_{\pi_2}, \ldots, y_{\pi_p})$, and we denote π^{-1} as the inverse of the permutation π. Then, the objective function can be written as

$$J(\pi) = \sum_{i \neq j} (\pi_i^{-1} - \pi_j^{-1})^2 s_{ij}. \tag{1}$$

The optimization of the objective function in Eq. (1) is not affected by a linear transformation. Assume $y_i = a\pi_i^{-1} + b$, where the two parameters a and b are determined such that $\sum_{i=1}^{p} y_i = 0$ and $\sum_{i=1}^{p} y_i^2 = 1$. Thus, we obtain

$$\begin{cases} a \sum_{i=1}^{p} \pi_i^{-1} + pb & = 0 \\ a^2 \sum_{i=1}^{p} (\pi_i^{-1})^2 + 2ab \sum_{i=1}^{p} \pi_i^{-1} + pb^2 & = 1 \end{cases}$$

or, equivalently,

$$\begin{cases} a \sum_{i=1}^{p} i + pb & = 0 \\ a^2 \sum_{i=1}^{p} i^2 - pb^2 & = 1 \end{cases}$$

From these equations, we can easily derive that $a^2 = \dfrac{12}{p^2 - 1}$ and $b = -\dfrac{p+1}{2a}$.

Instead of finding the best permutation π of the data dimensions by an exhaustive search or based on a heuristic, we introduce an asymptotically optimal algorithm to find the permutation π based on finding the continuous vector $\mathbf{y} = (y_1, y_2, \ldots, y_p)$. The permutation π is determined by the values of the continuous values $y_1, y_2, \ldots, y_p$ with

$$y_i < y_j \Leftrightarrow \pi_i^{-1} < \pi_j^{-1}.$$

Therefore, the permutation π is uniquely defined by the vector $\mathbf{y}$.

We compute a Laplace matrix L of the similarities matrix S [7], which is defined by $L = D - S$, where $D = (d_{ij})$ is a diagonal matrix with entries

$$d_{ij} = \begin{cases} \sum_{j=1}^{n} s_{ij} & i = j \\ 0 & i \neq j. \end{cases}$$

Hence, the Laplacian matrix $L = (l_{ij})$ has the entries

$$l_{ij} = \begin{cases} \sum_{j=1}^{n} s_{ij} & i = j \\ -s_{ij} & i \neq j. \end{cases}$$

Based on this definition, the objective function $J(\mathbf{y}) = \sum_{i \neq j} (y_i - y_j)^2 s_{ij}$ can be rewritten as

$$\begin{aligned} J(\mathbf{y}) &= \sum_{i \neq j} (y_i - y_j)^2 s_{ij} \\ &= \sum_{i \neq j} (y_i^2 + y_j^2 - 2y_i y_j) s_{ij} \\ &= 2 \sum_{i=1}^{n} \Big(\sum_{j \neq i} s_{ij}\Big) y_i^2 - 2 \sum_{i \neq j} y_i y_j s_{ij} \\ &= 2 \sum_{i=1}^{n} d_{ii} y_i^2 - 2 \sum_{i \neq j} y_i y_j s_{ij} \\ &= 2\mathbf{y}^T D\mathbf{y} - 2\mathbf{y}^T S\mathbf{y} \\ &= 2\mathbf{y}^T L\mathbf{y}. \end{aligned}$$

Moreover, we know that the Laplace matrix L is a semi-definite and symmetric matrix. The matrix L always has the smallest eigenvalue $\lambda = 0$ and a corresponding eigenvector $\mathbf{1} = (1, 1, \ldots, 1)$. To avoid the trivial solution, we find the second smallest eigenvalue λ and the corresponding eigenvector, which is called the Fiedler vector [15].

The Fiedler vector $\mathbf{y}$ can be determined as a solution to the optimization problem

$$J_1(\mathbf{y}) = \mathbf{y}^T L\mathbf{y} = \mathbf{y}^T(D - S)\mathbf{y} \longrightarrow \min, \quad (2)$$

under constraints $\mathbf{y}^T\mathbf{1} = 0, \mathbf{y}^T\mathbf{y} = 1$. Assuming that L has a unique eigenvalue 0, we find the Fiedler vector $\mathbf{y}$ and eigenvalue $\lambda \neq 0$ such that $L\mathbf{y} = \lambda\mathbf{y}$ or

$$\lambda = \frac{\mathbf{y}^T L\mathbf{y}}{\mathbf{y}^T\mathbf{y}}$$

is minimal, respectively. To avoid the matrix L to be singular, we find $\mathbf{y}$ and λ such that

$$\lambda + 1 = \frac{\mathbf{y}^T(L + I)\mathbf{y}}{\mathbf{y}^T\mathbf{y}}$$

is minimal or, equivalently, that

$$\mu = \frac{1}{\lambda + 1} = \frac{\mathbf{y}^T\mathbf{y}}{\mathbf{y}^T(L + I)\mathbf{y}}$$

is maximal. We assume that the matrix $L + I$ has a Cholesky decomposition [16] $L + I = U^T U$, where U is an upper triangular matrix. Let $\mathbf{z} = U\mathbf{y}$ or $\mathbf{y} = U^{-1}\mathbf{z}$, respectively, then we obtain

$$\mu = \frac{\mathbf{z}^T(U^{-1})^T U^{-1}\mathbf{z}}{\mathbf{z}^T\mathbf{z}}.$$

We apply the Power method [16] to find the largest eigenvalue μ and the corresponding eigenvector $\mathbf{z}$ of the matrix $(U^{-1})^T U^{-1}$. Hence, we have derived an algorithm for finding the Fiedler vector of the Laplacian matrix L. The algorithm is given as follows:

Algorithm 1

- Compute the Cholesky decomposition $L + I = U^T U$.
- Compute the inverse matrix for the upper triangular matrix U.
- Compute the largest eigenvalue μ and the corresponding eigenvector $\mathbf{z}$ of the matrix $(U^{-1})^T U^{-1}$ by the Power method.
- Compute $\mathbf{y} = U^{-1}\mathbf{z}$.
- Compute the Fiedler vector $\mathbf{y} \longleftarrow \frac{\mathbf{y}}{||\mathbf{y}||}$.

The constraints to avoid a trivial solution as defined above are not the only feasible solution. We also want to investigate a second choice of constraints.

Consequently, we propose an alternative optimal solution

$$J_2(\mathbf{y}) = \mathbf{y}^T L\mathbf{y} = \mathbf{y}^T (D - S)\mathbf{y} \longrightarrow \min, \tag{3}$$

for new constraints $\mathbf{y}^T D\mathbf{1} = 0, \mathbf{y}^T D\mathbf{y} = 1$. In the Algorithm 1, we consider all data dimensions equally. We propose a new constraint $\mathbf{y}^T D\mathbf{y} = 1$ for more emphasizing data dimensions more similarity with other data dimensions. To solve this, we compute eigenvalue λ and eigenvector $\mathbf{y}$ from $L\mathbf{y} = \lambda D\mathbf{y}$ or, equivalently, from $(L + D)\mathbf{y} = (1 + \lambda)D\mathbf{y}$. Again, instead of finding the minimal

$$\lambda = \frac{\mathbf{y}^T L\mathbf{y}}{\mathbf{y}^T D\mathbf{y}}$$

we find the maximal

$$\mu = \frac{1}{\lambda + 1} = \frac{\mathbf{y}^T D\mathbf{y}}{\mathbf{y}^T (L + D)\mathbf{y}}.$$

Analogously to Algorithm 1, we find the Cholesky decomposition [16] of the matrix $L + D = U^T U$ where U is an upper triangular matrix. Again, we define $\mathbf{z} = U\mathbf{y}$ or $\mathbf{y} = U^{-1}\mathbf{z}$, respectively, and obtain

$$\mu = \frac{\mathbf{z}^T (U^{-1})^T D U^{-1}\mathbf{z}}{\mathbf{z}^T \mathbf{z}}.$$

We apply the Power method [16] to find the largest eigenvalue μ and the corresponding eigenvector $\mathbf{z}$ of the matrix $(U^{-1})^T U^{-1}$. This leads to a second algorithm, where we find the Fiedler vector according to the changed set-up. The algorithm is given as follows:

Algorithm 2

- Compute the Cholesky decomposition $L + D = U^T U$.
- Compute the inverse matrix for the upper triangle matrix U.
- Compute the largest eigenvalue μ and the corresponding eigenvector z of the matrix $(U^{-1})^T D U^{-1}$ by the Power method.
- Compute $\mathbf{y} = U^{-1}\mathbf{z}$.
- Compute the Fiedler vector $\mathbf{y} \longleftarrow \frac{\mathbf{y}}{||\mathbf{y}||}$.

The asymptotic time complexity of the two algorithms is governed by the computation of the Cholesky decomposition and by the execution of the Power method. Both computations have complexity $O(p^3)$. Hence, the overall asymptotic time complexity of our algorithms is $O(p^3)$.

5 Experimental Results

First, we test our approach on some well-known data sets. The data set is the so-called Iris data set.[1] The Iris data set contains 150 data points, four attributes X_1 (sepal length), X_2 (sepal width), X_3 (petal length), and X_4 (petal width), and three classes Setosa (50 data points), Versicolour (50 data points), and Virginica (50 data points). Figure 1 shows the Iris data with original ordering of data dimensions. Each class is encoded by a different color.

First, we look into results when neglecting the classification, i.e., we treat the data as unclassified. The matrix of Pearson's correlation coefficient of the data dimensions for the Iris data set is given by

$$P = \begin{pmatrix} 1 & -0.1093 & 0.8717 & 0.8179 \\ -0.1093 & 1 & -0.4205 & -0.3565 \\ 0.8717 & -0.4205 & 1 & 0.9627 \\ 0.8179 & -0.3565 & 0.96275 & 1 \end{pmatrix}$$

The respective similarity matrix with Pearson's correlation coefficients for the Iris data set is then

$$S = \begin{pmatrix} 0 & 0.1093 & 0.8717 & 0.8179 \\ 0.1093 & 0 & 0.4205 & 0.3565 \\ 0.8717 & 0.4205 & 0 & 0.9627 \\ 0.8179 & 0.3565 & 0.96275 & 0 \end{pmatrix}$$

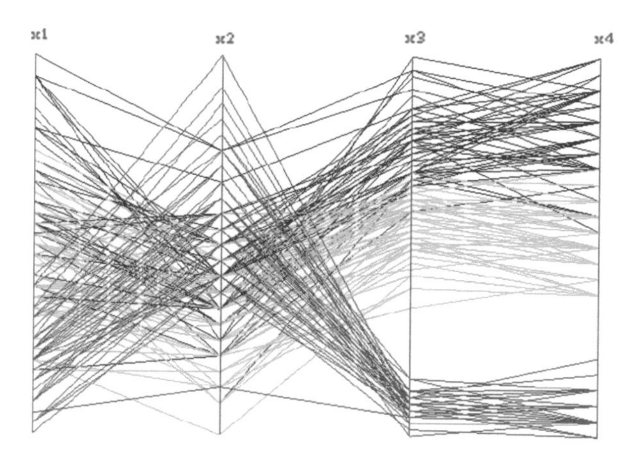

Fig. 1 Visualizing the Iris data set on parallel coordinates with original data dimension ordering

[1] http://archive.ics.uci.edu/ml/datasets/Iris.

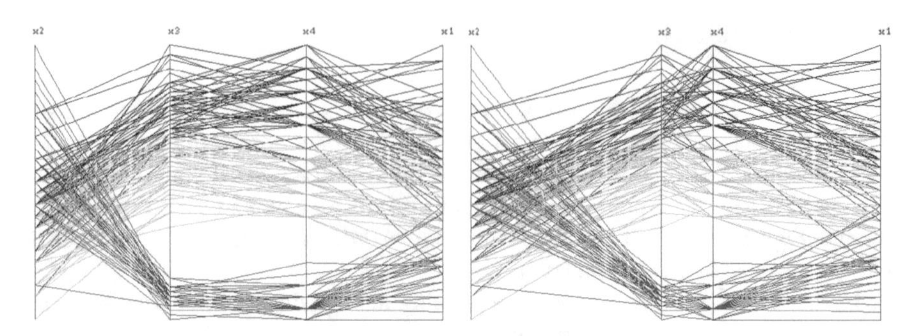

Fig. 2 Visualizing the Iris data set on parallel coordinates with Pearson's correlation similarity. *(Left)* The position of data dimensions are equidistant. *(Right)* The position of data dimensions are placed proportional to their similarity

Figure 2 shows the ordering of dimensions according to the Pearson's correlation similarities. Both, Algorithm 1 and Algorithm 2 produce the same order of data dimension. Figure 2 (left) shows the data dimensions placed equidistantly on the parallel coordinates. The pairwise correlations between the re-ordered adjacent dimensions is clearly higher than with the original ordering, which allows for immediate and intuitive interpretations. Figure 2 (right) shows the data dimensions place proportional to the values of vector **y**. It can easily be observed that the data dimensions X_3 and X_4 have highest similarity.

Next, we look into the results when considering the classification, i.e., we treat the data as classified. The similarity matrix based on CDC for the Iris data set is given by

$$S = \begin{pmatrix} 0 & 0.48 & 0.55 & 0.51 \\ 0.48 & 0 & 0.61 & 0.61 \\ 0.55 & 0.61 & 0 & 0.63 \\ 0.51 & 0.61 & 0.63 & 0 \end{pmatrix}.$$

Figure 3 shows the ordering of data dimensions based on the CDC similarity. Figure 3 (left) shows the order of data dimensions obtained by Algorithm 1, while Fig. 3 (right) shows the order obtained by Algorithm 2.

The optimal dimension ordering obtained by Algorithm 1 is X_3, X_1, X_2, and X_4. Hence, the three classes of the Iris data set is well classified in pairwise dimensions X_3 vs. X_1, X_1 vs. X_2, and X_2 vs. X_4 as displayed in Fig. 3 (left). Figure 3 (right) shows another optimal dimension ordering by Algorithm 2. Algorithm 1 only emphasizes the similarity between two pairwise data dimensions, while Algorithm 2 also emphasizes the similarity of each data dimension with other data dimensions. The weighted matrix has diagonal entries $d_{11} = 1.54$, $d_{22} = 1.70$, $d_{33} = 1.79$, and $d_{44} = 1.75$. Thus, the order of relationship degree is X_3, X_4, X_2, and X_1. Therefore, the optimal dimension ordering by Algorithm 2 is slightly better than that of Algorithm 1.

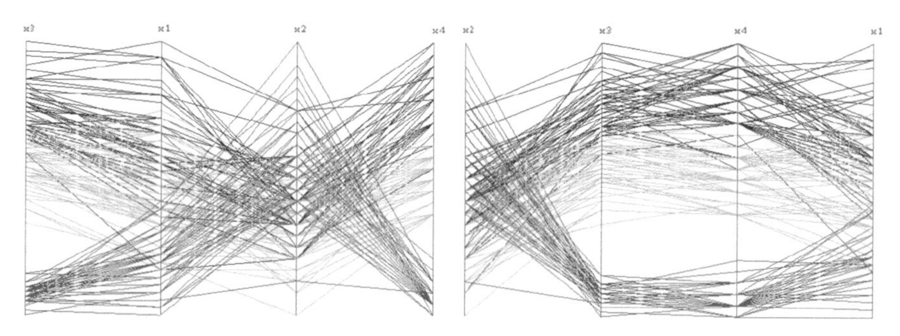

Fig. 3 The optimal dimension ordering based on class distance consistency (CDC) similarity. (*Left*) The optimal dimension ordering by Algorithm 1. (*Right*) The optimal dimension ordering by Algorithm 2

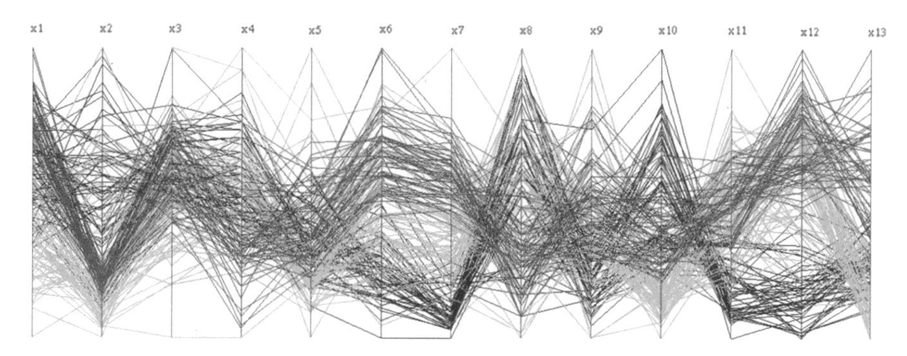

Fig. 4 Parallel coordinates of the Wine data set with original dimension ordering

The second well-known data set that we tested is the so-called Wine data set.[2] The Wine data includes 178 data points with 13 attributes X_1(Alcohol), X_2 (Malic acid), X_3 (Ash), X_4 (Alcalinity of ash), X_5 (Magnesium), X_6 (Total phenols), X_7 (Flavanoids), X_8 (Nonflavanoid phenols), X_8 (Proanthocyanins), X_{10} (Color intensity), X_{11} (Hue), X_{12} (OD280 / OD315 of diluted wines), and X_{13} (Proline). The Wine data set is classified into three classes: class 1 (59 data points), class 2 (71 data points), and class 3 (48 data points). Figure 4 shows the Wine data set with original dimension ordering. The different colors represent different classes of the Wine data set.

For the Wine data set, we use Algorithm 2 assuming both unclassified and classified data. For the exploration of the pairwise linear relationships of data dimensions, we use Pearson's correlation coefficient for measuring similarity. Figure 5 shows the optimal ordering with respect to the pairwise linear relationship in parallel coordinates. One can observe more consistent polygonal line directions between adjacent axes. Figure 6 shows the optimal dimension ordering assuming

[2]http://archive.ics.uci.edu/ml/datasets/Wine.

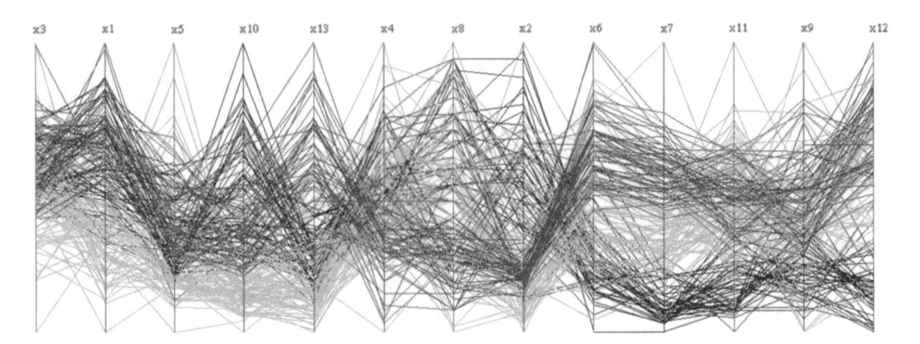

Fig. 5 Optimal dimension ordering for the Wine data set in the parallel coordinates assuming unclassified data

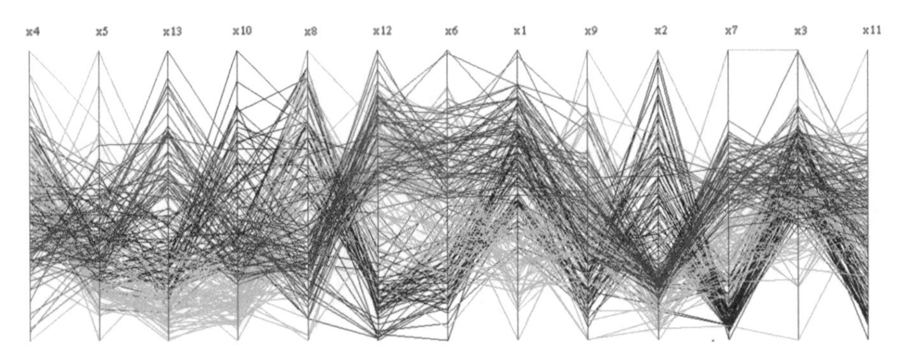

Fig. 6 Optimal dimension ordering for the Wine data set in the parallel coordinates assuming classified data

classified data. One can observe that there is a higher tendency to separating the classes.

The third data set we used for our experiments is a synthetic data set that is referred to as Sint data set. The Sint data set contains 14,831 data points with 20 attributes. It contains eight clusters. We add 20 noise attributes by using a uniform random noise distribution. Hence, we are dealing with 40 dimensions in total. The reasoning behind the adding of random noise dimensions is to test whether our algorithm is capable to identify clusters in a subspace. In this experiment, we only use the class distance consistency (CDC) similarity. Figure 7 shows the Sint data set without ordering dimensions. Each cluster is encoded by assigning a unique color. With the given order, it is not possible to visually detect a subspace that contains clusters. We emphasize the pairwise relationship of data dimensions that exhibit good clusters by replacing the CDC similarity s_{ij} by s_{ij}^k, where k is a control power parameter. Figure 8 shows the optimal ordering by Algorithm 2 with $k = 4$ and Fig. 9 shows the optimal ordering by Algorithm 2 with $k = 8$. When the parameter k increases, the noise attributes tend to group together (first 20 attributes in Fig. 9)

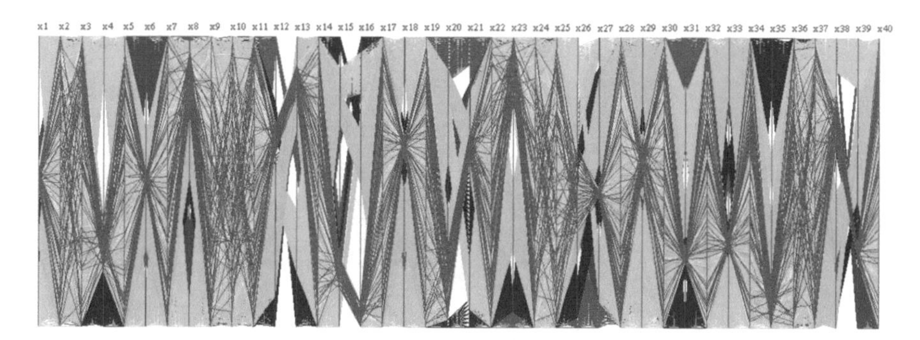

Fig. 7 Parallel coordinate plot for Sint data set without reordering dimensions

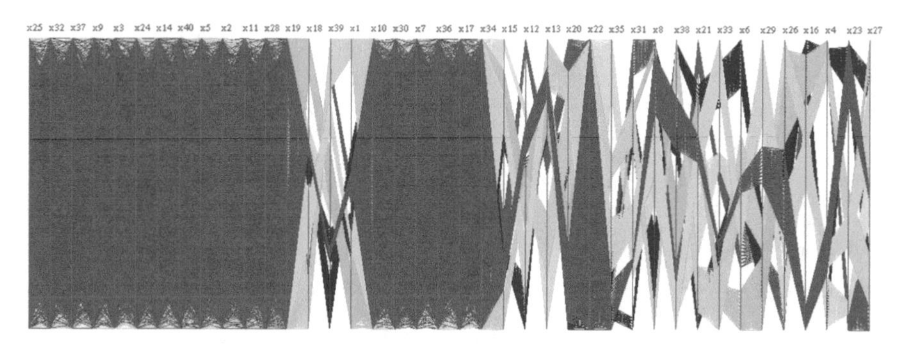

Fig. 8 Optimal ordering of dimensions in parallel coordinates with control parameter $k = 4$

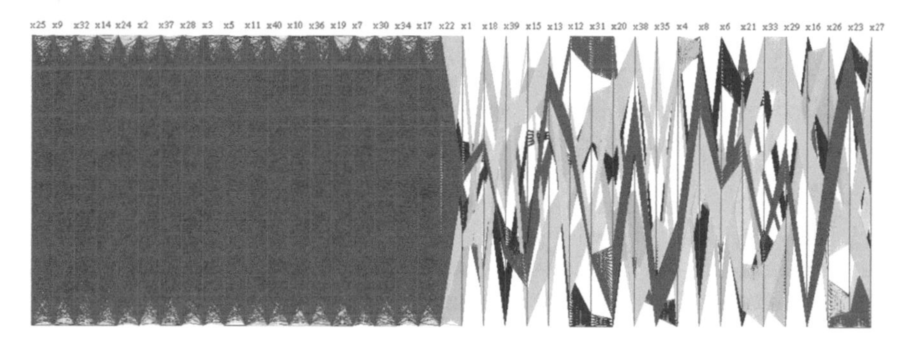

Fig. 9 Optimal ordering of dimensions in parallel coordinates with control parameter $k = 8$

and the subspace that contains the clusters of the data set (last 20 attributes in Fig. 9) becomes visible. Now, the structure that was hidden in Fig. 7 is visible.

Finally, we apply our method to the biological data referred to as Wisconsin Diagnostic Breast Cancer data set[3] (WDBC for short). Features are computed from

[3]http://archive.ics.uci.edu/ml/datasets.

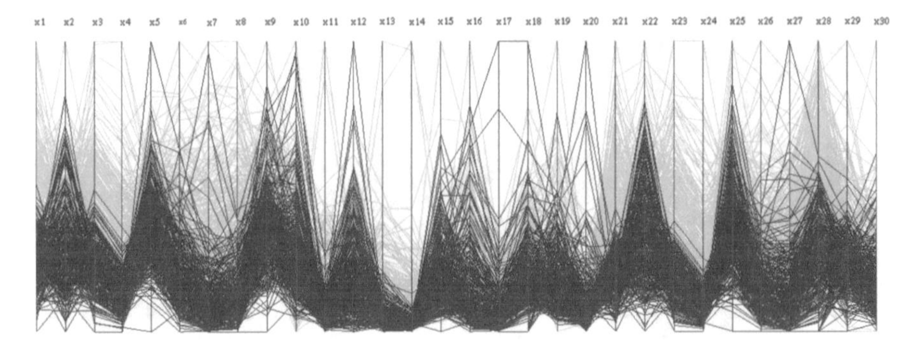

Fig. 10 Parallel coordinate plot of the WDBC data set without reordering dimension

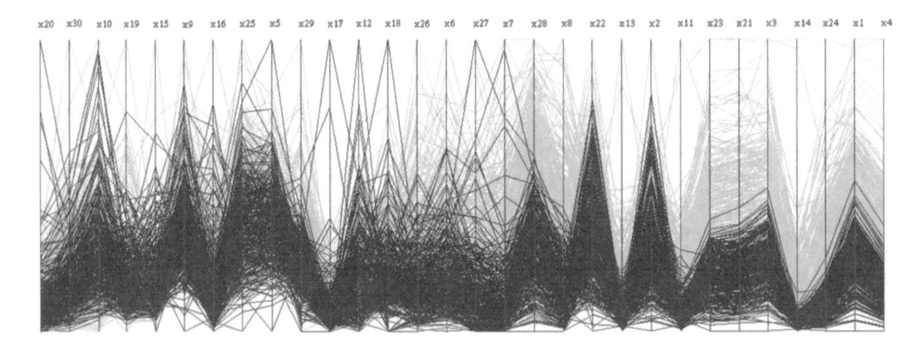

Fig. 11 Optimal ordering of dimensions in parallel coordinates for the WDBC data set based on Pearson's correlation coefficient with Algorithm 2

a digitized image of a fine needle aspirate (FNA) of a breast mass. They describe characteristics of the cell nuclei present in the image. Ten features are computed for each nucleus. The WDBC data set contains 569 data points and 30 attributes. The data set consists of the two classes malignant and benign. Figure 10 shows a parallel coordinate plot with original data dimension ordering. Color encodes the two classes. Figure 11 shows the parallel coordinate plot of the WDBC data set with an optimal reordering of data dimensions based on Pearson's correlation coefficient computed by Algorithm 2. Figure 12 shows the parallel coordinate plot of the WDBC data set with an optimal reordering of data dimensions based on CDC, again, computed by Algorithm 2. We observe again that Pearson's correlation coefficient strives for grouping pairs of dimensions with similar values, while the CDC produces higher visual separability of the classes.

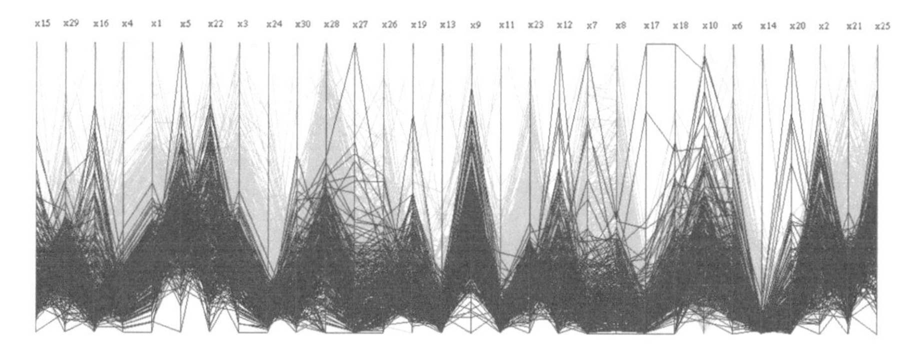

Fig. 12 Optimal ordering of dimensions in parallel coordinates for the WDBC data set based on class distance consistency with Algorithm 2

6 Comparison and Discussion

While the previous section showed results for visual inspection, this section aims at evaluating our results with respect to numerical quality measures and to compare our results to the literature. We compare our approach to the recently presented work by Albuquerque et al. [1]. They propose a method for dimension reordering in parallel coordinates based on sorting with respect to a quality metric for dimensions. The quality of the ith dimension is calculated by

$$Q_i = \sum_{j=1}^{n} \frac{n-i}{n} s_{ij}.$$

To quantify the quality of our reordering in comparison to the results by Albuquerque et al., we use the cost function introduced by Ankerst [4]. It computes the cost by summing the similarities of adjacent dimensions, i.e., the cost function is calculated by

$$S(\pi) = \sum_{i=1}^{n} s_{\pi_i,\pi_{i+1}}.$$

Consequently, the higher the cost function, the higher is the quality of the parallel coordinate plot.

We evaluate and compare our approach for six data sets including the ones presented in the previous section (Iris, Wine, Sint, and WDBC) plus two additional data sets[4] (Breast Tissue, Yeast). Yeast is another well-known multi-dimensional data set, while Breast Tissue represents impedance measurements of freshly excised

[4]http://archive.ics.uci.edu/ml/datasets.

Table 1 The cost function of dimension reordering based on CDC (higher values indicate higher quality)

Data set	Iris	Wine	Sint	WDBC	Breast tissue	Yeast
Algorithm in [1]	1.67	5.56	17.38	7.95	3.32	1.26
Algorithm 1	1.77	5.65	18.48	7.97	3.22	1.33
Algorithm 1	1.70	5.93	18.39	7.98	3.42	1.46

Table 2 The cost function of dimension reordering based on Pearson's correlation coefficient (higher values indicate higher quality)

Data set	Iris	Wine	Sint	WDBC	Breast tissue	Yeast
Algorithm in [1]	1.94	4.31	7.66	17.66	5.38	1.10
Algorithm 1	1.59	4.75	7.62	16.23	5.64	0.53
Algorithm 2	2.20	4.58	6.42	16.27	5.67	0.98

breast tissue taken at seven different frequencies (the six classes are carcinoma, fibro-adenoma, mastopathy, glandular, connective, and adipose).

Table 1 shows the quality of the parallel coordinate plots with different dimension reordering methods based on CDC similarity. We compare our two algorithms to the one by Albuquerque et al. [1]. In all six data set, the quality of our approach is rated higher than that of the algorithm in [1]. Only for the Breast Tissue data set the algorithm in [1] produced better results than our Algorithm 1. However, our Algorithm 2 always outperformed the algorithm in [1].

Table 2 shows the quality of the parallel coordinate plots with different dimension reordering methods based on Pearson's correlation coefficient. Here, our approaches were both better than the approach in [1] for the Iris, Wine, and Breast Tissue data sets, while we obtained lower quality for the Sint, WDBC, and Yeast data sets.

7 Conclusion and Future Work

We have presented a method for dimension reordering and dimension spacing in parallel coordinates based on data dimension similarity. For the dimension reordering we present two cost functions for measuring similarity, one for unclassified and one for classified data. We derived a scheme for solving the optimization problem with low computational complexity. We document the effectiveness of our method with various well-known data sets as well as biological data sets. We showed that our methods can improve the results and, therefore, support the detection of structures in multidimensional data sets. We also compared our method to a state-of-the-art approach in terms of a numerical quality measure. We showed that our approach outperforms the state-of-the-art approach for classified data and produces similar results for unclassified data.

For future work, we want to improve the display of our parallel coordinate plots to reduce clutter in the renderings. In particular, we would like to further enhance cluster structures in subspaces based on automatic dimension reordering and spacing in the parallel coordinates. Moreover, we want to use our approach for other multidimensional data visualization methods.

Acknowledgements This research is funded by Vietnam National Foundation for Science and Technology Development (NAFOSTED) under grant number 102.01-2012.04.

References

1. Albuquerque, G., Eisemann, M., Lehmann, D.J., Theisel, H., Magnor, M.A.: Quality-based visualization matrices. In: Proceedings of Vision, Modeling and Visualization (VMV), Braunschweig, pp. 341–350 (2009)
2. Albuquerque, G., Eisemann, M., Lehmann, D.J., Theisel, H., Magnor, M.: Improving the visual analysis of high-dimensional datasets using quality measures. In: IEEE Symposium on Visual Analytics Science and Technology (VAST), 2010, pp. 19–26. IEEE, Salt Lake City (2010)
3. Ankerst, M., Keim, D.A., Kriegel, H.-P.: Circle segments: a technique for visually exploring large multidimensional data sets. In: Proceedings of Visualization'96, Hot Topic Session. IEEE, San Francisco (1996)
4. Ankerst, M., Berchtold, S., Keim, D.A.: Similarity clustering of dimensions for an enhanced visualization of multidimensional data. In: Proceedings IEEE Symposium on Information Visualization, 1998, pp. 52–60. IEEE, Washington (1998)
5. Artero, A.O., de Oliveira, M.C.F., Levkowitz, H.: Enhanced high dimensional data visualization through dimension reduction and attribute arrangement. In The Tenth International Conference on Information Visualization, IV 2006, pp. 707–712. IEEE, London (2006)
6. Atkins, J.E., Boman, E.G., Hendrickson, B.: A spectral algorithm for seriation and the consecutive ones problem. SIAM J. Comput. **28**(1), 297–310 (1998)
7. Belkin, M., Niyogi, P.: Laplacian eigenmaps for dimensionality reduction and data representation. Neural Comput. **15**(6), 1373–1396 (2003)
8. Bertin, J.: Semiology of Graphics: Diagrams, Networks, Maps. University of Wisconsin press, Madison (1983)
9. Dasgupta, A., Kosara, R.: Pargnostics: screen-space metrics for parallel coordinates. IEEE Trans. Vis. Comput. Graph. **16**(6), 1017–1026 (2010)
10. De Leeuw, J., Michailidis, G.: Graph layout techniques and multidimensional data analysis. In: Game Theory. Optimal Stopping, Probability and Statistcs. Lecture Notes-Monograph Series, pp. 219–248. Institute of Mathematical Statistics, Beachwood (2000)
11. di Caro, L., Frias-martinez, V., Frias-martinez, E.: Analyzing the role of dimension arrangement for data visualization in radviz. In: Advances in Knowledge Discovery and Data Mining, pp. 125–132. Springer, Heidelberg (2010)
12. Ding, C., He, X.: Linearized cluster assignment via spectral ordering. In: Proceedings of the Twenty-First International Conference on Machine Learning, pp. 30–30. ACM, New York (2004)
13. Elmqvist, N., Dragicevic, P., Fekete, J.-D.: Rolling the dice: multidimensional visual exploration using scatterplot matrix navigation. IEEE Trans. Vis. Comput. Graph. **14**(6), 1539–1148 (2008)
14. Ferdosi, B.J., Roerdink, J.B.: Visualizing high-dimensional structures by dimension ordering and filtering using subspace analysis. In: Computer Graphics Forum, vol. 30, pp. 1121–1130. Wiley, Chichester (2011)

15. Fiedler, M.: A property of eigenvectors of nonnegative symmetric matrices and its application to graph theory. Czechoslov. Math. J. **25**(4), 619–633 (1975)
16. Golub, G.H., Van Loan, C.F.: Matrix Computations, vol. 3. Johns Hopkins University Press, Baltimore (1996)
17. Hoffman, P., Grinstein, G., Marx, K., Grosse, I., Stanley, E.: DNA visual and analytic data mining. In: Proceedings Visualization'97, pp. 437–441. IEEE, Los Alamitos (1997)
18. Inselberg, A.: The plane with parallel coordinates. Vis. Comput. **1**(2), 69–91 (1985)
19. Kandogan, E.: Star coordinates: a multi-dimensional visualization technique with uniform treatment of dimensions. In: Proceedings of the IEEE Information Visualization Symposium, vol. 650 (2000)
20. Keim, D.A., Ankerst, M., Kriegel, H.-P.: Recursive pattern: a technique for visualizing very large amounts of data. In: Proceedings of the 6th Conference on Visualization'95, pp. 279–286. IEEE, Washington (1995)
21. Koren, Y., Carmel, L.: Robust linear dimensionality reduction. IEEE Trans. Vis. Comput. Graph. **10**(4), 459–470 (2004)
22. Lehmann, D.J., Albuquerque, G., Eisemann, M., Magnor, M., Theisel, H.: Selecting coherent and relevant plots in large scatterplot matrices. In: Computer Graphics Forum, vol. 31, pp. 1895–1908. Wiley, Chichester (2012)
23. McDonnell, K.T., Mueller, K.: Illustrative parallel coordinates. In: Computer Graphics Forum, vol. 27, pp. 1031–1038. Wiley, Chichester (2008)
24. Peng, W., Ward, M.O., Rundensteiner, E.A.: Clutter reduction in multi-dimensional data visualization using dimension reordering. In: IEEE Symposium on Information Visualization, INFOVIS 2004, pp. 89–96. IEEE, Austin (2004).
25. Seo, J., Shneiderman, B.: A rank-by-feature framework for unsupervised multidimensional data exploration using low dimensional projections. In: IEEE Symposium on Information Visualization, INFOVIS 2004, pp. 65–72. IEEE, Austin (2004)
26. Siirtola, H.: Combining parallel coordinates with the reorderable matrix. In: Proceedings. International Conference on Coordinated and Multiple Views in Exploratory Visualization, 2003, pp. 63–74. IEEE, London (2003)
27. Sips, M., Neubert, B., Lewis, J.P., Hanrahan, P.: Selecting good views of high-dimensional data using class consistency. In: Computer Graphics Forum, vol. 28, pp. 831–838. Wiley, Chichester (2009)
28. Tatu, A., Albuquerque, G., Eisemann, M., Schneidewind, J., Theisel, H., Magnork, M., Keim, D.: Combining automated analysis and visualization techniques for effective exploration of high-dimensional data. In: IEEE Symposium on Visual Analytics Science and Technology, VAST 2009, pp. 59–66. IEEE, Atlantic City (2009)
29. Wang, J., Peng, W., Ward, M.O., Rundensteiner, E.A.: Interactive hierarchical dimension ordering, spacing and filtering for exploration of high dimensional datasets. In: IEEE Symposium on Information Visualization, INFOVIS 2003, pp. 105–112. IEEE, Washington (2003)

Extraction of Robust Voids and Pockets in Proteins

Raghavendra Sridharamurthy, Talha Bin Masood, Harish Doraiswamy, Siddharth Patel, Raghavan Varadarajan, and Vijay Natarajan

Abstract Voids and pockets in a protein, collectively called as cavities, refer to empty spaces that are enclosed by the protein molecule. Existing methods to compute, measure, and visualize the cavities in a protein molecule are sensitive to inaccuracies in the empirically determined atomic radii. This paper presents a topological framework that enables robust computation and visualization of these structures. Given a fixed set of atoms, cavities are represented as subsets of the weighted Delaunay triangulation of atom centres. A novel notion of (ε, π)-stable cavities helps identify cavities that are stable even after perturbing the atom radii by a small value. An efficient method is described to compute these stable cavities for a given input pair of values (ε, π). This approach is used to identify potential pockets and channels in protein structures.

1 Introduction

A cavity in a protein molecule refers to both voids (without openings) and pockets (with openings). These cavities play a key role in determining the stability and function of proteins. From the biologist's point of view, obtaining a stable protein is the starting point of many applications, from in-vitro studies of binding and interactions, to using the protein as an antigen or vaccine. Whereas surface pockets

R. Sridharamurthy (✉) • T.B. Masood
Department of Computer Science and Automation, Indian Institute of Science, Bangalore, India
e-mail: g.s.raghavendra@gmail.com; tbmasood@csa.iisc.ernet.in

H. Doraiswamy
Department of Computer Science and Engineering, NYU Polytechnic School of Engineering, Brooklyn, NY, USA
e-mail: harishd@nyu.edu

S. Patel • R. Varadarajan
Molecular Biophysics Unit, Indian Institute of Science, Bangalore, India
e-mail: spatel@mbu.iisc.ernet.in; varadar@mbu.iisc.ernet.in

V. Natarajan (✉)
Department of Computer Science and Automation, Department of Computational and Data Sciences, Indian Institute of Science, Bangalore 560012, India
e-mail: vijayn@csa.iisc.ernet.in

L. Linsen et al. (eds.), *Visualization in Medicine and Life Sciences III*, Mathematics and Visualization, DOI 10.1007/978-3-319-24523-2_15

often form part of the active site of enzymes or interacting sites for other proteins, internal voids are often relevant structurally as features that affect the overall thermodynamic stability of the protein. It is established that filling up internal voids improves the packing of the protein thus increasing stability. In this respect, detecting and visualizing structurally robust cavities inside the protein informs the biologist on which mutations to perform to improve internal packing and get a stable protein.

Related Work Several methods have been proposed to locate cavities in protein molecules. In this paper, we focus our attention on geometric methods. Edelsbrunner et al. [9, 10] and Liang et al. [20, 21] proposed a definition that is based on the theory of alpha shapes and discrete flows in Delaunay triangulations. Kim et al. [15, 16] proposes a definition of cavities based on an alternate representation of a set of atoms called beta shapes that faithfully captures proximity. Tools based on the above approach are available and widely used [5, 17, 18]. Till and Ullmann [31] employed a graph theoretic algorithm to identify cavities and compute their volume. Parulek et al. [27] used graph based methods on the implicit representation of molecular surfaces to identify pockets and potential binding sites. Varadarajan et al. [2] employed a Monte Carlo procedure to position water molecules together with a Voronoi region-based method to locate empty space. They discussed the importance of accurate identification of cavities for the study of protein structure and stability. Novel Voronoi diagram-based techniques for the extraction and visualization of cavities have also been developed from the viewpoint of studying and interactively exploring access paths to active sites [22, 23, 28, 29]. Krone et al. [19] presented a visualization tool for interactive exploration of protein cavities in dynamic data.

Motivation The input in the above-mentioned methods are protein structures determined from x-ray crystallography data or other lower resolution data. These cavity detection methods are sensitive to inaccuracies that are inherent in the crystallographic measurements. While the measurements may guarantee high resolution, it is important to note that even small inaccuracies may cause a difference in the reported number of cavities. Inaccuracies may also arise due to fundamental limitations such as the notion of radii of atoms, which is determined empirically. For example, as illustrated in Fig. 1, presence of such inaccuracies may result in a cavity detection method to report two distinct but large cavities in place of one, or report very small volume cavities. Figure 2 illustrates the problem as it occurs in a lysozyme protein.

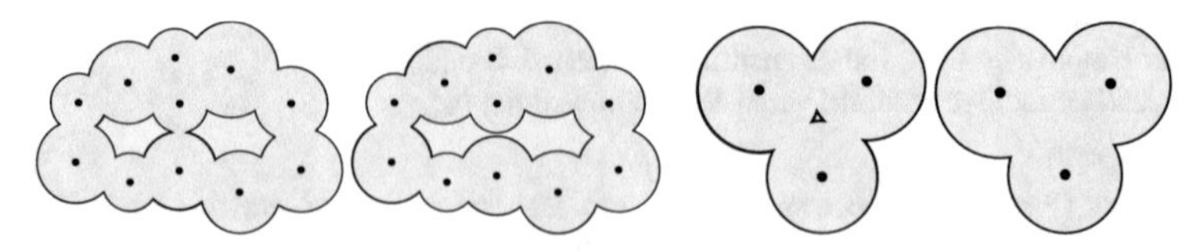

Fig. 1 *Left*: Two cavities that are apparently very near to each other may be a single cavity. *Right*: A very small cavity may be reported whereas no such cavity may exist

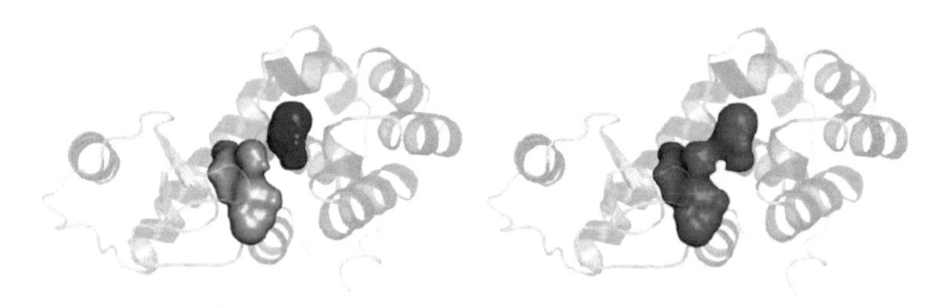

Fig. 2 *Left*: Two cavities that appear very near to each other in a lysozyme protein (PDB ID: 200L). The solid surface represents cavities while the protein is shown as cartoon for context. *Right*: The two cavities may be a single cavity

Contributions In this work, we aim to develop an interactive method to compute robust cavities in proteins. Our goal is to enable the user to reduce, if not completely eliminate, the inaccuracies mentioned earlier. In order to achieve this, we first provide a novel definition for robustness in the presence of inaccuracies in the measured radii.

We then propose a method for computing robust and stable cavities in proteins.[1] This is accomplished through the use of a simple and succinct structure called the alpha complex to represent protein molecules. The alpha complex is a simplicial complex that can be stored as a filtration, a series of simplicial complexes K^i with $K^i \subset K^{i+1}$. In order to identify the set of cavities that are stable with respect to small perturbations in the atom radii, our method symbolically modifies the radii of a select set of atoms by systematically processing and modifying the filtration. We show that this modification results in controlled changes in the number and properties of cavities and does not violate key properties of the filtration. The method is efficient in terms of running time performance and also supports the elimination of very small or insignificant voids as measured by the notion of topological persistence [11].

We develop software to visualize the stable cavities together with the molecule, and to calculate cavity volumes and surface areas. This software provides an interactive framework that a biologist can use to decide which cavities are more relevant and what mutations to perform. The software also supports exporting the detected cavities with the relevant biochemical context to enable their visualization in PyMOL [4].

Finally, we use this software to demonstrate the applicability of the notion of robust voids and pockets and apply it to detect potential channels and pockets in several proteins.

[1]A preliminary version of this work appeared as a short paper in the Proceedings of Eurographics Conference on Visualization [30].

2 Geometry Representation of Biomolecules

In this section, we briefly introduce the mathematical background required to define and represent the structure of biomolecules [7, 8, 26].

Simplicial Complex A *k-simplex* σ is the convex hull of $k+1$ affinely independent points. A vertex, edge, triangle, and tetrahedron are k-simplices of dimension $0-3$. A simplex τ is a *face* of σ, $\tau \leq \sigma$, if it is the convex hull of a non-empty subset of the $k+1$ points. A *simplicial complex* K is used to represent a topological space and is a finite collection of simplices such that (a) $\sigma \in K$ and $\tau \leq \sigma$ implies $\tau \in K$, and (b) $\sigma_1, \sigma_2 \in K$ implies $\sigma_1 \bigcap \sigma_2$ is either empty or a face of both σ_1 and σ_2. A *subcomplex* of K is a simplicial complex $L \subseteq K$.

Voronoi Diagram and Delaunay Triangulation Let $S \subseteq \mathbb{R}^d$ be a finite set of points. The *Voronoi cell* V_p, of a point $p \in S$, is the set of points in $\mathbb{R}^d$ whose Euclidean distance to p is smaller than or equal to any other point in S. The collection of Voronoi cells of all the points in S partitions $\mathbb{R}^d$, and is called the *Voronoi diagram* (Fig. 3a). The *Delaunay triangulation* D of S is the dual of the Voronoi diagram and partitions the convex hull of S, see Fig. 3b. The *weighted Voronoi diagram* and *weighted Delaunay triangulation* are similarly defined for a set of balls, which is considered as a set of weighted points. The weight is equal to the square of the

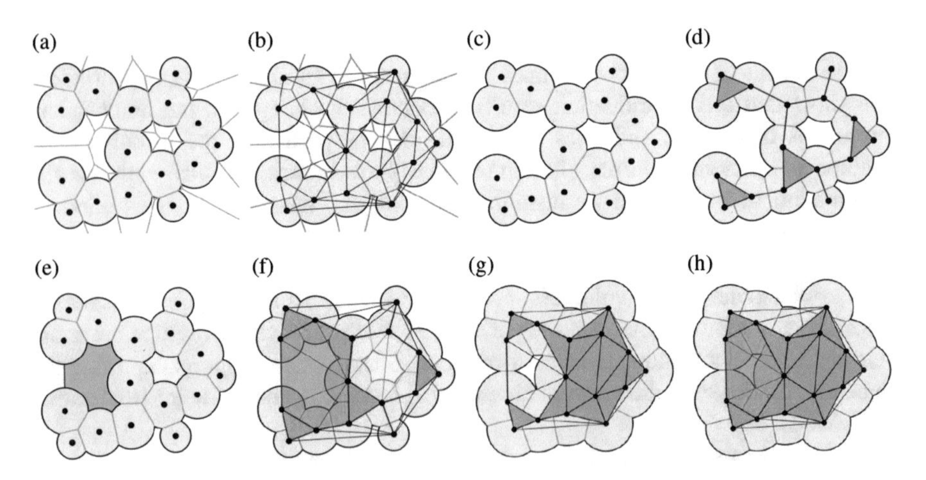

Fig. 3 (**a**) Voronoi diagram of a weighted point set in $\mathbb{R}^2$, Voronoi edges are in *green*. (**b**) The Delaunay complex is the dual of the Voronoi diagram. (**c**) Intersection of the weighted Voronoi diagram and the union of balls. (**d**) The dual complex is the dual of this partition of the union of balls that captures the incidence relationship. In this particular case, $\alpha = 0$. (**e**) Void and pocket in a collection of 2D balls. Void is shown in *yellow* and pocket is shown in *blue*. (**f**) Void and pocket shown as the connected components of the complement of alpha complex i.e., $D-K$. (**g**) The dual complex shown for some $\alpha > 0$ where the void has been filled up and original pocket has become a void. (**h**) The new void is highlighted using *blue*, Delaunay edges (in *black*) and alpha complex (in *red*) are also shown to provide context

radius of the ball, and the distance between a weighted point p with weight w_p and a point $x \in \mathbb{R}^d$ is given by the *power distance* $\|x - p\|^2 - w_p$.

Alpha Complex Molecules are often represented using a space-filling model such as a union of balls. The weighted Voronoi diagram helps represent the contribution from each atom to the union of balls. Consider an atom p. Define B_p as an open ball having the radius of the atom p. Let V_p be the weighted Voronoi cell corresponding to p. The contribution from each atom p is equal to $B_p \cap V_p$, the intersection between the ball corresponding to the atom and the weighted Voronoi cell of p, see Fig. 3c. The corresponding dual structure is a subcomplex of the weighted Delaunay triangulation and called the *dual complex*, see Fig. 3d.

Edelsbrunner et al. [6, 12, 13] consider a growth model, where the ball weights grow, and track the changes in the dual complex. The growth parameter, α, corresponds to a radius $\sqrt{r_p^2 \pm \alpha^2}$ for a ball centered at p with radius r_p. Positive values of α correspond to growing the balls and negative values correspond to shrinking the balls. The weight of the point $w(p)$ increases or decreases by α^2 and hence ranges between $-\infty$ and ∞. A negative weight corresponds to imaginary radius. Note that $\alpha = 0$ corresponds to no growth. The dual complex corresponding to a set of balls after they are grown by α is called the *alpha complex*.

Given a simplicial complex K, a finite sequence $\emptyset = K^0, K^1, \ldots, K^m = K$ of subcomplexes of K is a *filtration* if $K^0 \subset K^1 \subset \cdots \subset K^m$. Figure 3d, g show two subcomplexes (in red) which are part of a filtration. The *rank* of a subcomplex refers to its position in the filtration. The set of alpha complexes obtained by varying α from $-\infty$ to ∞ is a filtration of the Delaunay triangulation. In particular, we consider the filtration that is generated by inserting the simplices one at a time and if more than one simplex appear at the same value of α, we order them based on their dimension $(0 < 1 < 2 < 3)$. A vertex is inserted into the filtration when the weight of the ball becomes positive.

Voids and Pockets Let the alpha complex K represent a molecule at a given value α and D be the Delaunay complex of the weighted point set. A *cavity* is a maximally connected component of the complement $D - K$. *Voids* and *pockets* are cavities that are, respectively, bounded and not bounded by the union of balls [14]. Figure 3e, f illustrate a void and a pocket in 2D. Figure 3g, h show how they are affected by the growth model.

Topological Persistence The *boundary* of a triangle consists of its edge faces. The boundary of a collection of triangles is the formal sum of boundary edges of the individual triangles, where addition is performed modulo 2. A 2-*cycle* (two-dimensional cycle) is a collection of triangles whose boundary is empty. Cycles of other dimensions are defined similarly. A void is represented by a 2-cycle. The alpha complex K helps represent and track voids via the growth process. A void is said to be created when the last triangle in the 2-cycle is inserted into the filtration and it is destroyed when the volume that it occupies is filled by the last tetrahedron. *Topological persistence* of a 2-cycle measures its lifetime ($k \geq 0$) in a filtration [11]. It is equal to the difference between the α-values when the cycle is

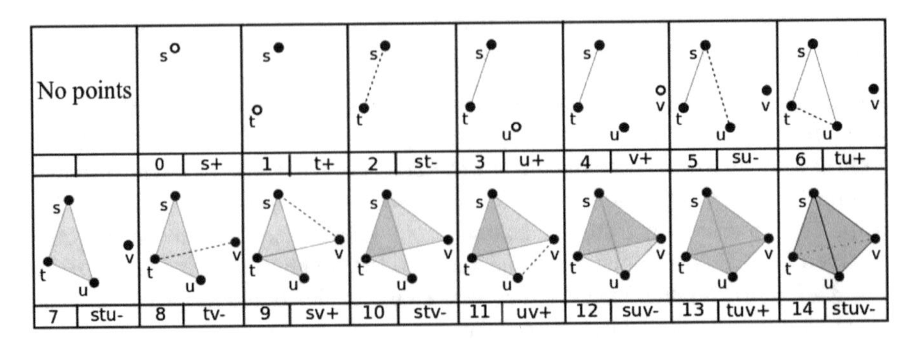

Fig. 4 A filtration generated by inserting simplices in a particular order. A k-simplex corresponds to an overlap of $k + 1$ balls. For each simplex, the box in the *bottom left* shows the rank/arrival time of the simplex, the box in the *bottom right* shows the simplex along with it's behaviour. The '+' implies that it is a positive simplex (creator) and the '−' implies that it is a negative simplex (destroyer). For example, the triangle '$tuv+$' creates a void and tetrahedron '$stuv-$' destroys it. The persistence of this void is therefore 1

created and destroyed. Given a filtration, the persistence of cycles can be computed efficiently [11]. The insertion of every simplex either creates a cycle or destroys a lower dimensional cycle. The persistence value associated with a simplex is equal to the topological persistence of the corresponding cycle. Figure 4 illustrates creation and destruction of cycles in a filtration of a small simplicial complex.

A void is represented by a 2-cycle and hence it has a well-defined creator and destroyer. However, this is not the case for a pocket, which may not necessarily be created as a result of a simplex insertion. This is because, initially $K = \emptyset$ and hence $D - K = D$ is a single connected component. This component corresponds to a pocket and has no creator. Hence, we cannot directly apply the notion of persistence to measure pockets. The notion of persistence of a void intuitively captures the volume of the void in terms of the range of α-values. To be consistent, we use a similar notion for pockets as well. We fix a value of $\alpha = \alpha_0$ and define the birth time of a pocket that has no creator to be equal to α_0. The α-value when the pocket interior is filled corresponds to its destruction time. Thus, similar to voids, the persistence of a pocket is equal to its lifetime and approximates the volume of the pocket in terms of the range of α-values.

3 Robust Cavities and Their Computation

We introduce a notion of robust cavities based on two parameters, one local and another global. The local parameter is referred to as stability and the global parameter is specified by topological persistence. In order to simplify the description, we assume that the cavities are computed for the α-complex corresponding to $\alpha = 0$. However, the proposed definitions, methods, and subsequent analysis are valid for all values of α.

3.1 ε-Stable and π-Persistent Cavities

Consider the interval $[-\varepsilon, \varepsilon]$ of α values, where $\varepsilon \geq 0$. A cavity is called an ε-*stable* cavity if it remains a single connected cavity within all α-complexes for α values in the range $[-\varepsilon, \varepsilon]$. In other words, using the lifetime terminology, the cavity is born, possibly split into multiple components, and destroyed at α-values that lie strictly outside of this interval. A cavity is π-*persistent* if its topological persistence is greater than π *i.e.*, the cavity size measured in terms of its lifetime is greater than π. The persistence of pockets is defined in this case by setting $\alpha_0 = -\varepsilon$. Combining the two notions of robustness, we call a cavity to be (ε, π)-*stable* if it is both ε-stable and π-persistent.

The above definitions help measure the stability of the cavities when the atom radii are perturbed by a small value. The local parameter considers perturbation within a small interval centered at the α-value of interest whereas the global parameter measures the size of the cavity in terms of its lifetime in the filtration. Cavities of interest may often not be stable with respect to both notions. For example, a large sized cavity (π-persistent for some large π) may be born within the interval $[-\varepsilon, \varepsilon]$. However, note that a small perturbation in the radii of atoms that line the surface of the cavity could result in an earlier birth time, hence making the cavity to be ε-stable. We aim to extract all cavities that are either stable as is or can be made stable via a small perturbation.

3.2 Computing (ε, π)-Cavities

The location of the atoms that constitute a protein molecule together with their van der Waals radii is obtained from the protein data bank in pdb format. Given ε and π, we compute the set of (ε, π)-stable cavities as follows.

1. Compute the weighted Delaunay triangulation of the input [18]. The atom centres form the set of points that are weighted using their van der Waals radii.
2. Build the alpha complex [6], which is a filtration of the weighted Delaunay triangulation.
3. **Modify the filtration based on the value of ε.**
4. Compute the set of (ε, π)-stable cavities by identifying all cavities [14, 20] of the modified filtration at $\alpha = 0$, and retaining only those cavities that have persistence greater than π.

The key idea in our proposed method is a modification of the filtration (Step 3) in order to compute the set of stable cavities. The filtration of the weighted Delaunay triangulation as defined by the α-value provides an explicit representation of the birth/death times of each cavity and its evolution. We propose to alter the birth/death times of the cavities by modifying the filtration instead of directly modifying the radii of atoms that line the surface of the cavity. While the latter approach follows directly from the definition, it is cumbersome and computationally inefficient. For example, varying the radii without explicit control may lead to changes in the triangulation and the alpha complex. These changes need to be explicitly tracked, else they may lead to inconsistencies between the alpha complex that represents the molecule and the space-fill model. Resolving such inconsistencies would necessitate the re-computation of all representations. On the other hand, the former approach is simpler and computationally efficient.

Modifying the Filtration We now describe this step in detail. One or more simplices are inserted to obtain a rank $i + 1$ simplicial complex from a rank i simplicial complex in the filtration. Higher ranks correspond to higher values of α. The topology of voids and pockets may change when the simplices are inserted. In particular, consider a triangle whose insertion changes the topology of a cavity. When this cavity is a void, the triangle splits the void into two voids (C1). In case of pockets, the insertion of the triangle could cause one of the following:

(C2) split the pocket into two pockets,
(C3) close one mouth of the pocket (for pockets with more than one mouth),
(C4) split the pocket into a pocket and a void, or
(C5) destroy the pocket and create a new void.

On the other hand, the insertion of a tetrahedron always destroys a void. These topology changes may be avoided by delaying the insertion of the simplices that cause a change in the topology of the cavities.

Let K^j and K^l be the alpha complexes corresponding to $\alpha = -\varepsilon$ and $\alpha = \varepsilon$ respectively. Consider the set of simplices, Σ, inserted into the filtration for values of α in the range $[-\varepsilon, \varepsilon]$. Let $\Sigma_t \subset \Sigma$ be a subset of the set of triangles that modifies the topology of a cavity and $\Sigma_T \subset \Sigma$ be the set of all tetrahedra in Σ. As mentioned earlier, our goal is to selectively alter the radii of atoms by altering the birth/death times of a cavity. In order to accomplish this, we delay the insertion of a select few simplices $\sigma_i \in \Sigma_t$ and all simplices in Σ_T such that $\sigma_i \notin K^j$ but $\sigma_i \in K^l$, where $K^j \subset K^l \subset D$. This delay corresponds to change in radii of the corresponding atoms enclosing the cavity.

Identifying the Set Σ_t The set Σ_t consists of all triangles that satisfies conditions C1, C2 and C3, while triangles that satisfy conditions C4 or C5 are optionally inserted into Σ_t. This is because a triangle that satisfies condition C4 or C5 creates a new void destroying the existing pocket. Depending on whether the perturbation decreases or increases the radii of the corresponding atoms, both the original pocket and the new void can be considered to be stable respectively.

Delayed Simplex Insertion Simplicial complexes in the filtration of the weighted Delaunay triangulation and the order of simplices that are inserted to generate the filtration satisfy several containment and incidence properties. These properties should be satisfied for the modified filtration as well. Towards this, we propose a conservative but computationally efficient approach to modify the filtration:

1. Move all tetrahedra in Σ_T to the end of the filtration. All such tetrahedra are present in D but not in any $K^i \subset D$.

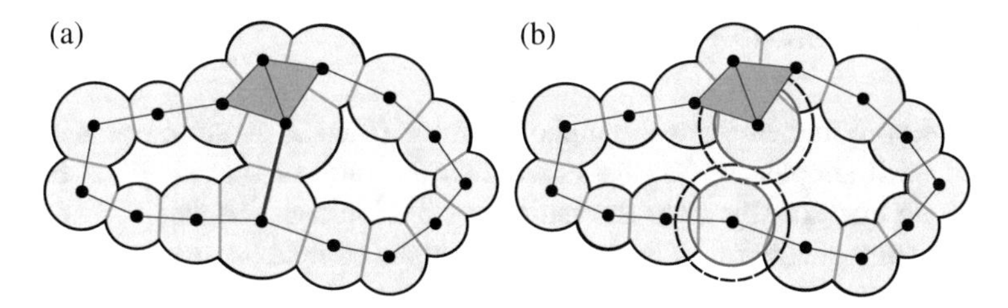

Fig. 5 2D illustration of simplex insertion causing a void to split. (**a**) Voids occur near to each other and the edge that splits the single void into two. (**b**) The two voids merge into one if the simplex insertion is delayed. The modified radii corresponding to the delay is highlighted in *red*

2. For each triangle in Σ_t, find its incident tetrahedra τ_1, τ_2.
3. Delay the insertion of the triangle and the two tetrahedra, τ_1 and τ_2, to the end of the filtration.

The above modification is illustrated using a 2D analogue in Fig. 5. Consider a void split into two as shown in Fig. 5a. Assume that the highlighted edge (triangle in 3D) is inserted into the alpha complex for α lying in the interval $[-\varepsilon, \varepsilon]$. Further, it also satisfies the criterion that it bounds two different voids. So it becomes a candidate for delayed insertion. We move the edge to the end of the filtration, which means that it does not belong to the alpha complex as shown in Fig. 5b. Also the radii of atoms centered at the end points of the edge (triangle in 3D) are decreased accordingly. Selective modification of the radii of a specific set of atoms is hence achieved in a controlled manner.

After the filtration is modified, we recompute the set of cavities at $\alpha = 0$, which correspond to the set of ε-stable cavities. From this set, we retain cavities having persistence greater than π to obtain the set of (ε, π)-stable cavities. Note that the persistence is computed with respect to the original filtration.

Discussion While computing stable cavities, our technique creates a single large stable cavity from two nearby smaller cavities. Larger cavities are more relevant for the study of stability of the molecules than smaller cavities. Biologists are therefore interested in identifying such structures in order to fill them. Therefore, in the presence of uncertainty, we choose to create a single large cavity instead of retaining the two smaller cavities.

Time Complexity Let m be the number of simplices in the Delaunay triangulation of the input protein having n atoms, $m = O(n^2)$. Computing the set of cavities takes $O(m\alpha(m))$ time using the union-find data structure. Here, α is the inverse Ackermann function. Given ε, Σ_t and Σ_T are computed in $O(m)$ time using a sequential search over the filtration. Identifying the set of tetrahedra incident on triangles in Σ_t, and moving all the simplices to the end of the filtration takes $O(m)$ time. Thus the time required to modify the filtration is $O(m\alpha(m))$.

3.3 Implementation Notes

The filtration obtained from the alpha complex (Step 2) is stored as a list. The index of each simplex in this list represents the rank of that simplex. For each triangle, we additionally store the indices of the incident tetrahedra. For a given ε, we first compute the ranks k_1 and k_2 of the alpha complex at $\alpha = -\varepsilon$ and $\alpha = \varepsilon$, respectively, and then we compute the set of pockets at rank k_2 [14]. Each pocket is stored as a set of tetrahedra and each tetrahedron has an entry that stores its parent pocket index.

Next, the set Σ_t and Σ_T are computed. The set Σ_T is simply the set of all tetrahedra σ_k having rank $k_1 \leq k \leq k_2$. In order to identify triangles that splits a void into two, it is sufficient to track the connected components of the complement of the α-complex. This is however not true for triangles that modify the topology of a pocket. The addition of such a triangle may not change the number of components present in the complement of the α-complex. In order to identify such triangles, starting from the set of pockets computed at rank k_2, we traverse the filtration in reverse from rank k_2 to rank k_1, and explicitly track the change in topology of the set of pockets. We also create and store the set $\Sigma_T^{'}$, which consists of the incident tetrahedra of all triangles present in Σ_t.

Let $\Sigma_S = \Sigma_t \bigcup \Sigma_T \bigcup \Sigma_T^{'}$. The simplices in Σ_S are sorted in the increasing order of their ranks. Instead of explicitly moving these simplices to the end of the filtration, we perform an implicit move. These simplicies are marked as invalid within the list that represents the original filtration. The new filtration is obtained by traversing the original filtration, ignoring the invalid simplicies, followed by traversing the simplices in Σ_S. An advantage of using this approach is that, when the value of ε is changed, it is easy to revert to the original filtration and recompute the new filtration.

4 Experimental Results

We have developed a software tool ROBUSTCAVITIES that interactively computes the set of stable cavities. The values of ϵ and π can be specified interactively by the user using a slider widget present in the tool. Following is a brief list of features supported in ROBUSTCAVITIES:

- Computation of stable cavities in a protein for specified values of ϵ, π and α.
- Computation of volume and surface area of cavities.
- Visualization and interactive exploration of cavities with support for multiple rendering modes and colormaps.
- Export cavities with the relevant biochemical context in order to be used in PyMOL [4]. In particular, we support skin mesh [3], union of balls, and tetrahedral representation of the detected cavities. We also provide Python scripts

which allow users to load these representations using different colormaps in PyMOL.

We first report experimental results that demonstrate the efficiency of our technique. We then present various examples of stable cavities present in different protein molecules. Finally, we demonstrate the utility of our technique in identifying potential channels and pockets in protein molecules. All experiments were performed on a workstation with a 8-core 2 GHz Intel Xeon processor, 16 GB RAM, and an NVidia GTX 600Ti graphics card.

4.1 Performance and Validation

Efficiency In order to test the efficiency of our technique, we computed the set of robust cavities of over 200 proteins having number of atoms ranging from 184 to 40,026. The value of $\alpha = 0$, $\epsilon = 1$ and $\pi = 0.01$ was used in these experiments.

We first measure the *interaction time*, which is equal to the time taken to modify the filtration and to identify the set of cavities from this modified filtration.This is essentially the time taken to update the set of cavities once the user changes the parameter values. Figure 6a plots the interaction time against the number of atoms in the protein. Note that even for very large proteins having 40,000 atoms, the set of robust cavities are computed within a second. Also note the near-linear behaviour of the interaction time.

Next, we measure the total time taken to compute the set of robust cavities. This includes time to compute the original filtration in addition to the interaction time.

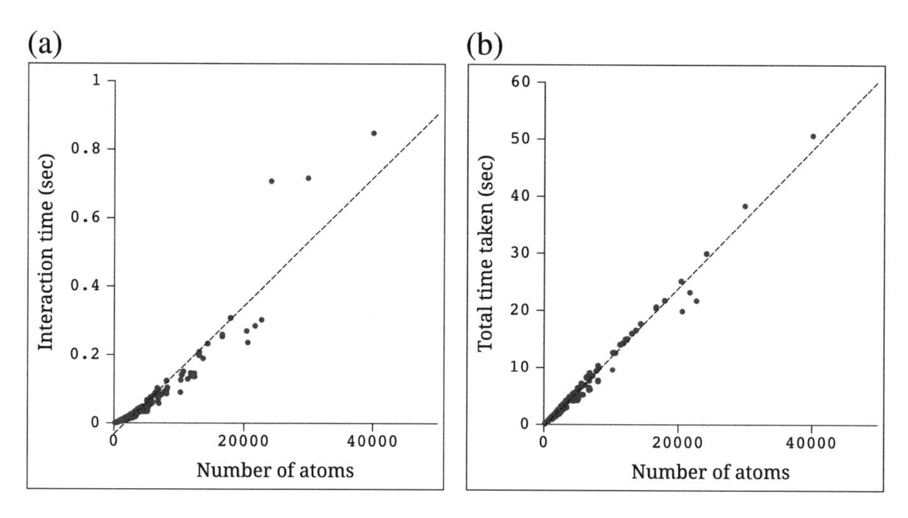

Fig. 6 (**a**) Graph showing variations in the interaction time with respect to varying number of atoms. (**b**) Graph showing variations in the total time with respect to varying number of atoms

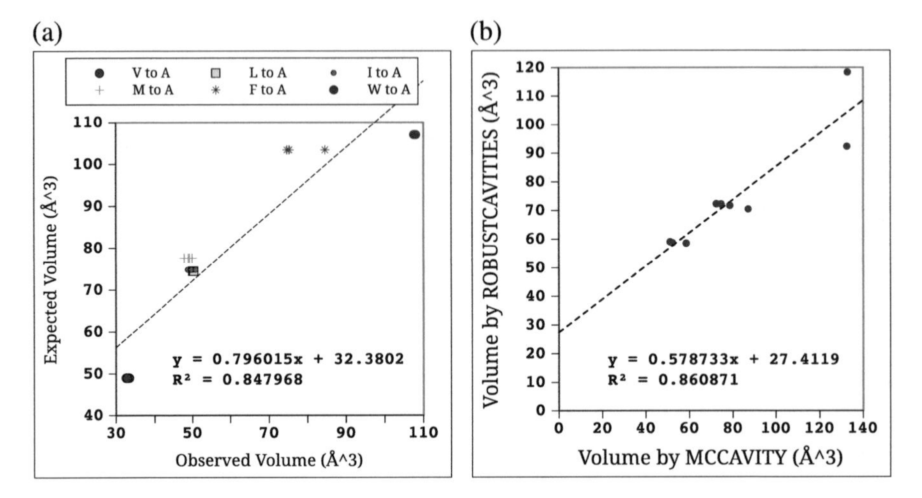

Fig. 7 (**a**) Plot of the computed and actual volumes of the artificial voids that were generated using mutant models. (**b**) Comparison of normalized volumes computed using ROBUSTCAVITIES and MCCAVITY

Figure 6b plots the variation of total time taken against the number of atoms in the protein. Even though the total time is significantly greater than the interaction time, it still requires only around 50 seconds for a protein with 40,000 atoms. Also, this performance is acceptable since the computation of the original filtration is a one-time operation done when loading the protein. Among the pre-processing steps, computing alpha complex is the most time consuming step, which can be improved further by employing the method proposed by Mach and Koehl [24].

Validation of Computed Volumes As mentioned earlier, ROBUSTCAVITIES also reports the volume and surface area of the cavities identified. Proteins adopt a variety of structures and it is known that the extent of packing of protein chain differs within the same structure as well as between different structures. Because of this reason the volume of the cavity which is created by converting a large residue to a small residue may not always be the same in different protein structures. To mitigate this problem we only examine completely buried cavities where it is known that proteins take up a close packed structure.

Different cavity volume computation methods may employ different molecular models resulting in a variation in the volumes that they report. We perform an additional normalization of the computed volumes using model mutants [2] to eliminate such variations. We use 28 different model mutants to create a set of artificial voids. We use the resulting volumes to compute a linear normalization function, as shown in Fig. 7a. The volumes computed using our computation are normalized as follows:

$$Volume = 0.8 \times ComputedVolume + 32.4$$

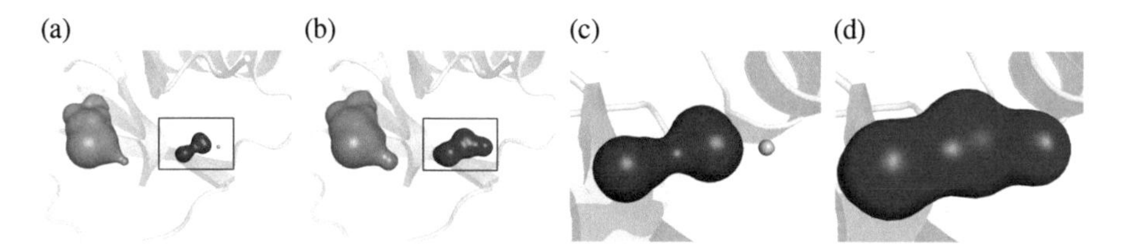

Fig. 8 Visualization of cavities in the protein 2CI2. (**a**) Cavities in the protein. (**b**) The set of $(1.0, 0.01)$-stable cavities. (**c**) Two of the nearby voids in the protein. (**d**) These cavities merge together resulting in a single stable void. Figures (**c**) and (**d**) are zoomed in for better context

The expected volume against which we compare our observed volume is the Voronoi volume of the cavity, which is created on a large to small residue substitution, averaged over many examples. Since we compare a specific large to small substitution in a protein with an averaged ideal value, there is a difference in the volumes, which is reflected as the constant in the above linear transformation.

In order to verify the correctness of the volumes computed by our software, we compare volume for some of these mutants to the volumes computed using MCCAVITY [2], see Fig. 7b. The graph shows normalized volumes and we observe that there is high correlation between the two sets of volumes. In the absence of an ideal normalizing function, the correlation coefficient helps determine if the volumes computed are consistent with data available from other methods.

4.2 Stable Cavities and Their Properties

We now illustrate different examples of robust cavities identified using our software. Unless otherwise specified, we use values $\alpha = 0$, $\epsilon = 1$ and $\pi = 0.01$ in following examples. Note that a value of $\varepsilon = 1$ is equivalent to a change of the radius of an atom by at most 0.2 Å, which is within the resolution at which the input data is available and hence within the tolerance threshold. In the following, we refer to a protein by specifying its PDB ID from the Protein Data Bank [1].

Figure 8 shows protein 2CI2, which has three cavities. Two of the cavities are voids, and are quite close as can be seen in the Fig. 8a. After modification, these two voids are detected as a single void as shown in Fig. 8b. Figure 8c, d show a close-up view of the two merging voids. Similarly, Figures 9 and 10 show stable cavities for proteins 4B87 and 1DKF, respectively. In both these proteins, we observe two significant pockets merging into a single stable pocket after filtration modification.

Properties of Stable Cavities Figures 11 and 12 plot the number and volume of (ε, π)-stable cavities for various values of ε. Note that increasing the value of ε implies that cavities from a wider range of α-values are considered. This could potentially increase the number of ε-stable cavities. However, such cavities usually have low persistence and are therefore not (ε, π)-stable. The total volume of all

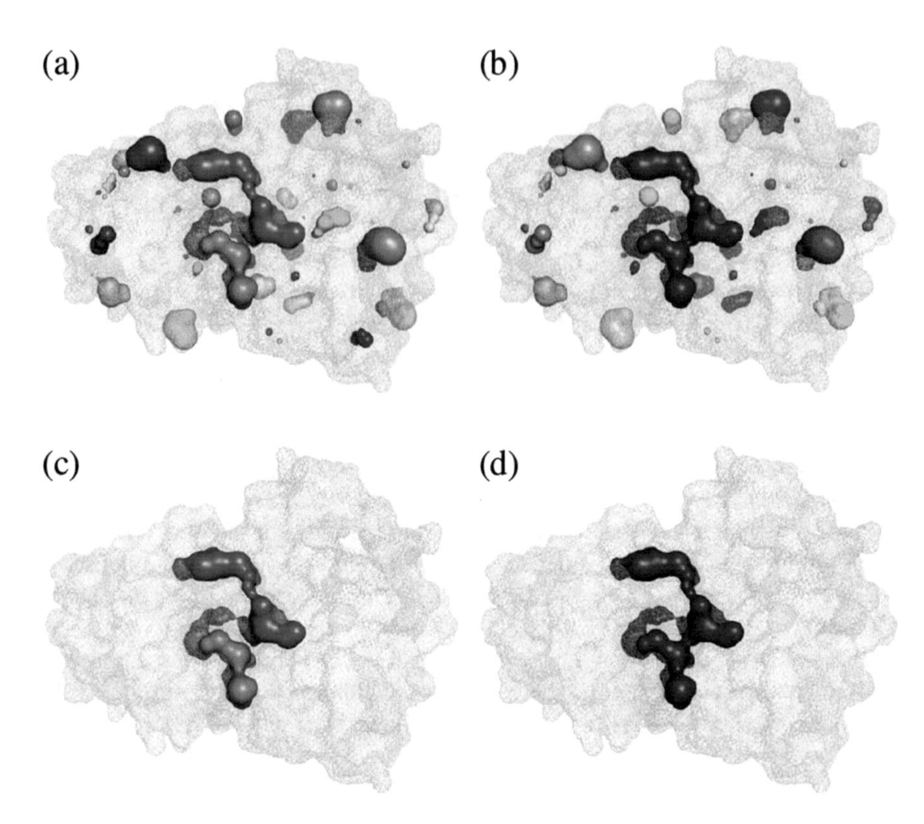

Fig. 9 Visualization of cavities in the protein 4B87. The molecular surface is shown for context. (**a**) The set of cavities in the protein. Number of cavities = 72. (**b**) The set of $(1.0, 0.01)$-stable cavities. Number of stable cavities = 70. (**c**) Two of the nearby pockets in the protein. (**d**) These pockets merge together resulting in a single stable pocket

stable cavities increases marginally ($< 1\,\%$) with increasing ε. The merging of two nearby cavities into a single stable cavity does not effect the total volume. However, volumes of individual cavities could change drastically. We have observed that the volume of a stable void is approximately equal to the sum of the volumes of the original voids.

Robustness of (ε, π)-Stable Cavities Figure 13 plots the number of (ε, π)-stable cavities and π-persistent cavities for various values of α for $\varepsilon = 1.0$ in case of 4HHB and 4B87 and $\varepsilon = 0.3$ in case of 2CI2. Note that the number of (ε, π)-stable cavities is mostly constant, while there is a significant variation in the number of π-persistent cavities. This is because, when using only persistence, even though small (noisy) voids are removed, a small change in the radius could change the number of voids. On the other hand, since our method adds in the additional constraint of stability, only the robust voids are retained.

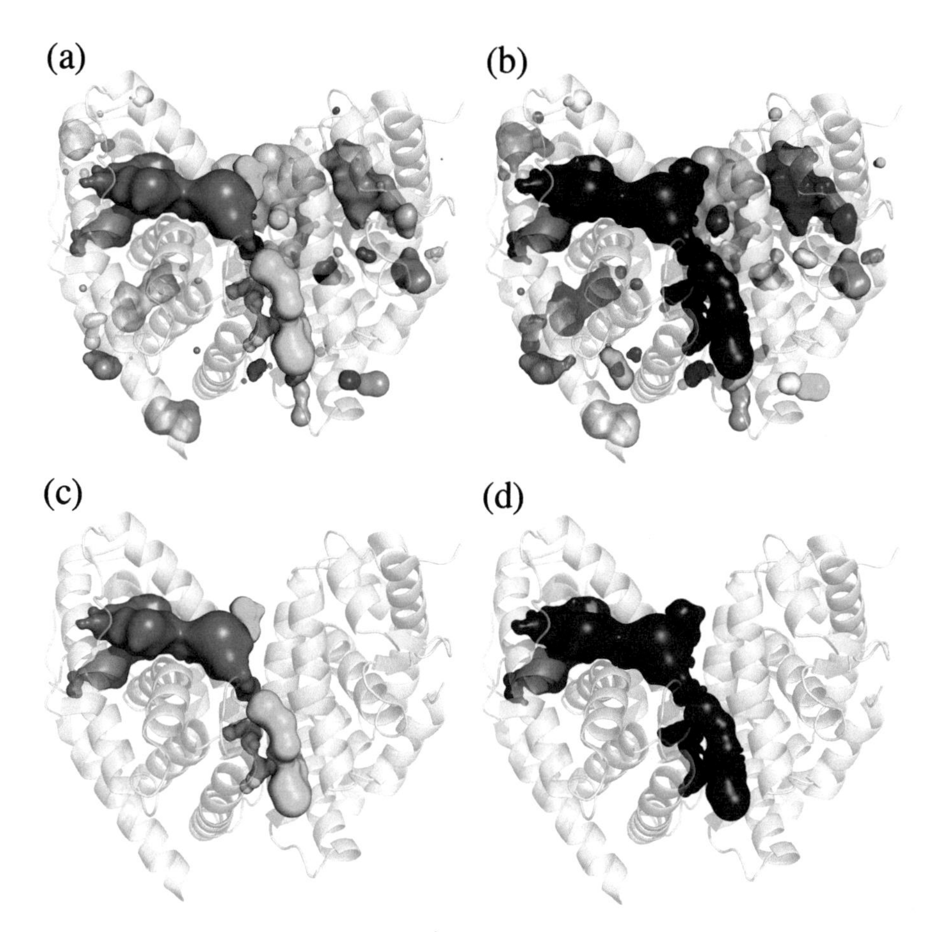

Fig. 10 Visualization of cavities in the protein 1DKF. A cartoon representation of the secondary structure is shown for context. (**a**) All cavities in the protein. Number of cavities = 56. (**b**) The set of (1.0, 0.01)-stable cavities. Number of stable cavities = 39. (**c**) Two of the nearby pockets in the protein. (**d**) These pockets merge together resulting in a single stable pocket

4.3 Detecting Potential Channels and Pockets in Proteins

Due to experimental errors or changed protein conformation, a true pocket could be labelled as a void, or a pore (also referred to as through-channels) may be labelled as a disconnected pocket by a cavity detection algorithm. Existing software used for finding cavities in proteins fail to identify such cavities. However, since our technique is robust to errors (specified using ϵ), it is possible to detect such potential pockets and pores.

While the case of identifying potential pores (or potential channels) is taken care of by conditions C2 and C3 (refer Sect. 3.2), in order to identify pockets that appear as voids, we additionally delay insertion of triangles satisfying conditions C4

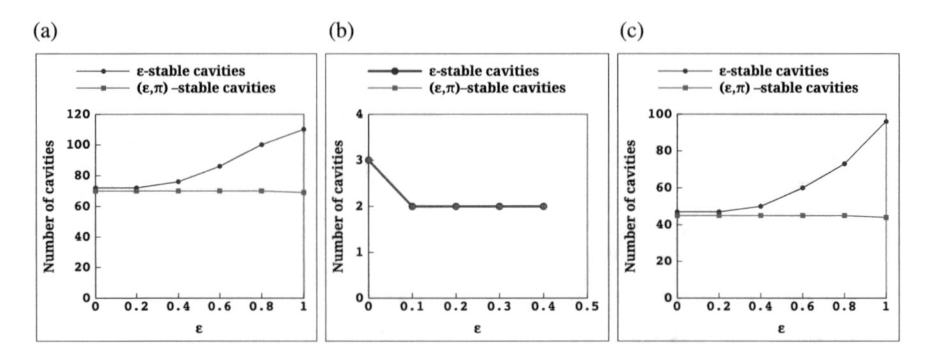

Fig. 11 Graphs showing the variation of the number of cavities with varying ε. Note that there is an increase in the number of ε-stable cavities as we consider a larger interval. But, the number of (ε, π) cavities is less than or equal to the original number of cavities. (**a**) Protein 4HHB. (**b**) Protein 2CI2. (**c**) Protein 4B87

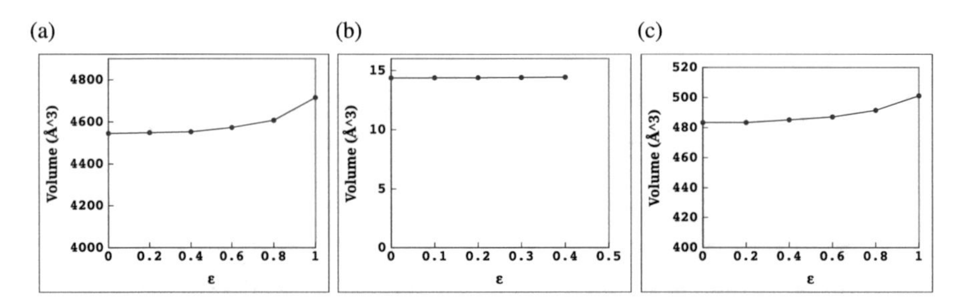

Fig. 12 Graphs showing the variation of the total volume of cavities with varying ε. The increase in total volume is insignificant. (**a**) Protein 4HHB. (**b**) Protein 2CI2. (**c**) Protein 4B87

and C5. Thus stability of pockets is given preference over stability of voids, since pockets usually correspond to functionally important regions of proteins as they are accessible from the outside environment.

Figure 14 illustrates an example where we detect a potential channel in protein 2OAR. This is a trans-membrane protein with a known ion-channel going though it. By default, two pockets and a small void are detected instead of the ion-channel. Using ROBUSTCAVITIES with $\epsilon = 1.4$ (which corresponds to maximum change of 0.41 Å in atomic radius), the void merges together with the two pockets to correctly identify this channel.

We now demonstrate the utility of our potential channel detection technique using the example of translocase SecY protein [32]. We consider three of its structures—1RHZ, 2YXQ and 2YXR. This unique transporter protein has its trans-membrane channel plugged in its wild type conformation (1RHZ). This plug only opens when specific molecules need to be transported. We tried detecting a channel through the wild type 1RHZ structure. Even after modifying the filtration, all cavities remained stably disconnected as shown in Fig. 15b. To probe the mechanics and regulation of this transporter, researchers created a half and full plug deletion

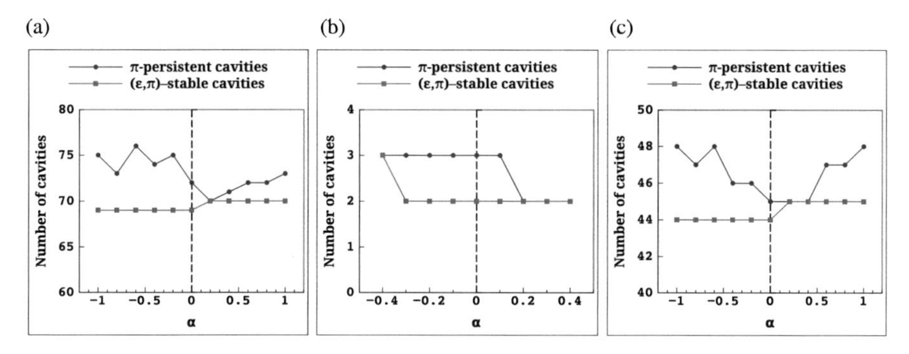

Fig. 13 Graphs showing the variation of the number of cavities for constant ε and varying α. The number of (ε, π)-stable cavities does not vary much but there is significant variation in the number of π-persistent cavities. (**a**) Protein 4HHB. (**b**) Protein 2CI2. (**c**) Protein 4B87

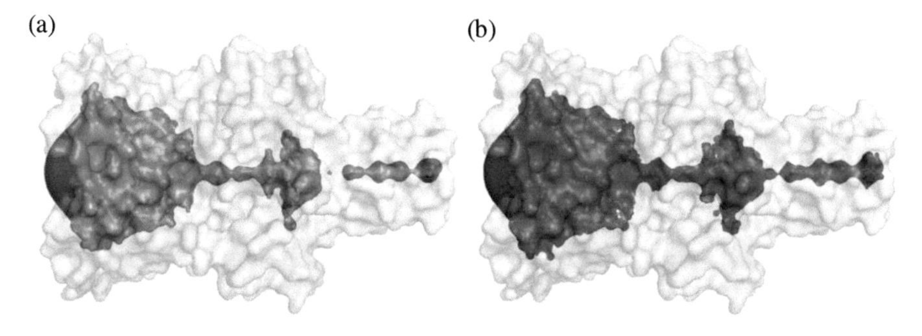

Fig. 14 Detection of potential channel in trans-membrane protein 2OAR. (**a**) At $\alpha = 0$, two pockets and a small void is detected. (**b**) After modification, these cavities merge together revealing the pore present in this trans-membrane protein

mutants of the same protein, labelled as 2YXQ and 2YXR respectively [25]. In 2YXQ, half of the plug region was deleted while in 2YXR the plug was deleted completely. Even after plug deletions, the protein compensates for the deletion and attains a tightly packed structure due to its dynamic nature; other methods still fail to detect a channel in this plug deletion mutant. However, after using ROBUSTCAVITIES with $\epsilon = 1.5$ (which corresponds to maximum change of 0.48 Å in atomic radius) to modify filtration, we are able to identify the potential channels in these mutants as is shown in Fig. 15d, f. In summary, we find that the trans-membrane pore is not present in the wild type structure but becomes progressively larger in the half plug and full plug deletions mutants. This is consistent with experimental data which show that plug deletion leads to increased translocation of proteins with defective signal sequences as well as small molecules and increase the propensity for the channel to adopt an open state.

In our final example shown in Fig. 16, we study cavity structures in low and high affinity states of the protein Hemoglobin. As shown in Fig. 16a, c, both low and high affinity structures consist of four heme sites surrounding a central cavity.

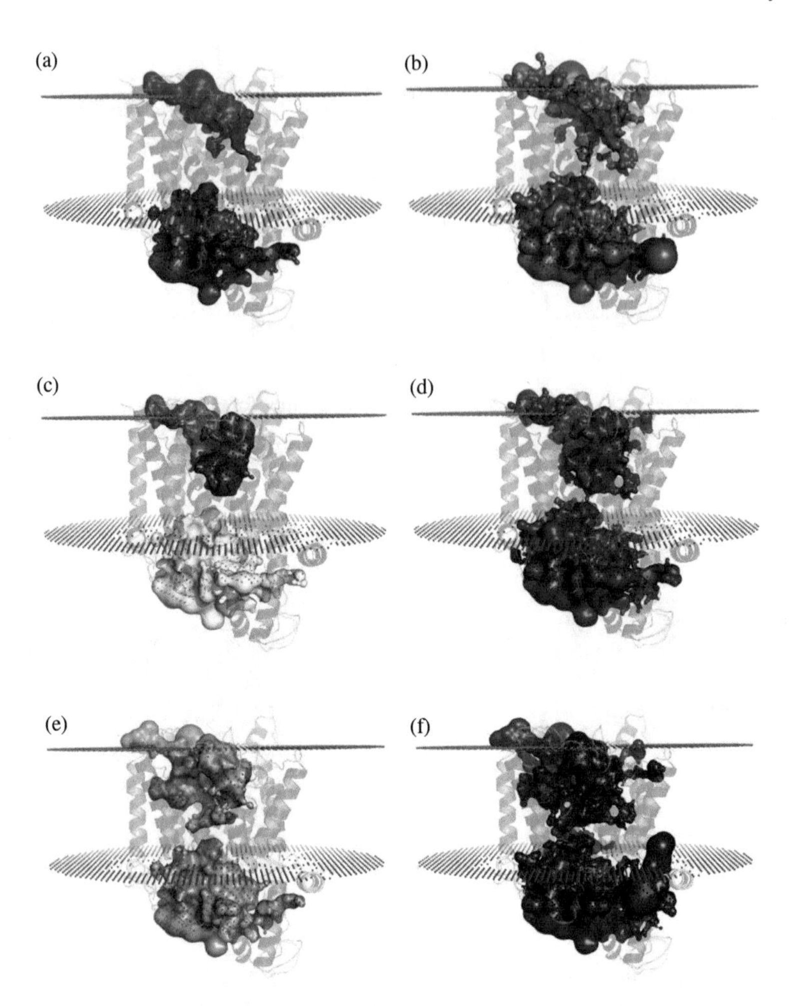

Fig. 15 The case study of protein translocase SecY. We show results for three structures of this protein viz. 1RHZ, 2YXQ and 2YXR. 1RHZ is the closed structure of this protein, while 2YXQ and 2YXR are mutants with half and full plug deletions respectively. In all the figures the membrane is shown as *blue* and *red planes*, where *blue plane* corresponds to intra-cellular region while *red plane* denotes extra-cellular region. (**a**) The two pockets in 1RHZ at $\alpha = 0$. (**b**) These two pockets remain disconnected even after modifying filtration. (**c**) The two pockets of interest in 2YXQ at $\alpha = 0$. Please note that these pockets are slightly larger than the pockets in closed structure 1RHZ. (**d**) The pockets merge to reveal transmembrane pore in this mutant. (**e**) The two relevant pockets of in 2YXR at $\alpha = 0$. It can be observed that these pockets are larger than the pockets detected in both 1RHZ and 2YXQ. (**f**) In this mutant too, the two pockets merge to reveal transmembrane connection

Also, all these cavities are disconnected at $\alpha = 0$ in both the structures. However, after modifying filtration, while the topology of cavities in low affinity structure remains unchanged (Fig. 16b), two heme sites in chains B and D of high affinity structure merge with central cavity (Fig. 16d). It is known that Oxygen binding to

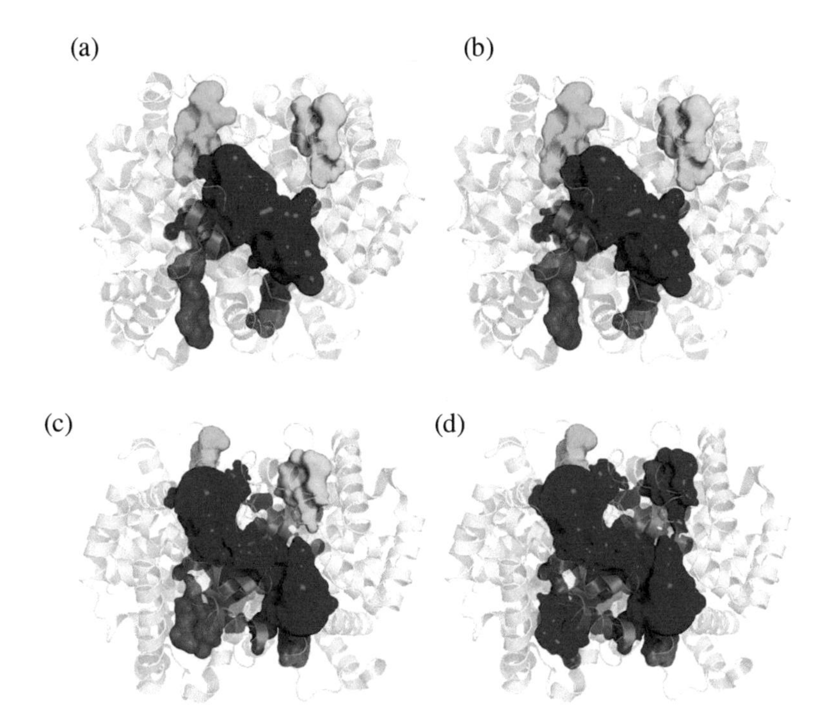

Fig. 16 Cavity structures in two states of Hemoglobin. In all the images, central cavity is shown in *blue color*, while *red*, *cyan*, *green* and *magenta colors* are used for heme sites in chains A, B, C and D, respectively. The *first row* in the figure is for 1HGA which is low affinity T state, while *second row* corresponds to high affinity R state (1BBB) of Hemoglobin. (**a**) The central cavity and four heme cavities in low affinity state of Hemoglobin. (**b**) Even after applying modification, the heme sites don't merge with central cavity. (**c**) The central cavity and four heme cavities in high affinity state of Hemoglobin. (**d**) The heme sites in chains B and D merge with the central cavity after application of ROBUSTCAVITIES

heme in hemoglobin causes a conformational change in the rest of the structure which leads to an increase in oxygen binding affinity. The binding results in the conformation transition from tense form (low affinity T state) to relaxed form (high affinity R state). This important conformational change is being correctly captured by the change in topology of the cavities of the R state (Fig. 16d).

5 Conclusions

We have defined a novel notion of robust cavities that is insensitive to the perturbation of the atomic radii. Robust cavities are computed via a controlled modification of the filtration that represents the molecule and its cavities. Identifying robust cavities is important so that the biologist only targets these cavities in tedious mutation-based experiments. The method addresses the inaccuracies in the measurements of the radii by selectively varying the radii for a specific set of atoms. However the positional uncertainties which arise due to the motion of the molecules is not addressed.

We show several examples which demonstrates using visual evidence that small perturbations in the radii results in a larger and robust cavities. The value of ε used in these experiments is lower than the typical experimental error in crystallographic measurements. We also show the efficiency of our method which allows for interactive exploration of robust cavities with varying ϵ. Finally, we use our technique to identify robust pockets and pores in different trans-membrane proteins.

In future, we plan to further investigate the relationship between the perturbation in the atom radii corresponding to the delayed simplex insertion and the structural and functional properties of the protein. Future work also includes generalizing the framework to use empirically determined intervals of radii for each atom type and addressing the issue of biological implications of the method.

Acknowledgements Talha Bin Masood was supported by Microsoft Corporation and Microsoft Research India under the Microsoft Research India PhD Fellowship Award. This work was supported in part by the Department of Science and Technology, India, under Grant SR/S3/EECE/0086/2012, the DST Center for Mathematical Biology, IISc, under Grant SR/S4/MS:799/12, the NYU School of Engineering, and NSF award CNS-1229185. We would like to thank Patrice Koehl for his suggestions and for sharing the source code of Proshape.

References

1. Berman, H., Westbrook, J., Feng, Z., Gilliland, G., Bhat, T., Weissig, H., Shindyalov, I., Bourne, P.: The protein data bank. Nucleic Acids Res. **28**(1), 235–242 (2000)
2. Chakravarty, S., Bhinge, A., Varadarajan, R.: A procedure for detection and quantitation of cavity volumes in proteins. J. Biol. Chem. **277**(35), 31345–31353 (2002)
3. Cheng, H., Shi, X.: Quality mesh generation for molecular skin surfaces using restricted union of balls. Comput. Geom. **42**(3), 196–206 (2009)
4. DeLano, W.: The pymol molecular graphics system. http://www.pymol.org (2002)
5. Dundas, J., Ouyang, Z., Tseng, J., Binkowski, A., Turpaz, Y., Liang, J.: CASTp: computed atlas of surface topography of proteins with structural and topographical mapping of functionally annotated residues. Nucleic Acids Res. **34**(2), W116–W118 (2006)
6. Edelsbrunner, H.: Weighted alpha shapes. University of Illinois at Urbana-Champaign, Department of Computer Science (1992)
7. Edelsbrunner, H.: Biological applications of computational topology. In: Goodman, J.E., O'Rourke, J. (eds.) Handbook of Discrete and Computational Geometry, pp. 1395–1412. CRC Press, Boca Raton (2004)
8. Edelsbrunner, H.: Computational Topology. An Introduction. American Mathematical Society, Providence (2010)
9. Edelsbrunner, H., Fu, P.: Measuring space filling diagrams and voids. Tech. rep., UIUC-BI-MB-94-01, Beckman Inst., Univ. Illinois, Urbana (1994)
10. Edelsbrunner, H., Koehl, P.: The geometry of biomolecular solvation. Comb. Comput. Geom. **52**, 243–275 (2005)

11. Edelsbrunner, H., Letscher, D., Zomorodian, A.: Topological persistence and simplification. Discrete Comput. Geom. **28**(4), 511–533 (2002)
12. Edelsbrunner, H., Mücke, E.: Three-dimensional alpha shapes. ACM Trans. Graph. **13**(1), 43–72 (1994)
13. Edelsbrunner, H., Kirkpatrick, D., Seidel, R.: On the shape of a set of points in the plane. IEEE Trans. Inf. Theory **29**(4), 551–559 (1983)
14. Edelsbrunner, H., Facello, M., Liang, J.: On the definition and the construction of pockets in macromolecules. Discrete Appl. Math. **88**(1), 83–102 (1998)
15. Kim, D.S., Sugihara, K.: Tunnels and voids in molecules via voronoi diagram. In: Proceedings of Symposium on Voronoi Diagrams in Science and Engineering (ISVD), pp. 138–143 (2012)
16. Kim, D.S., Cho, Y., Sugihara, K., Ryu, J., Kim, D.: Three-dimensional beta-shapes and beta-complexes via quasi-triangulation. Comput. Aided Des. **42**(10), 911–929 (2010)
17. Kim, D.S., Ryu, J., Shin, H., Cho, Y.: Beta-decomposition for the volume and area of the union of three-dimensional balls and their offsets. J. Comput. Chem. **33**(13), 1252–1273 (2012)
18. Koehl, P., Levitt, M., Edelsbrunner, H.: Proshape: understanding the shape of protein structures. Software at http://biogeometry.duke.edu/software/proshape (2004)
19. Krone, M., Falk, M., Rehm, S., Pleiss, J., Ertl, T.: Interactive exploration of protein cavities. Comput. Graph. Forum **30**, 673–682 (2011)
20. Liang, J., Edelsbrunner, H., Fu, P., Sudhakar, P., Subramaniam, S.: Analytical shape computation of macromolecules: II. Inaccessible cavities in proteins. Proteins Struct. Funct. Genet. **33**(1), 18–29 (1998)
21. Liang, J., Edelsbrunner, H., Woodward, C.: Anatomy of protein pockets and cavities. Protein Sci. **7**(9), 1884–1897 (1998)
22. Lindow, N., Baum, D., Hege, H.: Voronoi-based extraction and visualization of molecular paths. IEEE Trans. Vis. Comput. Graph. **17**(12), 2025–2034 (2011)
23. Lindow, N., Baum, D., Bondar, A., Hege, H.: Dynamic channels in biomolecular systems: path analysis and visualization. In: Proceedings of IEEE Symposium on Biological Data Visualization (BioVis), pp. 99–106 (2012)
24. Mach, P., Koehl, P.: Geometric measures of large biomolecules: surface, volume, and pockets. J. Comput. Chem. **32**(14), 3023–3038 (2011)
25. Minor, D.L.: Puzzle plugged by protein pore plasticity. Mol. Cell **26**(4), 459–460 (2007)
26. Munkres, J.: Elements of Algebraic Topology, vol. 2. Addison-Wesley, Menlo Park (1984)
27. Parulek, J., Turkay, C., Reuter, N., Viola, I.: Implicit surfaces for interactive graph based cavity analysis of molecular simulations. In: 2012 IEEE Symposium on Biological Data Visualization (BioVis), pp. 115–122 (2012)
28. Petřek, M., Otyepka, M., Banáš, P., Košinová, P., Koča, J., Damborský, J.: Caver: a new tool to explore routes from protein clefts, pockets and cavities. BMC Bioinf. **7**(1), 316 (2006)
29. Petřek, M., Košinová, P., Koča, J., Otyepka, M.: MOLE: A Voronoi diagram-based explorer of molecular channels, pores, and tunnels. Structure **15**(11), 1357–1363 (2007)
30. Sridharamurthy, R., Doraiswamy, H., Patel, S., Varadarajan, R., Natarajan, V.: Extraction of robust voids and pockets in proteins. In: EuroVis-Short Papers, pp. 67–71. The Eurographics Association, Geneve (2013)
31. Till, M.S., Ullmann, G.M.: Mcvol-a program for calculating protein volumes and identifying cavities by a monte carlo algorithm. J. Mol. Model. **16**(3), 419–429 (2010)
32. van den Berg, B., Clemons, W.M., Collinson, I., Modis, Y., Hartmann, E., Harrison, S.C., Rapoport, T.A.: X-ray structure of a protein-conducting channel. Nature **427**(6969), 36–44 (2003)

Author Index

L. Linsen et al. (eds.), *Visualization in Medicine and Life Sciences III*, Mathematics and Visualization, DOI 10.1007/978-3-319-24523-2

Printed in the United States
By Bookmasters